Progress in Mathematics
Volume 272

Luis Barreira

Dimension and Recurrence in Hyperbolic Dynamics

Birkhäuser
Basel · Boston · Berlin

Author:

Luis Barreira
Departamento de Matemática
Instituto Superior Técnico
1049-001 Lisboa
Portugal
e-mail: barreira@math.ist.utl.pt

2000 Mathematics Subject Classification: 37-02, 37C45, 37Dxx, 37Axx

Library of Congress Control Number: 2008929876

Bibliographic information published by Die Deutsche Bibliothek.
Die Deutsche Bibliothek lists this publication in the Deutsche Nationalbibliografie;
detailed bibliographic data is available in the Internet at http://dnb.ddb.de

ISBN 978-3-7643-8881-2 Birkhäuser Verlag AG, Basel · Boston · Berlin

© 2008 Birkhäuser Verlag AG
Basel · Boston · Berlin
P.O. Box 133, CH-4010 Basel, Switzerland
Part of Springer Science+Business Media
Printed on acid-free paper produced from chlorine-free pulp. TCF ∞
Printed in Germany

ISBN 978-3-7643-8881-2   e-ISBN 978-3-7643-8882-9

9 8 7 6 5 4 3 2 1   www.birkhauser.ch

Ferran Sunyer i Balaguer (1912–1967) was a self-taught Catalan mathematician who, in spite of a serious physical disability, was very active in research in classical mathematical analysis, an area in which he acquired international recognition. His heirs created the Fundació Ferran Sunyer i Balaguer inside the Institut d'Estudis Catalans to honor the memory of Ferran Sunyer i Balaguer and to promote mathematical research.

Each year, the Fundació Ferran Sunyer i Balaguer and the Institut d'Estudis Catalans award an international research prize for a mathematical monograph of expository nature. The prize-winning monographs are published in this series. Details about the prize and the Fundació Ferran Sunyer i Balaguer can be found at

http://www.crm.es/FSBPrize/ffsb.htm

**This book has been awarded the
Ferran Sunyer i Balaguer 2008 prize.**

The members of the scientific commitee of the 2008 prize were:

Hyman Bass
    University of Michigan

Antonio Córdoba
    Universidad Autónoma de Madrid

Paul Malliavin
    Université de Paris VI

Oriol Serra
    Universitat Politècnica de Catalunya, Barcelona

Alan Weinstein
    University of California at Berkeley

INSTITVT D'ESTVDIS CATALANS
MCMVII

Ferran Sunyer i Balaguer Prize winners since 1996:

To Claudia

Contents

Preface

The main objective of this book is to give a broad unified introduction to the study of *dimension* and *recurrence* in hyperbolic dynamics. It includes a discussion of the foundations, main results, and main techniques in the rich interplay of four main areas of research: *hyperbolic dynamics, dimension theory, multifractal analysis*, and *quantitative recurrence*. It also gives a panorama of several selected topics of current research interest. This includes topics on irregular sets, variational principles, applications to number theory, measures of maximal dimension, multifractal rigidity, and quantitative recurrence.

The book is directed to researchers as well as graduate students who wish to have a global view of the theory together with a working knowledge of its main techniques. It can also be used as a basis for graduate courses in dimension theory of dynamical systems, multifractal analysis (together with a discussion of several special topics), and pointwise dimension and recurrence in hyperbolic dynamics. I hope that the book may serve as a fast entry point to this exciting and active field of research, and also that it may lead to further developments.

The material is organized in four parts: dimension theory; multifractal analysis: core theory; multifractal analysis: further developments; and hyperbolicity and recurrence. With the exception of some basic well-known statements, all the results are included with detailed proofs, many of them simplified or rewritten specifically for the book. Furthermore, the text is self-contained. In particular, all the necessary notions and results from hyperbolic dynamics, symbolic dynamics, ergodic theory, dimension theory, and the thermodynamic formalism are recalled along the way, mostly without proofs but with appropriate references. I emphasize that each chapter can essentially be read independently.

Since the theory is so vast, in order to present a global view of the topics under discussion, but still keep the size of the book under control, I had to make a careful selection of material. Certainly, this selection also reflects a personal taste, undoubtedly biased towards my own interests. This causes some interesting topics to be barely mentioned, particularly when their study mostly requires techniques of a different nature from the ones consistently used in the book. Other topics are unfortunately not yet at a stage of development that makes it reasonable to include them in a monograph of this nature. I chose rather to present a sufficiently global view of the theory and to avoid introducing additional techniques that may well play an important role in the theory but as of now are still under development. The most notable example of this nature is the study of the dimension of invariant sets of nonconformal maps (both invertible and noninvertible) which, in spite of several important developments, still lacks today a completely satisfactory approach in its most general version. To include these topics would increase unreasonably the size of the book, even more when roughly two thirds of the material already appears here for the first time in book form. As a compromise, I added detailed notes about these and other topics at appropriate places in the book, together with references for further reading.

There are no words that can adequately express my gratitude to Claudia Valls for her help, patience, encouragement, and inspiration without which it would be impossible for this book to exist. I am also indebted to all my collaborators, and particularly Yakov Pesin, Benoît Saussol, Jörg Schmeling, and Christian Wolf, with whom I have obtained, in various combinations, several of the results in the book. I acknowledge the support of the Center for Mathematical Analysis, Geometry, and Dynamical Systems of Instituto Superior Técnico, and Fundação para a Ciência e a Tecnologia.

Luis Barreira
Barcelona, April 2008

Chapter 1

Introduction

We describe briefly in this chapter the research areas considered in the book. At this point, rather than giving a technical introduction we prefer to give a brief overview of the historical origins and main characteristics of each area. We also describe the contents of the book.

1.1 Dimension and recurrence in hyperbolic dynamics

Recurrence. The notion of nontrivial recurrence goes back to Poincaré in his study of the three-body problem. He proved in his celebrated memoir [122] of 1890 that, whenever a dynamical system preserves volume, almost all trajectories return infinitely often to any arbitrarily small neighborhood of their initial position. This is Poincaré's recurrence theorem. The memoir is the famous one that in its first version (printed in 1889, even having circulated shortly, and of which some copies still exist today) had the error that can be seen as the main cause for the study of chaotic behavior in the theory of dynamical systems. Incidentally, Poincaré's recurrence theorem was already present in the first printed version of the memoir and then again in [122]. Already after publication of [122], the following was observed by Poincaré about the complexity caused by the existence of homoclinic points in the restricted three-body problem (as quoted for example in [25, p. 162]):

> "One is struck by the complexity of this figure that I am not even attempting to draw. Nothing can give us a better idea of the complexity of the three-body problem and of all the problems of dynamics in general ..."

We recommend [25] for a detailed historical account.

Hyperbolicity. The study of hyperbolicity goes back to the seminal work of Hadamard [67] of 1898 concerning the geodesic flow in the unit tangent bundle of a surface with negative curvature, in particular revealing the instability of the flow with respect to the initial conditions. Hadamard observed (as quoted for example in [25, p. 209]) that:

> "...each stable trajectory can be transformed, by an infinitely small variation in the initial conditions, into a completely unstable trajectory extending to infinity, or, more generally, into a trajectory of any of the types given in the general discussion: for example, into a trajectory asymptotic to a closed geodesic."

It should be noted that the geodesic flow preserves volume and thus it exhibits a nontrivial recurrence that was also exploited by Hadamard. Considerable activity took place during the 1920s and 1930s, in particular with the important contributions of Hedlund and Hopf who established several topological and ergodic properties of geodesic flows, also in the case of manifolds with arbitrary negative sectional curvature. We refer to the survey [72] for details and further references. Also in [67], Hadamard laid the foundations of symbolic dynamics, subsequently developed by Morse and Hedlund and raised to a subject in its own right (see in particular their work [102] of 1938; incidentally, this is where the expression "symbolic dynamics" appeared for the first time).

Quantitative recurrence. It should be noted that even though Poincaré's recurrence theorem is a fundamental result in the theory of dynamical systems, it only provides information of a qualitative nature. In particular, it gives no information about the frequency of visits of each trajectory to a given set. This drawback was surmounted by Birkhoff [29, 30] and von Neumann [160] who in 1931 established independently the first versions of the ergodic theorem. Together with its variants and generalizations, the ergodic theorem is a fundamental result in the theory of dynamical systems and in particular in ergodic theory (incidentally, one of the first occurrences of the expression "ergodic theory" was in 1932 in joint work of Birkhoff and Koopman [31]). Nevertheless, the ergodic theorem considers only one aspect of the quantitative behavior of recurrence. In particular, it gives no information about the rate with which a given trajectory returns arbitrarily close to itself. More recently, there has been a growing interest in the area, starting with the work of Boshernitzan [34] and Ornstein and Weiss [107].

Dimension theory. In another direction, the dimension theory of dynamical systems progressively developed, during the last two decades, into an independent field of research. We emphasize that we are mostly concerned here with the study of dynamical systems from the point of view of dimension. The first monograph that clearly took this point of view was Pesin's book [115]. Roughly speaking, the main objective of the dimension theory of dynamical systems is to measure the complexity from the dimensional point of view of the objects that remain invariant under the dynamics, such as the invariant sets and measures. We note that the thermodynamic formalism developed by Ruelle in his seminal work [131] (see also [132]) has a privileged relation with the dimension theory of dynamical systems.

Multifractal analysis. The multifractal analysis of dynamical systems is a subfield of the dimension theory of dynamical systems. Roughly speaking, multifractal analysis studies the complexity of the level sets of invariant local quantities obtained from a dynamical system. For example, we can consider Birkhoff averages, Lyapunov exponents, pointwise dimensions, and local entropies. These functions are typically only measurable and thus their level sets are rarely manifolds. Hence, in order to measure their complexity it is appropriate to use quantities such as the topological entropy and the Hausdorff dimension. The concept of multifractal analysis was suggested by Halsey, Jensen, Kadanoff, Procaccia and Shraiman in [68]. The first rigorous approach is due to Collet, Lebowitz and Porzio in [44], for a class of measures invariant under one-dimensional Markov maps. In [94], Lopes considered the measure of maximal entropy for hyperbolic Julia sets, and in [128], Rand studied Gibbs measures for a class of repellers. We refer to the books by Pesin [115] and Falconer [55] for related discussions and further references. Multifractal analysis has also a privileged relation with the experimental study of dynamical systems. More precisely, the so-called multifractal spectra obtained from the topological entropy or the Hausdorff dimension of the level sets of a local quantity can be determined experimentally with considerable precision. Thus, we may expect to be able to recover some information about a dynamical system from the information contained in its multifractal spectra.

1.2 Contents of the book: a brief tour

Chapter 2 is of an introductory nature. It recalls in a pragmatic manner all the basic notions and results from dimension theory, ergodic theory, and the thermodynamic formalism that are needed in the book (symbolic dynamics is also used but at a very elementary level, and thus it is recalled only when needed). The chapter serves as a reference for the remaining chapters.

The text continues with four parts: I) dimension theory; II) multifractal analysis: core theory; III) multifractal analysis: further developments; IV) hyperbolicity and recurrence. Each part can essentially be read independently.

Part I is dedicated to the study of the dimension of invariant sets of dynamical systems. Chapter 3 initiates this study, including its relation with the thermodynamic formalism. We note that the dimension theory of invariant sets presents complications of a different nature from those in the dimension theory of invariant measures. In particular, the dimension of an invariant set may be affected by number-theoretical properties. The emphasis in Chapter 3 is on the study of the so-called geometric constructions which can be seen as models of invariant sets of dynamical systems. We first consider geometric constructions modeled by the full shift, in which case we can avoid the thermodynamic formalism. Chapter 4 is dedicated to the study of the dimension of invariant sets of a hyperbolic dynamics, both invertible and noninvertible. In particular, we present the dimension formulas for repellers and hyperbolic sets of a conformal map. We note that symbolic dynamics plays an important role. In particular, using Markov partitions we can model the invariant sets by geometric constructions. Chapter 5 establishes the existence of ergodic measures of maximal dimension for hyperbolic sets of conformal diffeomorphisms. This is a dimensional version of the existence of ergodic measures of maximal entropy. A crucial difference in this setting is that while the entropy map is upper semicontinuous, the Hausdorff dimension is neither upper semicontinuous nor lower semicontinuous.

The core of the theory of multifractal analysis is the main theme of Part II. Chapter 6 describes the multifractal analysis of repellers and hyperbolic sets of conformal maps. Chapter 7 introduces the general concept of multifractal analysis, considering in particular other invariant local quantities. This provides new multifractal spectra that can be seen as potential multifractal moduli, in the sense that they may contain nontrivial information about the dynamical system. In Chapter 8 we discuss the properties of the set of points for which the averages in Birkhoff's ergodic theorem do not converge. This set has zero measure with respect to any invariant measure, and thus it is very small from the point of view of measure theory. On the other hand, it is very large from the point of view of entropy and dimension. We then obtain a conditional variational principle in Chapter 9. In particular, this allows us to show that many spectra, including the so-called mixed spectra, are analytic in several contexts. On the other hand, we show that there are many nonconvex mixed spectra.

Part III is dedicated to several additional topics of the theory of multifractal analysis. Chapter 10 presents multidimensional versions of the conditional variational principle and gives applications to certain problems of number theory. It turns out that the multidimensional multifractal spectra exhibit several nontrivial phenomena that are absent in the one-dimensional case. In Chapter 11 we discuss how we can make rigorous a certain multifractal classification of dynamical systems. Namely, we consider the phenomenon of multifractal rigidity. Roughly speaking, it states that if two dynamical systems are topologically equivalent and some of their multifractal spectra coincide, then the systems must be equivalent (in some sense to be made precise). We also show that sometimes it is impossible to effect a multifractal classification. Chapter 12 is dedicated to the study of multifractal spectra obtained from considering simultaneously Birkhoff averages into the past and into the future in hyperbolic sets. We emphasize that the description of these spectra is not a direct consequence of the results in Chapter 6. The main difficulty is that even though the local product structure of the hyperbolic set is a Lipschitz homeomorphism with Lipschitz inverse, the level sets of the Birkhoff averages are never compact. This forces us to construct explicitly noninvariant measures concentrated on each product of level sets having the right pointwise dimension.

Finally, Part IV is dedicated to the interplay of hyperbolicity and recurrence. In Chapter 13, for repellers and hyperbolic sets of conformal maps, we establish explicit formulas for the pointwise dimension of an arbitrary invariant measure. These are expressed in terms of the local entropy and the Lyapunov exponents. This allows us to show that the Hausdorff dimension of a nonergodic invariant measure is equal to the essential supremum of the Hausdorff dimensions of the measures in an ergodic decomposition. In Chapter 14 we consider hyperbolic measures, that is, measures with all Lyapunov exponents nonzero, and we describe their almost product structure. It imitates the local product structure of a hyperbolic set, although its study is much more delicate. In Chapter 15 we study the problem of quantitative recurrence in a hyperbolic set, establishing a relation between the recurrence rate and the pointwise dimension. We also describe the almost product structure of the return time.

Chapter 2

Basic Notions

This chapter collects in a pragmatic manner all the notions and results from *dimension theory*, *ergodic theory*, and the *thermodynamic formalism* that are needed in the book. We emphasize that it is not intended to be an introduction to any of the three areas but instead to serve as a reference for the remaining chapters. Furthermore, it may be skipped without consequences, since whenever needed we shall refer in the main text of the book to the appropriate place in this chapter.

2.1 Dimension theory

We start by introducing the notions of Hausdorff dimension and of lower and upper box dimensions, both for sets and measures. We also introduce the notions of lower and upper pointwise dimensions, and we show how they can be used to compute or at least to estimate the dimension of measures. We refer to the books [56, 96, 115] for details.

We consider a set $X \subset \mathbb{R}^m$ for some $m \in \mathbb{N}$. Let d be the distance in X. We define the *diameter* of a set $U \subset X$ by

$$\operatorname{diam} U = \sup\{d(x, y) : x, y \in U\},$$

and the *diameter* of a collection $\mathcal{U}$ of subsets of X by

$$\operatorname{diam} \mathcal{U} = \sup\{\operatorname{diam} U : U \in \mathcal{U}\}.$$

Given $Z \subset X$ and $\alpha \in \mathbb{R}$, we define the α-*dimensional Hausdorff measure* of Z by

$$m(Z, \alpha) = \lim_{\varepsilon \to 0} \inf_{\mathcal{U}} \sum_{U \in \mathcal{U}} (\operatorname{diam} U)^\alpha, \tag{2.1}$$

where the infimum is taken over all finite or countable covers $\mathcal{U}$ of the set Z with $\operatorname{diam} \mathcal{U} \le \varepsilon$.

Definition 2.1.1. The *Hausdorff dimension* of $Z \subset X$ is defined by

$$\dim_H Z = \inf\{\alpha \in \mathbb{R} : m(Z, \alpha) = 0\}. \tag{2.2}$$

The *lower* and *upper box dimensions* of $Z \subset X$ are defined respectively by

$$\underline{\dim}_B Z = \liminf_{\varepsilon \to 0} \frac{\log N(Z, \varepsilon)}{-\log \varepsilon} \quad \text{and} \quad \overline{\dim}_B Z = \limsup_{\varepsilon \to 0} \frac{\log N(Z, \varepsilon)}{-\log \varepsilon},$$

where $N(Z, \varepsilon)$ denotes the least number of balls of radius ε that are needed to cover the set Z.

It is easy to show that

$$\dim_H Z \leq \underline{\dim}_B Z \leq \overline{\dim}_B Z. \tag{2.3}$$

In general these inequalities may be strict and the coincidence of the Hausdorff dimension and of the lower and upper box dimensions is a relatively rare phenomenon. Indeed, it occurs only in some particular situations such as those discussed in Chapters 3 and 4, even if they correspond to large classes of dynamics.

Now we introduce the corresponding notions for measures. Let μ be a finite measure in X.

Definition 2.1.2. The *Hausdorff dimension* and the *lower* and *upper box dimensions* of μ are defined respectively by

$$\dim_H \mu = \inf\{\dim_H Z : \mu(X \setminus Z) = 0\},$$

$$\underline{\dim}_B \mu = \lim_{\delta \to 0} \inf\{\underline{\dim}_B Z : \mu(Z) \geq \mu(X) - \delta\},$$

$$\overline{\dim}_B \mu = \lim_{\delta \to 0} \inf\{\overline{\dim}_B Z : \mu(Z) \geq \mu(X) - \delta\}.$$

One can easily show that

$$\dim_H \mu = \lim_{\delta \to 0} \inf\{\dim_H Z : \mu(Z) \geq \mu(X) - \delta\}. \tag{2.4}$$

Indeed, let c be the right-hand side of (2.4). Clearly,

$$\dim_H \mu \geq \inf\{\dim_H Z : \mu(Z) \geq \mu(X) - \delta\}$$

for every δ, and hence $\dim_H \mu \geq c$. On the other hand, there exists a sequence of sets Z_n with $\mu(Z_n) \to \mu(X)$ and $\dim_H Z_n \to c$ as $n \to \infty$. Therefore,

$$\dim_H \mu \leq \dim_H \bigcup_{n \in \mathbb{N}} Z_n = \sup_{n \in \mathbb{N}} \dim_H Z_n = c.$$

We note that in general the quantities $\dim_H \mu$, $\underline{\dim}_B \mu$, and $\overline{\dim}_B \mu$ do not coincide, respectively, with the Hausdorff dimension and the lower and upper box dimensions of the support of μ, and thus they contain additional information about the way in which the measure μ is distributed in its support. It follows from (2.3) and (2.4) that

$$\dim_H \mu \le \underline{\dim}_B \mu \le \overline{\dim}_B \mu. \tag{2.5}$$

As it happens with the inequalities in (2.3), in general the inequalities in (2.5) may also be strict. Nevertheless, in strong contrast with the dimension of sets, the coincidence of the three quantities in (2.5) for a given measure μ is a more common phenomenon (see Theorem 14.3.5).

The following quantities allow us to formulate a criterion for the coincidence of the three numbers in (2.5).

Definition 2.1.3. The *lower* and *upper pointwise dimensions* of the measure μ at the point $x \in X$ are defined by

$$\underline{d}_\mu(x) = \liminf_{r \to 0} \frac{\log \mu(B(x,r))}{\log r} \quad \text{and} \quad \overline{d}_\mu(x) = \limsup_{r \to 0} \frac{\log \mu(B(x,r))}{\log r}.$$

We have the following identities.

Proposition 2.1.4. *For each $a > 0$ and $x \in X$ we have*

$$\underline{d}_\mu(x) = \liminf_{n \to \infty} \frac{\log \mu(B(x, ae^{-n}))}{-n}$$

and

$$\overline{d}_\mu(x) = \limsup_{n \to \infty} \frac{\log \mu(B(x, ae^{-n}))}{-n}.$$

Proof. For each $r > 0$ sufficiently small there exists a unique $n = n(r) \in \mathbb{N}$ such that

$$ae^{-(n+1)} \le r < ae^{-n} < 1.$$

We have

$$\mu(B(x, ae^{-(n+1)})) \le \mu(B(x,r)) \le \mu(B(x, ae^{-n})),$$

and thus

$$\frac{\log \mu(B(x, ae^{-(n+1)}))}{\log(ae^{-n})} \le \frac{\log \mu(B(x,r))}{\log r} \le \frac{\log \mu(B(x, ae^{-n}))}{\log(ae^{-(n+1)})}.$$

As $r \to 0$ we have $n(r) \to \infty$, and all sufficiently large integers are attained by $n(r)$. Therefore,

$$\limsup_{n \to \infty} \frac{\log \mu(B(x, ae^{-n}))}{-n} = \limsup_{n \to \infty} \frac{\log \mu(B(x, ae^{-(n+1)}))}{\log(ae^{-n})}$$

$$\le \limsup_{r \to 0} \frac{\log \mu(B(x,r))}{\log r}$$

$$\le \limsup_{n \to \infty} \frac{\log \mu(B(x, ae^{-n}))}{\log(ae^{-(n+1)})}$$

$$= \limsup_{n \to \infty} \frac{\log \mu(B(x, ae^{-n}))}{-n},$$

with similar inequalities for the lower pointwise dimension.   □

The following statement relates the Hausdorff dimension with the lower pointwise dimension.

Theorem 2.1.5. *The following properties hold:*

1. *if $\underline{d}_\mu(x) \geq \alpha$ for μ-almost every $x \in X$, then $\dim_H \mu \geq \alpha$;*

2. *if $\underline{d}_\mu(x) \leq \alpha$ for every $x \in Z \subset X$, then $\dim_H Z \leq \alpha$;*

3. *we have*

$$\dim_H \mu = \operatorname{ess\,sup}\{\underline{d}_\mu(x) : x \in X\}. \tag{2.6}$$

Proof. Set

$$Y = \{x \in X : \underline{d}_\mu(x) \geq \alpha\}.$$

Given $\varepsilon > 0$, for each $x \in Y$ there exists $r(x) > 0$ such that if $r \in (0, r(x))$, then

$$\mu(B(x,r)) \leq (2r)^{\alpha-\varepsilon}. \tag{2.7}$$

Given $\rho > 0$, set

$$Y_\rho = \{x \in Y : r(x) \geq \rho\}.$$

Clearly,

$$Y_{\rho_1} \subset Y_{\rho_2} \quad \text{for} \quad \rho_1 \geq \rho_2, \quad \text{and} \quad Y = \bigcup_{\rho>0} Y_\rho.$$

Since $\mu(X \setminus Y) = 0$, there exists $\rho > 0$ such that $\mu(Y_\rho) \geq \mu(X)/2$. Now let $Z \subset Y$ be an arbitrary set of full μ-measure, and let $\mathcal{U}$ be a cover of $Z \cap Y_\rho$ by open balls. Without loss of generality we assume that $U \cap Y_\rho \neq \varnothing$ for every $U \in \mathcal{U}$. Then for each $U \in \mathcal{U}$, there exists $x_U \in U \cap Y_\rho$, and we consider the new cover

$$\mathcal{V} = \{B(x_U, \operatorname{diam} U) : U \in \mathcal{U}\}$$

of the set $Z \cap Y_\rho$. It follows from (2.7) that

$$\sum_{U \in \mathcal{U}} (\operatorname{diam} U)^{\alpha-\varepsilon} = 2^{\varepsilon-\alpha} \sum_{V \in \mathcal{V}} (\operatorname{diam} V)^{\alpha-\varepsilon}$$

$$\geq 2^{\varepsilon-\alpha} \sum_{V \in \mathcal{U}} \mu(V)$$

$$\geq 2^{\varepsilon-\alpha} \mu(Y_\rho) \geq 2^{\varepsilon-\alpha} \mu(X)/2.$$

Since $\mathcal{U}$ is arbitrary, we obtain

$$m(Z \cap Y_\rho, \alpha - \varepsilon) \geq 2^{\varepsilon-\alpha} \mu(X)/2.$$

This implies that $\dim_H(Z \cap Y_\rho) \geq \alpha - \varepsilon$, and by the arbitrariness of ε we have

$$\dim_H Z \geq \dim_H(Z \cap Y_\rho) \geq \alpha.$$

Now we establish the second property. For each $x \in Z$ and $\varepsilon > 0$ there exists a sequence $r_n = r_n(x, \varepsilon) \searrow 0$ as $n \to \infty$ such that

$$\mu(B(x, r_n)) \geq (2r_n)^{\alpha + \varepsilon} \tag{2.8}$$

for every $n \in \mathbb{N}$. Consider a cover

$$\mathcal{U} \subset \{B(x, r_n(x, \varepsilon)) : x \in Z \text{ and } n \in \mathbb{N}\}$$

of the set Z. Notice that its diameter can be made arbitrarily small. By Besicovitch's covering lemma (see, for example, [96, Theorem 2.7]) there exists a subcover $\mathcal{V} \subset \mathcal{U}$ of Z of finite multiplicity, that is, there exists $K > 0$ such that

$$\text{card}\{V \in \mathcal{V} : x \in V\} \leq K$$

for every $x \in Z$. Therefore, by (2.8),

$$\sum_{V \in \mathcal{V}} (\text{diam } V)^{\alpha + \varepsilon} \leq \sum_{V \in \mathcal{V}} \mu(B(x, r)) \leq K\mu(X),$$

and since the diameter of the cover $\mathcal{V}$ can be made arbitrarily small we obtain $m(Z, \alpha + \varepsilon) \leq K\mu(X)$. This implies that $\dim_H Z \leq \alpha + \varepsilon$, and since ε is arbitrary we obtain $\dim_H Z \leq \alpha$.

For the third property, let

$$\alpha = \text{ess sup}\{\underline{d}_\mu(x) : x \in X\} \quad \text{and} \quad Z = \{x \in X : \underline{d}_\mu(x) \leq \alpha\}.$$

We have $\mu(Z) = \mu(X)$ and by property 2, we have

$$\dim_H \mu \leq \dim_H Z \leq \alpha.$$

Given $\varepsilon > 0$, let

$$Z_\varepsilon = \{x \in X : \underline{d}_\mu(x) \geq \alpha - \varepsilon\}.$$

We have $\mu(Z_\varepsilon) > 0$, and it follows from property 1 that

$$\dim_H \mu \geq \dim_H (\mu | Z_\varepsilon) \geq \alpha - \varepsilon.$$

By the arbitrariness of ε we obtain $\dim_H \mu \geq \alpha$. $\qquad \square$

We also recall a criterion established by Young in [165] for the coincidence of the Hausdorff and box dimensions of a measure.

Theorem 2.1.6. *If μ is a finite measure in X and there exists $d \geq 0$ such that*

$$\lim_{r \to 0} \frac{\log \mu(B(x, r))}{\log r} = d \tag{2.9}$$

for μ-almost every $x \in X$, then

$$\dim_H \mu = \underline{\dim}_B \mu = \overline{\dim}_B \mu = d.$$

Proof. By Theorem 2.1.5 we have $\dim_H \mu \geq d$. Now we prove that $\overline{\dim}_B \mu \leq d$ and the result follows from (2.5). Set

$$Z = \{x \in X : \overline{d}_\mu(x) \leq d\}.$$

Given $\varepsilon > 0$, for each $x \in Z$ there exists $r(x) > 0$ such that if $r \in (0, r(x))$, then

$$\mu(B(x, r)) \geq (2r)^{d+\varepsilon}.$$

Given $\rho > 0$, we consider the set $Y_\rho = \{x \in Z : r(x) \geq \rho\}$. Clearly,

$$Y_{\rho_1} \subset Y_{\rho_2} \quad \text{for} \quad \rho_1 \geq \rho_2, \quad \text{and} \quad Z = \bigcup_{\rho > 0} Y_\rho.$$

Therefore, since $\mu(X \setminus Z) = 0$, we have $\mu(Y_\rho) \nearrow \mu(X)$ as $\rho \to 0$. For each $r < \rho$ the balls $B(x, r)$ form a cover $\mathcal{U}$ of the set Y_ρ. By Besicovitch's covering lemma (see, for example, [96, Theorem 2.7]) there exists a subcover $\mathcal{V} \subset \mathcal{U}$ of Y_ρ of finite multiplicity K. Therefore,

$$\sum_{V \in \mathcal{V}} (\operatorname{diam} V)^{d+\varepsilon} \leq \sum_{V \in \mathcal{V}} \mu(B(x, r)) \leq K\mu(X).$$

Since

$$\sum_{V \in \mathcal{V}} (\operatorname{diam} V)^{d+\varepsilon} = (2r)^{d+\varepsilon} \operatorname{card} \mathcal{V} \geq (2r)^{d+\varepsilon} N(Y_\rho, r),$$

we obtain

$$N(Y_\rho, r) \leq \frac{K\mu(X)}{(2r)^{d+\varepsilon}}$$

and hence,

$$\overline{\dim}_B Y_\rho = \limsup_{r \to 0} \frac{\log N(Y_\rho, r)}{-\log r} \leq d + \varepsilon.$$

Since $\mu(Y_\rho) \nearrow \mu(X)$ as $\rho \to 0$, we conclude that

$$\overline{\dim}_B \mu \leq \limsup_{\rho \to 0} \overline{\dim}_B Y_\rho \leq d + \varepsilon.$$

By the arbitrariness of ε we obtain $\overline{\dim}_B \mu \leq d$. This completes the proof of the theorem.   □

Definition 2.1.7. The limit in (2.9), when it exists, is called the *pointwise dimension* of μ at x, and we denote it by $d_\mu(x)$.

By Whitney's embedding theorem, the statements in this section also hold when X is a subset of a smooth manifold.

2.2 Ergodic theory

We recall in this section a few basic notions and results of ergodic theory, including Poincaré's recurrence theorem, Birkhoff's ergodic theorem, and the notion of Kolmogorov–Sinai entropy. We refer to the books [84, 95, 163] for details.

We first introduce the notion of invariant measure. Let X be a space with a σ-algebra.

Definition 2.2.1. Given a measurable transformation $T\colon X \to X$, we say that a measure μ in X is *T-invariant* if

$$\mu(T^{-1}A) = \mu(A)$$

for every measurable set $A \subset X$.

The study of transformations with an invariant measure is the main theme of ergodic theory. We denote by $\mathfrak{M}$ the set of all T-invariant probability measures in X. We say that a measure $\mu \in \mathfrak{M}$ is *ergodic* if every T-invariant measurable set $A \subset X$ (i.e., such that $T^{-1}A = A$) has measure $\mu(A) = 0$ or $\mu(A) = 1$. We denote by $\mathfrak{M}_E \subset \mathfrak{M}$ the subset of all ergodic measures.

Now we recall one of the basic but fundamental results of ergodic theory—Poincaré's recurrence theorem. It states that any measurable transformation preserving a finite measure exhibits a nontrivial recurrence in any set A with positive measure, in the sense that the orbit of almost every point in A returns infinitely often to A.

Theorem 2.2.2 (Poincaré's recurrence theorem). *Let $T\colon X \to X$ be a measurable transformation. Given $\mu \in \mathfrak{M}$, if $A \subset X$ is a measurable set with positive measure, then*

$$\operatorname{card}\{n \in \mathbb{N} : T^n x \in A\} = \infty$$

for μ-almost every point $x \in A$.

A slightly modified version of Theorem 2.2.2 was first established by Poincaré in his seminal memoir on the three-body problem [122] (we refer to [25] for a detailed historical account).

The following is another basic result of ergodic theory. We denote by $L^1(X, \mu)$ the space of all measurable functions $\varphi\colon X \to \mathbb{R}$ with $\int_X |\varphi|\, d\mu < \infty$.

Theorem 2.2.3 (Birkhoff's ergodic theorem). *Let $T\colon X \to X$ be a measurable transformation and let $\mu \in \mathfrak{M}$. For each $\varphi \in L^1(X, \mu)$ there exists the limit*

$$\lim_{n\to\infty} \frac{1}{n} \sum_{k=0}^{n-1} \varphi(T^k x)$$

for μ-almost every $x \in X$. If in addition $\mu \in \mathfrak{M}_E$, then

$$\lim_{n\to\infty} \frac{1}{n} \sum_{k=0}^{n-1} \varphi(T^k x) = \int_X \varphi\, d\mu$$

for μ-almost every $x \in X$.

For example, if $\varphi = \chi_A$ is the characteristic function of a measurable subset $A \subset X$, then by Theorem 2.2.3, if $\mu \in \mathcal{M}_E$, then

$$\lim_{n \to \infty} \frac{1}{n} \operatorname{card}\{0 \le k \le n - 1 : T^k x \in A\} = \mu(A)$$

for μ-almost every $x \in X$. This means that almost all orbits stay a proportion of time in a given set A equal to the measure of the set.

Now we introduce the notion of Kolmogorov–Sinai entropy. Let $\mu \subset \mathcal{M}$ and let ξ be a finite or countable partition of X into measurable subsets. This means that:

1. ξ is a finite or countable family of subsets of X, with $\mu(\bigcup_{C \in \xi} C) = 1$;

2. $\mu(C \cap D) = 0$ for every $C, D \in \xi$ with $C \neq D$.

We define the entropy of ξ with respect to μ by

$$H_\mu(\xi) = - \sum_{C \in \xi} \mu(C) \log \mu(C),$$

with the convention that $0 \log 0 = 0$.

Definition 2.2.4. We define the *Kolmogorov–Sinai entropy* of T with respect to the measure μ by

$$h_\mu(T) = \sup_\xi \lim_{n \to \infty} \frac{1}{n} H_\mu(\xi_n), \tag{2.10}$$

where ξ_n is the partition of the space X composed of the sets $\bigcap_{k=0}^{n-1} T^{-k} C_{i_{k+1}}$ with $C_{i_1}, \dots, C_{i_n} \in \xi$.

It can be shown that indeed there exists the limit in (2.10) when $n \to \infty$.

2.3   Thermodynamic formalism

We introduce in this section a few basic notions of the thermodynamic formalism, starting with topological pressure. It was introduced by Ruelle in [131] for expansive transformations and by Walters in [161] in the general case. For more details and further references about the thermodynamic formalism we refer to [38, 84, 86, 132, 163].

Let (X, d) be a compact metric space and let $T : X \to X$ be a continuous transformation. For each $n \in \mathbb{N}$ we define a new distance in X by

$$d_n(x, y) = \max \{d(T^k x, T^k y) : 0 \le k \le n - 1\}. \tag{2.11}$$

Given $\varepsilon > 0$, we say that a finite set $E \subset X$ is (n, ε)-*separated* if $d_n(x, y) > \varepsilon$ for every $x, y \in E$ with $x \neq y$.

Definition 2.3.1. The *topological pressure* of a continuous function $\varphi\colon X \to \mathbb{R}$ (with respect to T) is defined by

$$P(\varphi) = P_X(\varphi) = \lim_{\varepsilon \to 0} \limsup_{n \to \infty} \frac{1}{n} \log \sup_E \sum_{x \in E} \exp \sum_{k=0}^{n-1} \varphi(T^k x),$$

where the supremum is taken over all (n, ε)-separated sets $E \subset X$.

The notion of topological entropy is a particular case of topological pressure.

Definition 2.3.2. We define the *topological entropy* of the transformation T by $h(T) = P(0)$, that is,

$$h(T) = \lim_{\varepsilon \to 0} \limsup_{n \to \infty} \frac{1}{n} \log N(n, \varepsilon),$$

where $N(n, \varepsilon)$ is the largest cardinality of all (n, ε)-separated sets.

Now we present an equivalent description of topological pressure.

Theorem 2.3.3 (Variational principle of topological pressure). *If $T\colon X \to X$ is a continuous transformation in a compact metric space, and $\varphi\colon X \to \mathbb{R}$ is continuous, then*

$$P(\varphi) = \sup_{\mu \in \mathcal{M}} \left\{ h_\mu(T) + \int_X \varphi \, d\mu \right\}$$

$$= \sup_{\mu \in \mathcal{M}_E} \left\{ h_\mu(T) + \int_X \varphi \, d\mu \right\}. \tag{2.12}$$

For example, setting $\varphi = 0$ in Theorem 2.3.3 we obtain the variational principle of topological entropy, namely

$$h(T) = \sup_{\mu \in \mathcal{M}} h_\mu(T) = \sup_{\mu \in \mathcal{M}_E} h_\mu(T).$$

Definition 2.3.4. A measure $\mu \in \mathcal{M}$ is called an *equilibrium measure* for φ (with respect to T) if the first supremum in (2.12) is attained by this measure, that is, if

$$P(\varphi) = h_\mu(T) + \int_X \varphi \, d\mu.$$

Part I

Dimension Theory

Chapter 3

Dimension Theory and Thermodynamic Formalism

We start in this chapter our study of the dimension theory of dynamical systems and its relation with the thermodynamic formalism (see Section 2.3). We first consider geometric constructions modeled by the full shift in which case we can avoid the thermodynamic formalism without too much pain. On the other hand, this should make the reader appreciate the simplification and unification that are allowed by the thermodynamic formalism in the following chapters.

3.1 Dimension theory of geometric constructions

There are important differences between the dimension theory of invariant *sets* and the dimension theory of invariant *measures*. In particular, while virtually all dimensional characteristics of invariant hyperbolic *measures* coincide, the study of the dimension of invariant hyperbolic *sets* reveals that the dimensional characteristics frequently depend on other properties, and in particular on number-theoretical properties. This is another reason for our interest in simpler models. The theory of geometric constructions precisely provides such a class of models.

Figure 3.1: Geometric construction in the real line

We start with the description of a particular geometric construction in the real line. We consider constants λ_1, ..., $\lambda_p \in (0,1)$ and disjoint closed intervals

$\Delta_1, \ldots, \Delta_p \subset \mathbb{R}$ with length $\lambda_1, \ldots, \lambda_p$ (see Figure 3.1). For each $k = 1, \ldots, p$, we choose again p disjoint closed intervals $\Delta_{k1}, \ldots, \Delta_{kp} \subset \Delta_k$ with length $\lambda_k \lambda_1$, $\ldots, \lambda_k \lambda_p$. Iterating this procedure, for each $n \in \mathbb{N}$ we obtain p^n disjoint closed intervals $\Delta_{i_1 \cdots i_n}$ with length $\prod_{k=1}^{n} \lambda_{i_k}$. We then define the limit set

$$F = \bigcap_{n=1}^{\infty} \bigcup_{i_1 \cdots i_n} \Delta_{i_1 \cdots i_n}. \qquad (3.1)$$

The following result was essentially proved by Moran in [101] (more precisely, he simply did not consider the box dimensions).

Theorem 3.1.1. *We have*

$$\dim_H F = \underline{\dim}_B F = \overline{\dim}_B F = s$$

and $0 < m(F, s) < \infty$, where s is the unique real number such that

$$\sum_{k=1}^{p} \lambda_k{}^s = 1. \qquad (3.2)$$

Proof. To verify that there is a unique real number s satisfying (3.2) it is sufficient to observe that the function $L(s) = \sum_{k=1}^{p} \lambda_k{}^s$ is strictly decreasing, with

$$\lim_{s \to -\infty} L(s) = +\infty \quad \text{and} \quad \lim_{s \to +\infty} L(s) = 0.$$

For the remaining properties, we first observe that for each $n \in \mathbb{N}$ the sets $\Delta_{i_1 \cdots i_n}$ form a cover of F. Furthermore,

$$\sum_{i_1 \cdots i_n} (\operatorname{diam} \Delta_{i_1 \cdots i_n})^s = \sum_{i_1 \cdots i_n} (\lambda_{i_1} \cdots \lambda_{i_n})^s = \left(\sum_{k=1}^{p} \lambda_k^s \right)^n = 1, \qquad (3.3)$$

and since

$$\operatorname{diam} \Delta_{i_1 \cdots i_n} \le \left(\max_{k=1,\ldots,p} \lambda_k \right)^n \to 0 \quad \text{as} \quad n \to \infty,$$

it follows from (3.3) that $m(F, s) \le 1$. In particular, $\dim_H F \ge s$.

To obtain a lower bound for the Hausdorff measure we define a probability measure μ in the limit set F by requiring that

$$\mu(\Delta_{i_1 \cdots i_n}) = (\lambda_{i_1} \cdots \lambda_{i_n})^s$$

for every $i_1, \ldots, i_n \in \{1, \ldots, p\}$. It follows from (3.3) that $\mu(F) = 1$. Now we construct a special cover of F, that we call a *Moran cover*. For each sequence $\omega = (i_1 i_2 \cdots) \in \{1, \ldots, p\}^{\mathbb{N}}$ and $r \in (0, 1)$, we consider the unique integer $n = n(\omega, r)$ such that

$$\lambda_{i_1} \cdots \lambda_{i_n} < r \le \lambda_{i_1} \cdots \lambda_{i_{n-1}}. \qquad (3.4)$$

We can easily verify that for each fixed r the sets

$$\Delta(\omega, r) = \Delta_{i_1 \cdots i_{n(\omega, r)}} \tag{3.5}$$

are pairwise disjoint, that is,

$$\text{if} \quad \Delta(\omega, r) \cap \Delta(\omega', r) \neq \varnothing \quad \text{then} \quad \Delta(\omega, r) = \Delta(\omega', r). \tag{3.6}$$

Furthermore, they form a cover of F. Notice that for each $\omega \in \{1, \ldots, p\}^{\mathbb{N}}$ we have

$$r/c \leq \operatorname{diam} \Delta(\omega, r) < r, \tag{3.7}$$

where $c = 1/\min_{k=1,\ldots,p} \lambda_k$. This implies that for any interval I of length r there is at most a number c of sets $\Delta(\omega, r)$ that intersect I. Therefore,

$$\mu(I) \leq \sum_{\Delta(\omega, r) \cap B \neq \varnothing} \mu(\Delta(\omega, r)) < \sum_{\Delta(\omega, r) \cap B \neq \varnothing} r^s \leq cr^s.$$

Since each set $U \subset F$ is contained in an interval of length $\operatorname{diam} U$, we have

$$\mu(U) \leq c(\operatorname{diam} U)^s.$$

Therefore, for any countable cover $\mathcal{U}$ of F we have

$$1 = \mu(F) = \sum_{U \in \mathcal{U}} \mu(U) \leq c \sum_{U \in \mathcal{U}} (\operatorname{diam} U)^s,$$

and it follows from (2.1) that $m(F, s) \geq 1/c$.

It remains to establish an upper bound for the upper box dimension. Given $r \in (0, 1)$, we consider again the Moran cover of F formed by the sets $\Delta(\omega, r)$. Let $\tilde{\Delta}_1, \ldots, \tilde{\Delta}_{N(r)}$ be these sets (it follows from (3.6), (3.7), and the compactness of F that $N(r) < \infty$). By (3.7) we have $\operatorname{diam} \tilde{\Delta}_i < r$ for $i = 1, \ldots, N(r)$. Furthermore, since

$$(\operatorname{diam} \Delta_{i_1 \cdots i_m})^s = \sum_{i_{m+1} \cdots i_n} (\operatorname{diam} \Delta_{i_1 \cdots i_n})^s,$$

it follows from (3.3) that

$$\sum_{i=1}^{N(r)} (\operatorname{diam} \tilde{\Delta}_i)^s = 1. \tag{3.8}$$

On the other hand, again by (3.7), we have $\operatorname{diam} \tilde{\Delta}_i \geq r/c$ for $i = 1, \ldots, N(r)$. Since the sets $\tilde{\Delta}_i$ are pairwise disjoint, it follows from (3.8) that $N(r) \leq (c/r)^s$. Therefore,

$$N(F, r) \leq N(r) \leq (c/r)^s,$$

and

$$\overline{\dim}_B F \leq \limsup_{r \to 0} \frac{s \log(c/r)}{-\log r} = s.$$

This completes the proof of the theorem. $\qquad \square$

It is remarkable that the Hausdorff and box dimensions of the limit set F do not depend on the location of the intervals $\Delta_{i_1 \cdots i_n}$ but only on their length.

Now we introduce the general notion of geometric construction modeled by an arbitrary symbolic dynamics. Given an integer $p \in \mathbb{N}$, we consider the space of sequences $\Sigma_p^+ = \{1, \ldots, p\}^{\mathbb{N}}$ and we equip it with the distance

$$d(\omega, \omega') = \sum_{k=1}^{\infty} e^{-k} |i_k - i_k'|, \tag{3.9}$$

where $\omega = (i_1 i_2 \cdots)$ and $\omega' = (i_1' i_2' \cdots)$. With this distance Σ_p^+ becomes a compact metric space. We also consider the *shift map* $\sigma \colon \Sigma_p^+ \to \Sigma_p^+$ defined by $\sigma(i_1 i_2 \cdots) = (i_2 i_3 \cdots)$.

Definition 3.1.2. A *geometric construction* in $\mathbb{R}^m$ consists of:

1. a compact subset $\Sigma \subset \Sigma_p^+$ such that $\sigma^{-1}\Sigma \supset \Sigma$, for some $p \in \mathbb{N}$;

2. a decreasing sequence of compact sets $\Delta_{i_1 \cdots i_n} \subset \mathbb{R}^m$ for each $\omega \in \Sigma$, with diameter $\operatorname{diam} \Delta_{i_1 \cdots i_n} \searrow 0$ as $n \to \infty$, and such that whenever $(i_1 \cdots i_n) \neq (j_1 \cdots j_n)$ we have

$$\operatorname{int} \Delta_{i_1 \cdots i_n} \cap \operatorname{int} \Delta_{j_1 \cdots j_n} \neq \varnothing. \tag{3.10}$$

We say that the geometric construction is *modeled* by Σ.

For example, to model repellers and hyperbolic sets, we can consider geometric constructions modeled by topological Markov chains (see Chapter 4 for details).

Definition 3.1.3. The *limit set* of a geometric construction modeled by Σ is the compact set F defined by (3.1), with the union taken over all $i_1, \ldots, i_n \in \{1, \ldots, p\}$ such that $(i_1 \cdots i_n) = (j_1 \cdots j_n)$ for some sequence $(j_1 j_2 \cdots) \in \Sigma$.

We point out that to determine or even to estimate the dimension of the limit set F, sometimes it is not sufficient to know the geometric shape of the sets $\Delta_{i_1 \cdots i_n}$, in strong contrast with what happens in Theorem 3.1.1. For example, the dimension can be affected by certain number-theoretical properties. We shall not make any detailed discussion but we give an example. Namely, consider a geometric construction in $\mathbb{R}^2$ modeled by Σ_2^+ such that the sets

$$\Delta_{i_1 \cdots i_n} = (f_{i_1} \circ \cdots \circ f_{i_n})([0, 1] \times [0, 1])$$

are rectangles with sides of length a^n and b^n, obtained from the composition of the functions

$$f_1(x, y) = (ax, by) \quad \text{and} \quad f_2(x, y) = (ax - a + 1, by - b + 1),$$

for some fixed constants $a \in (0,1)$ and $b \in (0,1/2)$ (see Figure 3.2). In particular, the projection of $\Delta_{i_1 \cdots i_n}$ on the horizontal axis is an interval with right endpoint given by

$$a^n + \sum_{k=0}^{n-1} j_k a^k, \tag{3.11}$$

where

$$j_k = \begin{cases} 0 & \text{if } i_k = 1, \\ 1-a & \text{if } i_k = 2. \end{cases}$$

Now we assume that $a = (\sqrt{5}-1)/2$. In this case we have $a^2 + a = 1$, and thus for each $n > 2$ there is more than one vector $(i_1 \cdots i_n)$ for which we obtain the same value in (3.11). This causes a larger concentration of the sets $\Delta_{i_1 \cdots i_n}$ in certain regions of the limit set F. Therefore, to compute the Hausdorff dimension, when we take an open cover of F it may happen that it is possible to replace, in the regions of larger concentration of the sets $\Delta_{i_1 \cdots i_n}$, several elements of the cover by a single element. This may cause F to have a smaller Hausdorff dimension than expected, with respect to a certain "generic" value obtained by Falconer in [51]. It was established by Neunhäuserer in [104] that indeed this happens when $a = (\sqrt{5}-1)/2$. See also [124] for former related results of Przytycki and Urbański. We note that the constant $(\sqrt{5}-1)/2$ is only an example among many other possible values that lead to a similar phenomenon. Additional complications can occur when f_1 and f_2 are replaced by functions that are not affine.

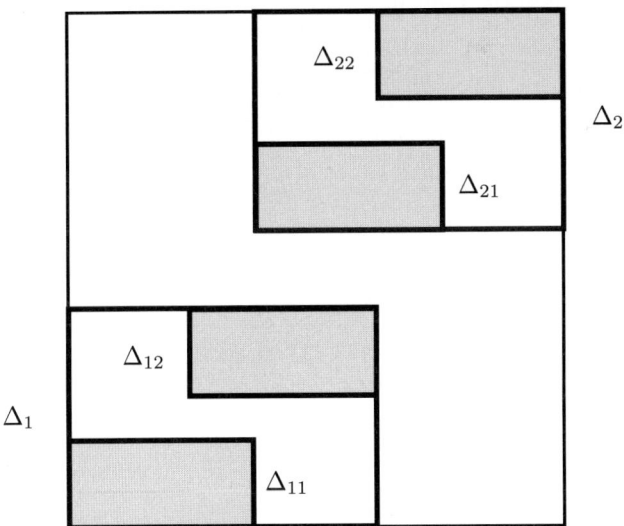

Figure 3.2: Number-theoretical properties and dimension theory

3.2   Thermodynamic formalism and dimension theory

In [118] Pesin and Weiss extended the result of Moran in Theorem 3.1.1 to an arbitrary symbolic dynamics, with the help of the thermodynamic formalism. Before presenting their result we consider the thermodynamic formalism in the particular case of symbolic dynamics. We also introduce the notion of Gibbs measure and we illustrate the relation of dimension theory with the thermodynamic formalism, taking advantage of the simplification given by symbolic dynamics.

3.2.1   Thermodynamic formalism for symbolic dynamics

One can easily show that in the case of symbolic dynamics the limit in Definition 2.3.1 when $\varepsilon \to 0$ is not needed.

Proposition 3.2.1. *Given a compact set $\Sigma \subset \Sigma_p^+$ such that $\sigma^{-1}\Sigma \supset \Sigma$ and a continuous function $\varphi \colon \Sigma \to \mathbb{R}$, the topological pressure of φ (with respect to σ) is given by*

$$P(\varphi) = P_\Sigma(\varphi) = \lim_{n \to \infty} \frac{1}{n} \log \sum_{i_1 \cdots i_n} \exp \sup_{C_{i_1 \cdots i_n}} \left(\sum_{k=0}^{n-1} \varphi \circ \sigma^k \right), \qquad (3.12)$$

where

$$C_{i_1 \cdots i_n} = \{(j_1 j_2 \cdots) \in \Sigma : (j_1 \cdots j_n) = (i_1 \cdots i_n)\}. \qquad (3.13)$$

The existence of the limit in (3.12) follows from the fact that the sequence

$$a_n = \log \sum_{i_1 \cdots i_n} \exp \sup \left(\sum_{k=0}^{n-1} \varphi \circ \sigma^k \right)$$

is subadditive. This means that $a_{n+m} \leq a_n + a_m$ for every $m, n \in \mathbb{N}$. One can easily show that it is possible to replace the supremum in (3.12) by the infimum of the same expression without changing the value of the limit. In particular, one can take any value for the second sum in (3.12) between the infimum and the supremum of $\sum_{k=0}^{n-1} \varphi \circ \sigma^k$ in $C_{i_1 \cdots i_n}$.

Definition 3.2.2. Each set $C_{i_1 \cdots i_n}$ in (3.13) is called a *cylinder set* of length n, and we denote its length by $|C_{i_1 \cdots i_n}| = n$.

By Definition 2.3.2, the topological entropy of $\sigma|\Sigma$ is given by $h(\sigma|\Sigma) = P(0)$, that is,

$$h(\sigma|\Sigma) = \lim_{n \to \infty} \frac{1}{n} \log N(\Sigma, n),$$

where $N(\Sigma, n)$ is the number of vectors $(i_1 \cdots i_n)$ such that $(i_1 \cdots i_n) = (j_1 \cdots j_n)$ for some sequence $(j_1 j_2 \cdots) \in \Sigma$.

We also need the notion of Gibbs measure.

Definition 3.2.3. We say that a σ-invariant probability measure μ in Σ is a *Gibbs measure* for the function $\varphi \colon \Sigma \to \mathbb{R}$ if there exist constants D_1, $D_2 > 0$ such that

$$D_1 \le \frac{\mu(C_{i_1 \cdots i_n})}{\exp\left(-nP(\varphi) + \sum_{k=0}^{n-1} \varphi(\sigma^k \omega)\right)} \le D_2 \tag{3.14}$$

for every $n \in \mathbb{N}$ and $\omega \in C_{i_1 \cdots i_n}$.

It is easy to show that any Gibbs measure is an equilibrium measure (see Definition 2.3.4). The converse does not hold but, for example, for a topologically mixing topological Markov chain (see Section 4.1 for the definitions), if φ is Hölder continuous, then its unique equilibrium measure is a Gibbs measure.

To illustrate the relation between dimension theory and the thermodynamic formalism we consider numbers $\lambda_1, \dots, \lambda_p \in (0,1)$ and we define the function $\varphi \colon \Sigma \to \mathbb{R}$ by

$$\varphi(i_1 i_2 \cdots) = \log \lambda_{i_1}. \tag{3.15}$$

We have

$$
\begin{aligned}
P(s\varphi) &= \lim_{n \to \infty} \frac{1}{n} \log \sum_{i_1 \cdots i_n} \exp\left(s \sum_{k=1}^{n} \log \lambda_{i_k}\right) \\
&= \lim_{n \to \infty} \frac{1}{n} \log \sum_{i_1 \cdots i_n} \prod_{k=1}^{n} \lambda_{i_k}{}^{s} \\
&= \lim_{n \to \infty} \frac{1}{n} \log \left(\sum_{i=1}^{p} \lambda_i{}^{s}\right)^{n} \\
&= \log \sum_{i=1}^{p} \lambda_i{}^{s}.
\end{aligned}
\tag{3.16}
$$

Therefore, equation (3.2) is equivalent to the new equation

$$P(s\varphi) = 0 \tag{3.17}$$

involving the topological pressure. Equation (3.17) was introduced by Bowen in [40] (in his study of quasi-circles) and is usually called *Bowen's equation*. It is also appropriate to call it *Bowen–Ruelle's equation*, taking into account the fundamental role of the thermodynamic formalism developed by Ruelle, and of his article [134]. Equation (3.17) establishes the connection between the thermodynamic formalism and dimension theory. Taking into consideration that the topological pressure and the Hausdorff dimension may be defined in a similar manner (see (2.1)–(2.2) and Section 7.2), namely, both can be obtained as Carathéodory dimension characteristics (see [115] for details)—the relation between the two is in fact very natural. It happens that equation (3.17) has a rather universal character: indeed, virtually all known equations used to compute or to estimate the dimension of invariant sets of dynamical systems are particular cases of this equation or of appropriate generalizations.

Before proceeding with the extension of the result of Moran to arbitrary symbolic dynamics, we present an equivalent description of the topological pressure. Given $\alpha \in \mathbb{R}$ and $N \in \mathbb{N}$, we set

$$M(\alpha, N) = \inf_{\mathcal{C}} \sum_{C \in \mathcal{C}} \exp\left(-\alpha |C| + \sup_{C} \sum_{k=0}^{|C|-1} \varphi \circ \sigma^k\right) \qquad (3.18)$$

where the infimum is taken over all finite covers $\mathcal{C}$ of Σ by cylinder sets in the collection

$$\mathcal{H}_N = \{C_{i_1 \cdots i_n} : n \geq N \text{ and } (i_1 i_2 \cdots) \in \Sigma\}. \qquad (3.19)$$

Theorem 3.2.4. *Given a compact set $\Sigma \subset \Sigma_p^+$ such that $\sigma^{-1}\Sigma \supset \Sigma$ and a continuous function $\varphi \colon \Sigma \to \mathbb{R}$ we have*

$$P(\varphi) = \inf\left\{\alpha : \lim_{N \to \infty} M(\alpha, N) = 0\right\}.$$

Proof. Clearly

$$M(\alpha, N) \leq \sum_{i_1 \cdots i_N} \exp\left(-\alpha N + \sup_{C_{i_1 \cdots i_N}} \sum_{k=0}^{N-1} \varphi \circ \sigma^k\right). \qquad (3.20)$$

On the other hand, given $\varepsilon > 0$ there exists $C > 0$ such that for every $N \in \mathbb{N}$,

$$\sum_{i_1 \cdots i_N} \exp\left(\sup_{C_{i_1 \cdots i_N}} \sum_{k=0}^{N-1} \varphi \circ \sigma^k\right) \leq Ce^{(P(\varphi)+\varepsilon)N}.$$

Therefore, provided that $\alpha > P(\varphi)$ and ε is sufficiently small, it follows from (3.20) that

$$M(\alpha, N) \leq e^{-\alpha N} Ce^{(P(\varphi)+\varepsilon)N} \to 0$$

as $N \to \infty$. Hence,

$$p := \inf\left\{\alpha : \lim_{N \to \infty} M(\alpha, N) = 0\right\} \leq P(\varphi).$$

Now let $\alpha > p$. There exist $N \in \mathbb{N}$ and a finite cover $\mathcal{C} \subset \mathcal{H}_N$ of Σ by cylinder sets $\tilde{C}_1, \ldots, \tilde{C}_q$ such that

$$N(\alpha, \mathcal{C}) := \sum_{C \in \mathcal{C}} \exp\left(-\alpha |C| + \sup_{C} \sum_{k=0}^{|C|-1} \varphi \circ \sigma^k\right) < 1. \qquad (3.21)$$

Let $I_1, \ldots, I_q$ be the finite sequences such that

$$\tilde{C}_i = C_{I_i} \quad \text{for} \quad i = 1, \ldots, q. \qquad (3.22)$$

For each $n \in \mathbb{N}$ we consider the cover $\mathcal{C}_n$ of Σ formed by the cylinder sets $C_{I_{i_1} \cdots I_{i_n}}$ with $i_1, \ldots, i_n \in \{1, \ldots, q\}$, and we write $\Gamma_{i_1 \cdots i_n} = C_{I_{i_1} \cdots I_{i_n}}$. Since

$$\sup_{\Gamma_{i_1 \cdots i_n}} \sum_{k=0}^{|\Gamma_{i_1 \cdots i_n}|-1} \varphi \circ \sigma^k \leq \sum_{j=1}^{n} \sup_{\Gamma_{i_j \cdots i_n}} \sum_{k=0}^{|\Gamma_{i_j}|-1} \varphi \circ \sigma^k$$

$$\leq \sum_{j=1}^{n} \sup_{\Gamma_{i_j}} \sum_{k=0}^{|\Gamma_{i_j}|-1} \varphi \circ \sigma^k,$$

it follows from (3.22) that

$$N(\alpha, \mathcal{C}_n) \leq \prod_{k=1}^{n} \sum_{C \in \mathcal{C}} \exp\left(-\alpha |C| + \sup_C \sum_{j=0}^{|C|-1} \varphi \circ \sigma^j\right) \leq N(\alpha, \mathcal{C})^n.$$

Therefore, by (3.21),

$$N(\alpha, \mathcal{C}_\infty) \leq \sum_{n \in \mathbb{N}} N(\alpha, \mathcal{C})^n < \infty,$$

where $\mathcal{C}_\infty = \bigcup_{n \in \mathbb{N}} \mathcal{C}_n$. Now let M be the maximal length of the elements of $\mathcal{C}$ (recall that $\mathcal{C}$ is a finite cover). For each $n \in \mathbb{N}$ and $\omega \in \Sigma$, there exists a cylinder set $C_I \in \mathcal{C}_\infty$ such that

$$\omega \in C_I \quad \text{and} \quad n \leq |C_I| < n + M.$$

Let $\mathcal{C}^*$ be the collection of all cylinder sets C_J such that J is a finite sequence consisting of the first n elements of some finite sequence I such that $C_I \in \mathcal{C}_\infty$. We have

$$N(\alpha, \mathcal{C}^*) \leq N(\alpha, \mathcal{C}_\infty) e^{M \sup |\varphi|} \max\{1, e^{\alpha M}\} < \infty.$$

Since

$$N(\alpha, \mathcal{C}^*) = \sum_{i_1 \cdots i_n} e^{-\alpha n} \exp\left(\sup_{C_{i_1 \cdots i_n}} \sum_{k=0}^{n-1} \varphi \circ \sigma^k\right),$$

we obtain $P(\varphi) - \alpha \leq 0$, and thus $p \geq P(\varphi)$. $\qquad \square$

3.2.2 Dimension of limit sets of geometric constructions

We present in this section the result of Pesin and Weiss in [118] that extends Theorem 3.1.1 to arbitrary symbolic dynamics. We consider the general case of geometric constructions whose sets $\Delta_{i_1 \cdots i_n}$ are balls (see Figure 3.3). The value of the dimension is again given by Bowen's equation in (3.17).

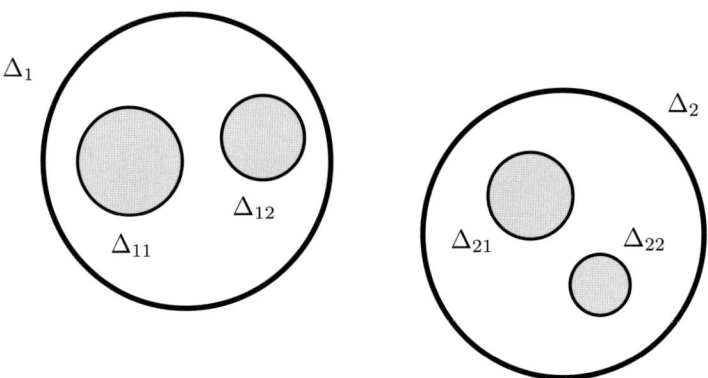

Figure 3.3: Geometric construction with balls

Theorem 3.2.5 (Dimension of the limit set). *For a geometric construction modeled by $\Sigma \subset \Sigma_p^+$ such that the sets $\Delta_{i_1 \cdots i_n} \subset \mathbb{R}^m$ are balls of diameter $\prod_{k=1}^{n} \lambda_{i_k}$, for some numbers $\lambda_1, \ldots, \lambda_p \in (0,1)$, we have*

$$\dim_H F = \underline{\dim}_B F = \overline{\dim}_B F = s, \tag{3.23}$$

where s is the unique real number satisfying $P(s\varphi) = 0$ with the function $\varphi \colon \Sigma \to \mathbb{R}$ as in (3.15).

Proof. Given $t \geq s$ we have

$$s \sum_{k=0}^{n-1} \varphi \circ \sigma^k \leq t \sum_{k=0}^{n-1} \varphi \circ \sigma^k + (s-t) n \inf \varphi,$$

and thus,

$$P(s\varphi) \leq P(t\varphi) + (s-t) \inf \varphi. \tag{3.24}$$

On the other hand,

$$t \sum_{k=0}^{n-1} \varphi \circ \sigma^k \leq s \sum_{k=0}^{n-1} \varphi \circ \sigma^k + (t-s) n \sup \varphi,$$

and thus,

$$P(t\varphi) \leq P(s\varphi) + (s-t) \sup \varphi. \tag{3.25}$$

Putting together the inequalities in (3.24) and (3.25) we obtain

$$(s-t) \sup \varphi \leq P(s\varphi) - P(t\varphi) \leq (s-t) \inf \varphi. \tag{3.26}$$

Since $\varphi < 0$ this implies that the function $s \mapsto P(s\varphi)$ is strictly decreasing. It also follows from (3.26) that

$$\lim_{s \to -\infty} P(s\varphi) = +\infty \quad \text{and} \quad \lim_{t \to +\infty} P(t\varphi) = -\infty.$$

Therefore, there exists a unique real number s satisfying $P(s\varphi) = 0$.

Now we establish (3.23). The argument can be considered an elaboration of the proof of Theorem 3.1.1, although now using the thermodynamic formalism and for an arbitrary symbolic dynamics. We start with the construction of a Moran cover. Given $\omega = (i_1 i_2 \cdots) \in \Sigma$ and $r \in (0,1)$, let $n = n(\omega, r)$ be the unique integer such that (3.4) holds. We also consider the sets $\Delta(\omega, r)$ in (3.5). By (3.7) we have

$$\operatorname{diam} \Delta(\omega, r) \geq r/c, \tag{3.27}$$

where $c = 1/\min_{k=1,\dots,p} \lambda_k$. On the other hand, by (3.10) the sets $\operatorname{int} \Delta(\omega, r)$ are pairwise disjoint. Therefore, by (3.27) and elementary geometry, there exists a constant $C > 0$ (independent of r) such that for each $x \in \mathbb{R}^m$ the ball $B(x, r)$ intersects at most a number C of sets $\Delta(\omega, r)$. This is a crucial property of the Moran cover.

Given $\varepsilon, r, \theta > 0$ there exists a countable cover $\mathcal{U}$ of the limit set F with $\operatorname{diam} \mathcal{U} < r$ such that

$$\sum_{U \in \mathcal{U}} (\operatorname{diam} U)^{\dim_H F + \varepsilon} < \theta. \tag{3.28}$$

For each $U \in \mathcal{U}$ we consider the family

$$\mathcal{U}_U = \big\{ \Delta(\omega, \operatorname{diam} U) : \Delta(\omega, \operatorname{diam} U) \cap U \neq \varnothing \text{ and } \omega \in \Sigma \big\},$$

which is a cover of U. Then

$$\mathcal{V} = \{ V : V \in \mathcal{U}_U \text{ and } U \in \mathcal{U} \} \tag{3.29}$$

is a cover of F of diameter at most r, and by (3.7) and (3.28) we obtain

$$\begin{aligned} \sum_{V \in \mathcal{V}} (\operatorname{diam} V)^{\dim_H F + \varepsilon} &\leq \sum_{U \in \mathcal{U}} \sum_{V \in \mathcal{U}_U} (\operatorname{diam} V)^{\dim_H F + \varepsilon} \\ &\leq \sum_{U \in \mathcal{U}} \sum_{V \in \mathcal{U}_U} (\operatorname{diam} U)^{\dim_H F + \varepsilon} \\ &\leq \sum_{U \in \mathcal{U}} C(\operatorname{diam} U)^{\dim_H F + \varepsilon} < C\theta. \end{aligned} \tag{3.30}$$

Since r and θ can be made arbitrarily small, using the notation in (3.18) we obtain

$$\lim_{N \to \infty} M(0, N) = 0,$$

and by Theorem 3.2.4,

$$P((\dim_H F + \varepsilon)\varphi) \geq 0.$$

Since $t \mapsto P(t\varphi)$ is strictly decreasing we obtain $\dim_H F + \varepsilon \geq s$, and it follows from the arbitrariness of ε that $\dim_H F \geq s$.

To estimate the upper box dimension we consider again the Moran cover constructed above. Let $\tilde{\Delta}_j = \Delta(\omega_j, r)$ be the elements of the Moran cover for some sequences $\omega_j \in \Sigma$. Since

$$\operatorname{diam} \Delta_{i_1 \cdots i_n} = \prod_{k=1}^{n} \lambda_{i_k}, \tag{3.31}$$

we have

$$n(\omega, r) \leq -\frac{\log r}{\log c} + 1, \tag{3.32}$$

where $c = \max_{k=1,\ldots,p} \lambda_k$. Since $\operatorname{diam} \tilde{\Delta}_j < r$ for each j, we have

$$\sum_{m \in \mathbb{N}} \operatorname{card}\{j : n(\omega_j, r) = m\} \geq N(F, r). \tag{3.33}$$

Hence, there exists $m = m(r)$ such that

$$\operatorname{card}\{j : n(\omega_j, r) = m\} \geq \frac{N(F, r)}{-\log r / \log c + 1}. \tag{3.34}$$

On the other hand, for each $\theta > 0$ there exists a sequence $r_n \searrow 0$ such that

$$N(F, r_n) > r_n^{\theta - \overline{\dim}_B F} \tag{3.35}$$

for all sufficiently large $n \in \mathbb{N}$. Set $m_n = m(r_n) - 1$. For each $\alpha < \overline{\dim}_B F - 2\theta$ we obtain

$$p \sum_{i_1 \cdots i_{m_n}} (\operatorname{diam} \Delta_{i_1 \cdots i_{m_n}})^\alpha \geq \sum_{i_1 \cdots i_{m(r_n)}} (\operatorname{diam} \Delta_{i_1 \cdots i_{m_n}})^\alpha$$

$$\geq r_n^\alpha \frac{N(F, r_n)}{-\log r_n / \log c + 1} \tag{3.36}$$

$$\geq \frac{r_n^{\alpha + \theta - \overline{\dim}_B F}}{-\log r_n / \log c + 1} \geq 1$$

for all sufficiently large $n \in \mathbb{N}$ (we note that $1 / - \log r \geq r^\theta$ for all sufficiently small $r > 0$). Furthermore, again by (3.31) we have

$$n(\omega, r) \geq -\frac{\log r}{\log c'}, $$

where $c' = \min_{k=1,\ldots,p} \lambda_k$, and hence $\inf_{\omega \in \Sigma} n(\omega, r) \to \infty$ as $r \to 0$. This implies that

$$\min_j n(\omega_j, r) \to \infty \quad \text{as} \quad r \to 0,$$

and hence $m_n \to \infty$. Therefore, by (3.36) we obtain

$$P(\alpha\varphi) = \lim_{n\to\infty} \frac{1}{n} \log \sum_{i_1 \cdots i_n} (\operatorname{diam} \Delta_{i_1 \cdots i_n})^\alpha \geq 0.$$

We conclude that $\alpha \leq s$, and hence

$$\overline{\dim}_B F - 2\theta \leq s.$$

It follows from the arbitrariness of θ that $\overline{\dim}_B F \leq s$.   □

3.3   Nonstationary geometric constructions

It turns out that the classical topological pressure is not adapted well to all geometric constructions. For example, consider again a geometric construction such that all sets $\Delta_{i_1 \cdots i_n}$ are balls, although now without any type of multiplicativity assumption for the diameters. On the contrary, in Theorems 3.1.1 and 3.2.5 we have

$$\operatorname{diam} \Delta_{i_1 \cdots i_n} = \prod_{k=1}^{n} \lambda_{i_k} \qquad (3.37)$$

for every set $\Delta_{i_1 \cdots i_n}$. We call the geometric constructions for which (3.37) fails *nonstationary geometric constructions*. It happens that the thermodynamic formalism is of no help in this situation. However, the nontrivial generalization given by the so-called *nonadditive thermodynamic formalism* can still be used with success. The main idea is to replace the sequence of functions

$$\varphi_n = \sum_{k=0}^{n-1} \varphi \circ \sigma^k \qquad (3.38)$$

in (3.12) by an arbitrary sequence ψ_n. We note that while the functions φ_n satisfy the identity

$$\varphi_{n+m} = \varphi_n + \varphi_m \circ \sigma^n,$$

the functions ψ_n may have no similar property, and thus the expression *nonadditive*. Due to technical problems related with the existence of the limit in (3.12) when the sequence φ_n in (3.38) is replaced by an arbitrary sequence ψ_n, Barreira used a different approach in [3] to introduce the notion of nonadditive topological pressure. It is based on the theory of Carathéodory dimension characteristics developed by Pesin (see [115] for references and full details).

3.3.1   Nonadditive thermodynamic formalism

We first introduce the notion of nonadditive topological pressure. Let (X, d) be a compact metric space and let $f\colon X \to X$ be a continuous transformation. Given a finite open cover $\mathcal{U}$ of X, we denote by $\mathcal{W}_n(\mathcal{U})$ the collection of vectors $\mathbf{U} = (U_0, \ldots, U_n)$ with $U_0, \ldots, U_n \in \mathcal{U}$, and we write $m(\mathbf{U}) = n$. For each $\mathbf{U} \in \mathcal{W}_n(\mathcal{U})$ we define the open set

$$X(\mathbf{U}) = \bigcap_{k=0}^{n} f^{-k} U_k.$$

Now we consider a sequence of continuous functions $\Phi = \{\varphi_n \colon X \to \mathbb{R}\}_{n \in \mathbb{N}}$. For each $n \in \mathbb{N}$ we define

$$\gamma_n(\Phi, \mathcal{U}) = \sup \left\{ |\varphi_n(x) - \varphi_n(y)| : x, y \in X(\mathbf{U}) \text{ for some } \mathbf{U} \in \mathcal{W}_n(\mathcal{U}) \right\}.$$

We always assume that

$$\limsup_{\operatorname{diam} \mathcal{U} \to 0} \; \limsup_{n \to \infty} \frac{\gamma_n(\Phi, \mathcal{U})}{n} = 0 \qquad (3.39)$$

(we note that since X is compact it has finite covers with diameter as small as desired).

Example 3.3.1. In the additive case, that is, when the sequence Φ is composed of continuous functions obtained from a given function φ as in (3.38), condition (3.39) is automatically satisfied. More precisely, given a continuous function $\varphi \colon X \to \mathbb{R}$ we consider the sequence of functions $\varphi_n = \sum_{k=0}^{n-1} \varphi \circ f^k$. We have

$$|\varphi_n(x) - \varphi_n(y)| \le \sum_{k=0}^{n-1} |\varphi(f^k x) - \varphi(f^k y)|$$

and hence,

$$\frac{\gamma_n(\Phi, \mathcal{U})}{n} \le \frac{1}{n} \sum_{k=0}^{n-1} \sup \left\{ |\varphi(x) - \varphi(y)| : x, y \in f^k(X(\mathbf{U})) \text{ for } \mathbf{U} \in \mathcal{W}_n(\mathcal{U}) \right\} \qquad (3.40)$$

$$\le \sup \{ |\varphi(x) - \varphi(y)| : x, y \in U \text{ for some } U \in \mathcal{U} \},$$

since $f^k(X(\mathbf{U})) \subset U_{k+1}$ for $k = 0, \ldots, n-1$. By the uniform continuity of φ in X, the last supremum goes to 0 as $\operatorname{diam} \mathcal{U} \to 0$. Hence, in the additive case condition (3.39) is automatically satisfied.

Given $\mathbf{U} \in \mathcal{W}_n(\mathcal{U})$, we set

$$\varphi(\mathbf{U}) = \begin{cases} \sup_{X(\mathbf{U})} \varphi_n & \text{if } X(\mathbf{U}) \ne \varnothing, \\ -\infty & \text{otherwise.} \end{cases}$$

Given $Z \subset X$ and $\alpha \in \mathbb{R}$, we define the function

$$M(Z, \alpha, \Phi, \mathcal{U}) = \lim_{n \to \infty} \inf_{\Gamma} \sum_{\mathbf{U} \in \Gamma} \exp \left(-\alpha m(\mathbf{U}) + \varphi(\mathbf{U}) \right), \qquad (3.41)$$

where the infimum is taken over all finite or countable collections $\Gamma \subset \bigcup_{k \geq n} \mathcal{W}_k(\mathcal{U})$ such that $\bigcup_{U \in \Gamma} X(\mathbf{U}) \supset Z$. We also set

$$P_Z(\Phi, \mathcal{U}) = \inf\{\alpha \in \mathbb{R} : M(Z, \alpha, \Phi, \mathcal{U}) = 0\}.$$

The following result was established by Barreira in [3].

Theorem 3.3.2 (Nonadditive topological pressure). *The following properties hold:*

1. *there exists the limit*

$$P_Z(\Phi) := \lim_{\mathrm{diam}\, \mathcal{U} \to 0} P_Z(\Phi, \mathcal{U}); \tag{3.42}$$

2. *if there exist constants c_1, $c_2 < 0$ such that $c_1 n \leq \varphi_n \leq c_2 n$ for every $n \in \mathbb{N}$, and the topological entropy $h(f|X)$ is finite, then there exists a unique number $s \in \mathbb{R}$ such that*

$$P_Z(s\Phi) = 0.$$

Proof. This is a simple modification of the proof of Proposition 2.8 in [38]. Let $\mathcal{V}$ be a finite open cover of X with diameter smaller than the Lebesgue number of $\mathcal{U}$. Then each element $V \in \mathcal{V}$ is contained in some element $U(V) \in \mathcal{U}$. We write

$$\mathbf{U}(\mathbf{V}) = (U(V_1), \dots, U(V_n))$$

for each $\mathbf{V} \in \mathcal{W}_n(\mathcal{V})$, and we observe that if $\Gamma \subset \bigcup_{k \in \mathbb{N}} \mathcal{W}_k(\mathcal{V})$ is a cover of Z, then

$$\{\mathbf{U}(\mathbf{V}) : \mathbf{V} \in \Gamma\} \subset \bigcup_{k \in \mathbb{N}} \mathcal{W}_k(\mathcal{U})$$

is also a cover of Z. Set

$$\gamma(\mathcal{U}) = \limsup_{n \to \infty} \frac{\gamma_n(\Phi, \mathcal{U})}{n}.$$

Given $\varepsilon > 0$, we have $\gamma_n(\Phi, \mathcal{U})/n \leq \gamma(\mathcal{U}) + \varepsilon$ for all sufficiently large n. Hence,

$$\varphi(\mathbf{U}(\mathbf{V})) \leq \varphi(\mathbf{V}) + n(\gamma(\mathcal{U}) + \varepsilon)$$

for each $\mathbf{V} \in \mathcal{W}_n(\mathcal{V})$, and

$$M(Z, \alpha, \Phi, \mathcal{U}) \leq M(Z, \alpha - \gamma(\mathcal{U}) - \varepsilon, \Phi, \mathcal{V}).$$

Therefore,

$$P_Z(\Phi, \mathcal{U}) \leq P_Z(\Phi, \mathcal{V}) + \gamma(\mathcal{U}) + \varepsilon,$$

and

$$P_Z(\Phi, \mathcal{U}) - \gamma(\mathcal{U}) - \varepsilon \leq \liminf_{\mathrm{diam}\, \mathcal{V} \to 0} P_Z(\Phi, \mathcal{V}).$$

By condition (3.39) we have that $\gamma(\mathcal{U}) \to 0$ as $\operatorname{diam} \mathcal{U} \to 0$. Since ε is arbitrary, we conclude that

$$\limsup_{\operatorname{diam} \mathcal{U} \to 0} P_Z(\Phi, \mathcal{U}) \leq \liminf_{\operatorname{diam} \mathcal{V} \to 0} P_Z(\Phi, \mathcal{V})$$

and thus the number $P_Z(\Phi)$ in (3.42) is well-defined.

For the second property we observe that by (3.41), for each $s \geq 0$ we have

$$M(Z, \alpha - sc_1, 0, \mathcal{U}) \leq M(Z, \alpha, s\Phi, \mathcal{U}) \leq M(Z, \alpha - sc_2, 0, \mathcal{U}),$$

and hence

$$P_Z(0) + sc_1 \leq P_Z(s\Phi) \leq P_Z(0) + sc_2. \tag{3.43}$$

Similarly, we can show that for each $s \leq 0$,

$$P_Z(0) + sc_2 \leq P_Z(s\Phi) \leq P_Z(0) + sc_1. \tag{3.44}$$

Since

$$P_Z(0) \leq P_X(0) = h(f|X) < \infty,$$

it follows from (3.43) and (3.44) that the number $P_Z(s\Phi)$ is finite for every s. On the other hand, given $t \geq s$ we have

$$c_1(t - s)n \leq (t - s)\varphi_n \leq c_2(t - s)n,$$

and hence,

$$M(Z, \alpha - (t - s)c_1, s\Phi, \mathcal{U}) \leq M(Z, \alpha, t\Phi, \mathcal{U}) \leq M(Z, \alpha - (t - s)c_2, s\Phi, \mathcal{U}).$$

Therefore,

$$P_Z(s\Phi, \mathcal{U}) + (t - s)c_1 \leq P_Z(t\Phi, \mathcal{U}) \leq P_Z(s\Phi, \mathcal{U}) + (t - s)c_2,$$

and

$$(t - s)c_1 \leq P_Z(t\Phi) - P_Z(s\Phi) \leq (t - s)c_2.$$

This shows that the function $s \mapsto P_Z(s\Phi)$ is Lipschitz and strictly decreasing, with

$$\lim_{t \to +\infty} P_Z(t\Phi) = -\infty \quad \text{and} \quad \lim_{s \to -\infty} P_Z(s\Phi) = +\infty.$$

In particular, there exists a unique $s \in \mathbb{R}$ such that $P_Z(s\Phi) = 0$. $\square$

Definition 3.3.3. The number $P_Z(\Phi)$ is called the *nonadditive topological pressure* of Φ in the set Z (with respect to f).

We note that Z need not be compact nor f-invariant. The nonadditive topological pressure is a generalization of the notion of topological pressure (see Definition 2.3.1), and contains as a particular case the subadditive version introduced by Falconer in [52]. In the additive case we recover the notion of topological pressure introduced by Pesin and Pitskel' in [116]. Moreover, the quantity $P_Z(0)$ coincides with the notion of topological entropy for noncompact sets introduced in [116], and can be shown to be equivalent to the notion of topological entropy introduced earlier by Bowen in [37] (see [115] for details).

The following statement follows easily from the definitions (see [3] for details).

Theorem 3.3.4. *The following properties hold:*

1. *if $Z_1 \subset Z_2$, then $P_{Z_1}(\Phi) \leq P_{Z_2}(\Phi)$;*

2. *if $Z = \bigcup_{i \in I} Z_i$ is a countable union of sets, then $P_Z(\Phi) = \sup_{i \in I} P_{Z_i}(\Phi)$;*

3. *if $\varphi_n \leq \psi_n$ for all sufficiently large n, then $P_Z(\Phi) \leq P_Z(\Psi)$.*

A variational principle for the nonadditive topological pressure is also established in [3], among other properties. We formulate it without proof.

Theorem 3.3.5 (Nonadditive variational principle). *If there exists a continuous function $\psi \colon X \to \mathbb{R}$ such that*

$$\varphi_{n+1} - \varphi_n \circ f \to \psi \quad uniformly,$$

then

$$P_X(\Phi) = \sup_{\mu \in \mathcal{M}} \left\{ h_\mu(f) + \int_X \psi \, d\mu \right\}.$$

3.3.2 Dimension of limit sets of nonstationary constructions

The following result gives a formula for the dimension of the limit sets of a class of *nonstationary* geometric constructions, in terms of the nonadditive topological pressure for the symbolic dynamics. It was obtained by Barreira in [3].

Theorem 3.3.6 (Dimension of the limit set). *Consider a geometric construction modeled by $\Sigma \subset \Sigma_p^+$ such that the sets $\Delta_{i_1 \cdots i_n}$ are balls of diameter $r_{i_1 \cdots i_n} < 1$. If there is a constant $\delta \in (0,1)$ such that*

$$r_{i_1 \cdots i_n} \geq \delta r_{i_1 \cdots i_{n-1}} \quad and \quad r_{i_1 \cdots i_{n+m}} \leq r_{i_1 \cdots i_n} r_{i_{n+1} \cdots i_m} \tag{3.45}$$

for every $(i_1 i_2 \cdots) \in \Sigma$ and $n, m \in \mathbb{N}$, then

$$\dim_H F = \underline{\dim}_B F = \overline{\dim}_B F = s, \tag{3.46}$$

where s is the unique real number satisfying the identity $P_\Sigma(s\Phi) = 0$, where Φ is the sequence of functions $\varphi_n \colon \Sigma \to \mathbb{R}$ defined by

$$\varphi_n(i_1 i_2 \cdots) = \log \operatorname{diam} \Delta_{i_1 \cdots i_n}.$$

Proof. The argument is an elaboration of the proof of Theorem 3.2.5. We first show that there exists the limit

$$Q(s) = \lim_{n\to\infty} \frac{1}{n} \log \sum_{i_1\cdots i_n} r_{i_1\cdots i_n}{}^s \tag{3.47}$$

for every $s \in \mathbb{R}$. It follows from the second inequality in (3.45) that

$$\sum_{i_1\cdots i_{n+m}} r_{i_1\cdots i_{n+m}}{}^s \le \sum_{i_1\cdots i_n} r_{i_1\cdots i_n}{}^s \sum_{i_{n+1}\cdots i_{n+m}} r_{i_{n+1}\cdots i_{n+m}}{}^s.$$

Therefore, the sequence

$$a_n = \log \sum_{i_1\cdots i_n} r_{i_1\cdots i_n}{}^s$$

is subadditive, that is,

$$a_{n+m} \le a_n + a_m \quad \text{for each} \quad n, m \in \mathbb{N},$$

and hence there exists the limit of a_n/n when $n \to \infty$. Furthermore, given $t \ge s$ we have

$$\lim_{n\to\infty} \frac{1}{n} \log \sum_{i_1\cdots i_n} r_{i_1\cdots i_n}{}^t \le \lim_{n\to\infty} \frac{1}{n} \log \sum_{i_1\cdots i_n} r_{i_1\cdots i_n}{}^s c^{n(t-s)},$$

where $c = \max_{k=1,\ldots,p} r_k < 1$. Hence,

$$Q(t) \le Q(s) + (t-s)\log c,$$

and Q is strictly decreasing. On the other hand, it follows from the first inequality in (3.45) that $r_{i_1\cdots i_n} \ge \theta^n$, where $\theta = \min\{\delta, r_1, \ldots, r_p\} < 1$. Therefore

$$Q(s) \ge \lim_{n\to\infty} \frac{1}{n} \sum_{i_1\cdots i_n} \theta^{sn} = \log p + s\log\theta,$$

and we obtain

$$\lim_{s\to-\infty} Q(s) = +\infty \quad \text{and} \quad \lim_{s\to+\infty} Q(s) = -\infty.$$

This ensures that there exists a unique real number s satisfying $Q(s) = 0$.

Given $t \ge 0$, $\alpha \in \mathbb{R}$, and a finite cover $\mathcal{C}$ of Σ by cylinder sets, we define

$$N_t(\alpha, \mathcal{C}) = \sum_{C_I \in \mathcal{C}} r_I{}^t e^{-\alpha|I|},$$

where $|I|$ is the length of the finite sequence I. We also set

$$p_t(\alpha) = \lim_{n\to\infty} \inf_{\mathcal{C}} N_t(\alpha, \mathcal{C}), \tag{3.48}$$

where the infimum is taken over all finite covers $\mathcal{C} \subset \mathcal{H}_m$ of Σ (see (3.19)). We can easily verify that

$$P(t) := P_\Sigma(t\Phi) = \inf\{\alpha \in \mathbb{R} : p_t(\alpha) = 0\}. \tag{3.49}$$

The following is a nonadditive version of Theorem 3.2.4.

Lemma 3.3.7. *We have $P(t) = Q(t)$ for every $t \in \mathbb{R}$.*

Proof of the lemma. If $\alpha > P(t)$, then for each sufficiently large $m \in \mathbb{N}$, there exists a finite cover $\mathcal{C} \subset \mathcal{H}_m$ of Σ such that $N_t(\alpha, \mathcal{C}) < 1$. For each $n \in \mathbb{N}$ we consider the (finite) cover $\mathcal{C}_n$ of Σ formed by the cylinder sets $C_{I_{i_1}\cdots I_{i_n}}$ such that $C_{I_j} \in \mathcal{C}$ for $j = 1, \ldots, n$. By the second inequality in (3.45) and (3.22) we obtain

$$N_t(\alpha, \mathcal{C}_n) \le \prod_{k=1}^{n} \sum_{C_{I_{i_k}} \in \mathcal{C}} r_{I_{i_k}}{}^t e^{-\alpha|I_{i_k}|} = N_t(\alpha, \mathcal{C})^n. \tag{3.50}$$

It follows from (3.50) that

$$N_t(\alpha, \mathcal{C}_\infty) \le \sum_{n \in \mathbb{N}} N_t(\alpha, \mathcal{C}_n) < \infty$$

where

$$\mathcal{C}_\infty = \bigcup_{n \in \mathbb{N}} \mathcal{C}_n.$$

Now let M be the maximal length of the finite sequences I such that $C_I \in \mathcal{C}$. As in the proof of Theorem 3.2.4, given $n \in \mathbb{N}$ let $\mathcal{C}^*$ be the collection of all cylinder sets C_J such that J is a finite sequence consisting of the first n elements of some finite sequence I such that $C_I \in \mathcal{C}_\infty$. We obtain $r_J \le r_I \delta^{-M}$ and

$$N_t(\alpha, \mathcal{C}^*) \le N_t(\alpha, \mathcal{C}_\infty)\delta^{-tM} \max\{1, e^{\alpha M}\}.$$

Therefore,

$$N_t(\alpha, \mathcal{C}^*) = \sum_{i_1 \cdots i_n} r_{i_1 \cdots i_n}{}^t e^{-\alpha n} < \infty,$$

and

$$\lim_{n \to \infty} \frac{1}{n} \log \sum_{i_1 \cdots i_n} r_{i_1 \cdots i_n}{}^t \le \alpha.$$

This shows that $P(t) \ge Q(t)$.

On the other hand, if $\mathcal{D}_n$ is the cover of Σ by the cylinder sets $C_{i_1 \cdots i_n}$, it follows from (3.48) that

$$p_t(\alpha) \le \liminf_{n \to \infty} N_t(\alpha, \mathcal{D}_n) = \liminf_{n \to \infty} \sum_{i_1 \cdots i_n} r_{i_1 \cdots i_n}{}^t e^{-\alpha n}.$$

By (3.47), for each $\varepsilon > 0$ there exists $C > 0$ such that

$$\sum_{i_1 \cdots i_n} r_{i_1 \cdots i_n}{}^t \leq C e^{n(Q(t)+\varepsilon)}.$$

Therefore, taking ε sufficiently small so that $\alpha > Q(t)+\varepsilon$ we obtain that $p_t(\alpha) = 0$ whenever $\alpha > Q(t)$. It follows from (3.49) that $P(t) \leq Q(t)$. This completes the proof of the lemma. $\qquad\square$

By Lemma 3.3.7, if s is the unique real number satisfying $Q(s) = 0$ then $P(s) = 0$.

Now we establish the identities in (3.46). We start by constructing a Moran cover. The construction is analogous to the one in the proof of Theorem 3.1.1. Namely, for each $\omega = (i_1 i_2 \cdots) \in \Sigma$ and $r \in (0,1)$, we consider the unique integer $n = n(\omega, r)$ such that

$$r_{i_1 \cdots i_n} < r \leq r_{i_1 \cdots i_{n-1}} \tag{3.51}$$

(we note that $r_{i_1 \cdots i_n}$ is strictly decreasing in n). Then the sets $\Delta(\omega, r)$ given by (3.5) are pairwise disjoint for each given r. We denote them by $\tilde{\Delta}_j = \Delta(\omega_j, r)$ for some sequences $\omega_j \in \Sigma$. Proceeding as in the proof of Theorem 3.2.5 (see (3.29)–(3.30)) we show that $p_{\dim_H F + \varepsilon}(0) = 0$ and hence,

$$P(\dim_H F + \varepsilon) \leq 0 = P(s).$$

Since the function $t \mapsto P(t)$ is decreasing, we have $\dim_H F + \varepsilon \geq s$ for every $\varepsilon > 0$, and thus $\dim_H F \geq s$.

Now we consider the box dimension. It follows from (3.51) that for $n = n(\omega, r)$ we have

$$r \leq r_{i_1 \cdots i_{n-1}} \leq c^{n-1} \quad \text{where} \quad c = \max_{k=1,\dots,p} r_k,$$

and hence (3.32) holds. Proceeding as in the proof of Theorem 3.2.5 we show that there exist $m = m(r)$ and for each $\theta > 0$ a sequence $r_n \searrow 0$ satisfying (3.34) and (3.35) for all sufficiently large $n \in \mathbb{N}$. Set $m_n = m(r_n) - 1$. For each $\alpha < \overline{\dim}_B F - 2\theta$ we obtain

$$p \sum_{i_1 \cdots i_{m_n}} r_{i_1 \cdots i_{m_n}}{}^\alpha \geq \sum_{i_1 \cdots i_{m(r_n)}} r_{i_1 \cdots i_{m_n}}{}^\alpha$$

$$\geq r_n{}^\alpha \frac{N(F, r_n)}{-\log r_n / \log c + 1}$$

$$\geq \frac{r_n{}^{\alpha + \theta - \overline{\dim}_B F}}{-\log r_n / \log c + 1} \geq 1$$

for all sufficiently large n. This shows that

$$Q(\alpha) \geq 0 \quad \text{for every} \quad \alpha < \overline{\dim}_B F - 2\theta,$$

and hence,

$$Q(\overline{\dim}_B F - 2\theta) \geq 0 = Q(s).$$

Since the function $t \mapsto Q(t)$ is decreasing this implies that $\overline{\dim}_B F - 2\theta \leq s$ for every $\theta > 0$, and hence $\overline{\dim}_B F \leq s$. This completes the proof of the theorem. □

The equation $P_\Sigma(s\Phi) = 0$ is a nonadditive version of Bowen's equation in (3.17). We observe that Theorem 3.3.6 contains the results of Moran in [101] and of Pesin and Weiss in [118]. In these two cases we have

$$r_{i_1 \cdots i_n} = \prod_{k=1}^{n} \lambda_{i_k}. \tag{3.52}$$

For example, when $\Sigma = \Sigma_p^+$ it follows from (3.52) that

$$\sum_{i_1 \cdots i_n} r_{i_1 \cdots i_n}{}^s = \sum_{i_1 \cdots i_n} \left(\prod_{k=1}^{n} \lambda_{i_k} \right)^s = \left(\sum_{i=1}^{p} \lambda_i{}^s \right)^n,$$

and thus the equation $P_\Sigma(s\Phi) = 0$ is equivalent to equation (3.2). The value of the dimensions in Theorem 3.3.6 is also independent of the location of the sets $\Delta_{i_1 \cdots i_n}$.

Chapter 4

Repellers and Hyperbolic Sets

We start in this chapter the study of the dimension of hyperbolic invariant sets of *conformal* dynamical systems (both invertible and noninvertible). As we observed in Section 3.1, one of the motivations for the study of geometric constructions is precisely the study of the dimension of invariant sets of hyperbolic dynamics. We show in this chapter that indeed a similar approach can be effected for repellers and hyperbolic sets of conformal maps, using Markov partitions and essentially following the arguments for geometric constructions in Chapter 3.

4.1 Dimension of repellers of conformal maps

We first consider the case of repellers, that is, invariant sets of a noninvertible expanding dynamics.

Let $f\colon M \to M$ be a differentiable map of a smooth manifold. We consider a compact f-invariant set $J \subset M$, that is, a compact set $J \subset M$ such that $f^{-1}J = J$.

Definition 4.1.1. We say that J is a *repeller* of f and that f is *expanding* on J if there exist constants $c > 0$ and $\beta > 1$ such that

$$\|d_x f^n v\| \geq c\beta^n \|v\| \tag{4.1}$$

for every $n \in \mathbb{N}$, $x \in J$, and $v \in T_x M$.

Let J be a repeller of the differentiable map f.

Definition 4.1.2. A finite cover of J by nonempty closed sets $R_1, \ldots, R_p$ is called a *Markov partition* of J (with respect to f) if:

1. $\overline{\operatorname{int} R_i} = R_i$ for each i;

2. $\operatorname{int} R_i \cap \operatorname{int} R_j = \varnothing$ whenever $i \neq j$;

3. $f(R_i) \supset R_j$ whenever $f(\operatorname{int} R_i) \cap \operatorname{int} R_j \neq \varnothing$.

We note that in Definition 4.1.2 the interior of each set R_i is computed with respect to the induced topology on J. Any repeller has Markov partitions with diameter

$$\operatorname{diam} \mathcal{R} := \max\{\operatorname{diam} R_i : i = 1, \ldots, p\}$$

as small as desired (see [134]).

Now we explain how Markov partitions can be used to model repellers by geometric constructions. Let R_1, ..., R_p be the elements of a Markov partition of a repeller J. We define a $p \times p$ matrix $A = (a_{ij})$ with entries

$$a_{ij} = \begin{cases} 1 & \text{if } f(\operatorname{int} R_i) \cap \operatorname{int} R_j \neq \varnothing, \\ 0 & \text{if } f(\operatorname{int} R_i) \cap \operatorname{int} R_j = \varnothing. \end{cases} \tag{4.2}$$

We also consider the space of sequences $\Sigma_p^+ = \{1, \ldots, p\}^{\mathbb{N}}$ and the shift map $\sigma \colon \Sigma_p^+ \to \Sigma_p^+$ (see Section 3.1 for the definition).

Definition 4.1.3. The restriction of σ to the set

$$\Sigma_A^+ = \{(i_1 i_2 \cdots) \in \Sigma_p^+ : a_{i_n i_{n+1}} = 1 \text{ for every } n \in \mathbb{N}\}$$

is called a *(one-sided) topological Markov chain* with *transition matrix A*.

We recall that a transformation f is *topologically mixing* on a set J if given open sets U and V with nonempty intersection with J there exists $n \in \mathbb{N}$ such that

$$f^m(U) \cap V \cap J \neq \varnothing \quad \text{for every} \quad m > n.$$

One can easily show that if f is topologically mixing on J, then there exists $k \in \mathbb{N}$ such that A^k has only positive entries (this means that after k steps one can make a transition between any two symbols in $\{1, \ldots, p\}$).

$$
\begin{array}{ccc}
\Sigma_A^+ & \xrightarrow{\ \sigma\ } & \Sigma_A^+ \\
\chi \downarrow & & \downarrow \chi \\
J & \xrightarrow{\ f\ } & J
\end{array}
$$

Figure 4.1: Symbolic coding of a repeller

Using (4.1), it is easy to show that one can define a *coding map* $\chi \colon \Sigma_A^+ \to J$ of the repeller J by

$$\chi(i_1 i_2 \cdots) = \bigcap_{k=0}^{\infty} f^{-k} R_{i_{k+1}}. \tag{4.3}$$

The map χ is surjective and satisfies

$$\chi \circ \sigma = f \circ \chi \tag{4.4}$$

(i.e., the diagram in Figure 4.1 is commutative). In addition, χ is Hölder continuous, with the distance in Σ_p^+ given by (3.9). In general the map χ is not invertible (although one can show that card $\chi^{-1}x \le p^2$ for every $x \in J$). Nevertheless, the identity in (4.4) still allows one to see χ as a dictionary that transfers the symbolic dynamics $\sigma|\Sigma_A^+$ and often the results at this level to the dynamics of f on J and its corresponding results. In particular, the coding map χ allows one to see each repeller as a geometric construction (see Section 3.1) defined by the sets

$$\Delta_{i_1 \cdots i_n} = \bigcap_{k=0}^{n-1} f^{-k} R_{i_{k+1}}. \tag{4.5}$$

Definition 4.1.4. We say that f is *conformal* on J if $d_x f$ is a multiple of an isometry for every $x \in J$.

We give two examples of repellers of conformal maps.

Example 4.1.5. In [152], Takens introduced the class of geometric constructions defined by:

1. p disjoint closed intervals $\Delta_1, \ldots, \Delta_p \subset \mathbb{R}$;

2. a C^1 map $f \colon U \to \mathbb{R}$, where U is an open neighborhood of $\Delta = \bigcup_{i=1}^p \Delta_i$.

We require that f be topologically mixing and expanding on U, and that $f(\partial \Delta) \subset \partial \Delta$ and $\Delta_i \subset f(\Delta_j)$ whenever $\partial \Delta_i \cap \partial f(\Delta_j) \ne \varnothing$. The map f is conformal, since it is defined in a subset of $\mathbb{R}$. We define a $p \times p$ matrix $A = (a_{ij})$ by

$$a_{ij} = \begin{cases} 1 & \text{if } \Delta_i \cap f^{-1}\Delta_j \ne \varnothing, \\ 0 & \text{if } \Delta_i \cap f^{-1}\Delta_j = \varnothing. \end{cases}$$

We consider a geometric construction modeled by Σ_A^+, defined by the sets

$$\Delta_{i_1 \cdots i_n} = \bigcap_{j=1}^n f^{-j+1} \Delta_{i_j}$$

for $(i_1 i_2 \cdots) \in \Sigma_A^+$. The corresponding limit set is a repeller of f.

Example 4.1.6 (Hyperbolic Julia sets). Let S be the Riemann sphere, and let $R \colon S \to S$ be a rational map of degree greater than 1. Since R is holomorphic, it is conformal. We say that a periodic point z of R with $R^n z = z$ is *repelling* if $|(R^n)'(z)| > 1$. The *Julia set* J of R is the closure of the set of repelling periodic points of R. It is well known that for any nonempty domain U intersecting J, we have $R^n(U) \supset J$ for all sufficiently large $n \in \mathbb{N}$. This implies that R is topologically mixing on J. We say that J is *hyperbolic* if R is expanding on J. Hence, each hyperbolic Julia set is a repeller of a conformal map.

Roughly speaking, the conformality of f on J allows one to show that even if the sets $\Delta_{i_1 \cdots i_n}$ are not balls they essentially behave as if they were. This allows one to reproduce with some changes the proof of Theorem 3.2.5 to obtain a formula for the dimension of repellers of conformal maps. Consider the function $\varphi \colon J \to \mathbb{R}$ defined by

$$\varphi(x) = -\log \|d_x f\|.$$

Theorem 4.1.7 (Dimension of repellers of conformal maps). *If J is a repeller of a $C^{1+\varepsilon}$ transformation f, for some $\varepsilon > 0$, such that f is conformal on J, then*

$$\dim_H J = \underline{\dim}_B J = \overline{\dim}_B J = s, \qquad (4.6)$$

where s is the unique real number such that $P(s\varphi) = 0$.

Proof. The uniqueness of the number s in the theorem follows immediately from the strict monotonicity of the function $s \mapsto P(s\varphi)$. When f is topologically mixing on J, this function is analytic, and using (4.1) and the conformality of f on J we obtain

$$
\begin{aligned}
\frac{d}{ds} P(s\varphi) &= \int_J \varphi \, d\mu \\
&= -\int_J \lim_{n \to +\infty} \frac{1}{n} \log \|d_x f^n\| \, d\mu(x) \le -\log \beta < 0,
\end{aligned}
\qquad (4.7)
$$

where μ is the unique equilibrium measure of $s\varphi$ (see [132] for details), and thus the function $s \mapsto P(s\varphi)$ is strictly decreasing. In the general case (that is, when f is not necessarily topologically mixing) we can proceed as in the proof of Theorem 3.2.5 to establish the strict monotonicity of the function $s \mapsto P(s\varphi)$.

The remaining arguments follow the proof of Theorem 3.2.5, with some appropriate modifications. The main difference is that now the function φ is not constant in each set R_i although it still satisfies a bounded distortion property. We first construct a Moran cover of the repeller. Let $R_1, \ldots, R_p$ be the elements of a Markov partition of J. Given $\omega = (i_1 i_2 \cdots) \in \Sigma_A^+$ and $r \in (0, 1)$, we consider the unique integer $n = n(\omega, r)$ such that

$$\|d_{\chi(\omega)} f^n\|^{-1} < r \le \|d_{\chi(\omega)} f^{n-1}\|^{-1}, \qquad (4.8)$$

and we set

$$\Delta(\omega, r) = \bigcap_{k=0}^{n-1} f^{-k} R_{i_{k+1}}. \qquad (4.9)$$

We note that these sets are pairwise disjoint for each given r. Since f is of class $C^{1+\varepsilon}$ the function $x \mapsto \log \|d_x f\|$ is Hölder continuous with Hölder exponent ε.

Therefore, for each $n \in \mathbb{N}$ and $x, y \in \Delta_{i_1 \cdots i_n}$, using (4.1) we obtain

$$
\begin{aligned}
\frac{\|d_x f^n\|}{\|d_y f^n\|} &= \prod_{k=0}^{n-1} \frac{\|d_{f^k x} f\|}{\|d_{f^k y} f\|} \\
&\leq \prod_{k=0}^{n-1} \left(1 + \frac{\|d_{f^k x} f - d_{f^k y} f\|}{\|d_{f^k y} f\|}\right) \\
&\leq \prod_{k=0}^{n-1} \left(1 + C_1 \|f^k x - f^k y\|^\varepsilon\right) \\
&\leq \prod_{k=0}^{n-1} \left(1 + C_2 \|f^n x - f^n y\|^\varepsilon \beta^{\varepsilon(k-n)}\right),
\end{aligned}
\tag{4.10}
$$

for some constants $C_1, C_2 > 0$. Since

$$
\|f^n x - f^n y\| \leq \max_{k=1,\dots,p} \operatorname{diam} R_k =: d,
$$

we conclude that

$$
\begin{aligned}
\frac{\|d_x f^n\|}{\|d_y f^n\|} &\leq \prod_{k=0}^{n-1} \left(1 + C_2 d^\varepsilon \beta^{\varepsilon(k-n)}\right) \\
&\leq \prod_{j=1}^{\infty} \left(1 + C_2 d^\varepsilon \beta^{-\varepsilon j}\right) =: D < \infty.
\end{aligned}
\tag{4.11}
$$

Now let h be the local inverse of f^n restricted to the set $\Delta(\omega, r)$, where $n = n(\omega, r)$. We obtain

$$
\begin{aligned}
\operatorname{diam} \Delta(\omega, r) &= \sup_{x,y \in \Delta(\omega,r)} \|x - y\| \\
&\leq \sup_{x,y \in \Delta(\omega,r)} \|f^n x - f^n y\| \cdot \sup_{z \in R_{i_n}} \|d_z h\|.
\end{aligned}
$$

Since $(d_x f^n)^{-1} = d_{f^n x} h$ and f is conformal on J, we have

$$
\|(d_x f^n)^{-1}\| = \|d_x f^n\|^{-1}.
$$

Hence, by (4.11),

$$
\operatorname{diam} \Delta(\omega, r) \leq dD \|d_x f^n\|^{-1} < r
\tag{4.12}
$$

for every $x \in \Delta(\omega, r)$, provided that d is sufficiently small, that is, provided that the diameter of the Markov partition is sufficiently small. Furthermore, since each set R_i is the closure of its interior, there exists $\rho > 0$ such that R_i contains a ball

B_i of radius ρ for $i = 1, \ldots, p$. Therefore, for each $x, y \in B_{i_n}$ and the same inverse map h,

$$\|hx - hy\| \geq \|x - y\| \cdot \inf_{z \in R_{i_n}} \|d_z h\|$$
$$\geq 2\rho D^{-1} \|d_w f^n\|^{-1}$$

for every $w \in \Delta_{i_1 \cdots i_n}$. This shows that

$$\text{each set} \quad \Delta_{i_1 \cdots i_n} \quad \text{contains a ball of radius} \quad \rho D^{-1} \|d_w f^n\|^{-1} \tag{4.13}$$

for some point $w \in \Delta_{i_1 \cdots i_n}$. By (4.8) we conclude that there exists a constant $\kappa \in (0, 1)$ (independent of w and r) such that each set $\Delta(w, r)$ contains a ball of radius κr. Since the sets $\Delta(w, r)$ are pairwise disjoint it follows from elementary geometry that there exists a constant $C > 0$ (independent of r) such that each ball $B(x, r) \subset \mathbb{R}^m$ intersects at most a number C of the sets $\Delta(w, r)$.

Now we establish (4.6). Given $\delta, r, \theta > 0$ there exists a countable cover $\mathcal{U}$ of J with $\text{diam}\, \mathcal{U} < r$ such that

$$\sum_{U \in \mathcal{U}} (\text{diam}\, U)^{\dim_H J + \delta} < \theta.$$

In a similar manner to that in the proof of Theorem 3.2.5, for each $U \in \mathcal{U}$ we consider the family

$$\mathcal{U}_U = \left\{ \Delta(\omega, \text{diam}\, U) : \Delta(\omega, \text{diam}\, U) \cap U \neq \varnothing \text{ and } \omega \in \Sigma_A^+ \right\}.$$

Then the family $\mathcal{V}$ in (3.29) is a cover of J, and proceeding as in (3.30) we show that

$$\sum_{V \in \mathcal{V}} (\text{diam}\, V)^{\dim_H J + \delta} < C\theta. \tag{4.14}$$

Now, given $\alpha \in \mathbb{R}$ and $N \in \mathbb{N}$ we consider the function $M(\alpha, N)$ in (3.18). It follows from (4.14) that

$$\lim_{N \to \infty} M(0, N) = 0,$$

and we can apply Theorem 3.2.4 (with $\Sigma = \Sigma_A^+$) to conclude that

$$P((\dim_H J + \delta)\varphi) \geq 0.$$

Since the function $t \mapsto P(t\varphi)$ is strictly decreasing we obtain $\dim_H J + \delta \geq s$, and it follows from the arbitrariness of δ that $\dim_H J \geq s$.

Now we consider the upper box dimension. For each $\delta > 0$ there exists $r > 0$ as small as desired such that $N(J, r) > r^{\delta - \overline{\dim}_B J}$. Furthermore, it follows from (4.8) that

$$n(\omega, r) \leq \frac{\log r}{-\log \beta} + 1$$

(with β as in (4.1)). Proceeding as in (3.33) we conclude that there exists $m = m(r)$ such that

$$\text{card}\{j : n(\omega_j, r) = m\} \geq \frac{r^{\delta - \overline{\dim}_B J}}{-\log r / \log \beta + 1}.$$

Setting $m = m(r)$ and using (4.8) we obtain

$$p \sum_{i_1 \cdots i_{m-1}} \sup_{x \in \Delta_{i_1 \cdots i_{m-1}}} \|d_x f^{m-1}\|^{-\alpha} \geq \sum_{i_1 \cdots i_m} \sup_{x \in \Delta_{i_1 \cdots i_{m-1}}} \|d_x f^{m-1}\|^{-\alpha}$$

$$\geq r^{-\alpha} \frac{r^{\delta - \overline{\dim}_B J}}{-\log r / \log \beta + 1} \geq r^{2\delta + \alpha - \overline{\dim}_B J}$$

for all sufficiently small r. Therefore, when $\alpha < \overline{\dim}_B J - 2\delta$ we obtain

$$\sum_{i_1 \cdots i_{m-1}} \exp \sup_{x \in \Delta_{i_1 \cdots i_{m-1}}} \left(\alpha \sum_{k=0}^{m-2} \varphi \circ f^k \right) \geq 1 \tag{4.15}$$

for all sufficiently small r. Furthermore, it follows from (4.8) that

$$n(\omega, r) \geq \frac{\log r}{-\log \beta'}, \tag{4.16}$$

where

$$\beta' = \sup\{\|d_x f\| : x \in J\}.$$

We notice that $\beta' > 1$. Otherwise we would have $\|d_x f^n v\| \leq \|v\|$ for every $n \in \mathbb{N}$, $x \in J$, and $v \in T_x M$, which contradicts (4.1). It follows from (4.16) that

$$\min_j n(\omega_j, r) \to \infty \quad \text{as} \quad r \to 0,$$

and thus $m(r) \to \infty$ as $r \to 0$. Since r can be chosen as small as desired, it follows from (4.15) that $P(\alpha \varphi) \geq 0$ and $\alpha \leq s$ whenever $\alpha < \overline{\dim}_B J - 2\delta$. This implies that $\overline{\dim}_B J - 2\delta \leq s$, and by the arbitrariness of δ we obtain that $\overline{\dim}_B J \leq s$. $\square$

Ruelle showed in [134] that $\dim_H J = s$ (under the additional assumption that f is topologically mixing on J). The equality between the Hausdorff dimension and the lower and upper box dimensions is due to Falconer [53]. Theorem 4.1.7 was extended independently to expanding maps of class C^1 by Gatzouras and Peres in [65] and by Barreira in [3], using different approaches.

It was also shown by Ruelle in [134] that if μ is the unique equilibrium measure of $s\varphi$ (again assuming that f is topologically mixing on J), then

$$\dim_H J = \dim_H \mu. \tag{4.17}$$

His proof consists of showing that μ is equivalent to the s-dimensional Hausdorff measure on J (in fact with Radon–Nikodym derivative bounded and bounded away from zero). More precisely, we have the following statement.

Theorem 4.1.8 (Measure of maximal dimension). *If J is a repeller of a $C^{1+\varepsilon}$ transformation f, for some $\varepsilon > 0$, such that f is conformal and topologically mixing on J, then setting $\dim_H J = s$:*

1. *the equilibrium measure μ of $s\varphi$ is equivalent to the s-dimensional Hausdorff measure m, with Radon–Nikodym derivative bounded and bounded away from zero;*

2. $0 < m(J, s) < \infty$;

3. $\dim_H \mu = \underline{\dim}_B \mu = \overline{\dim}_B \mu = s$.

Proof. Since f is of class $C^{1+\varepsilon}$, the function $s\varphi$ is Hölder continuous and thus μ is a Gibbs measure (see, for example, [38] for details). Hence, by (3.14), there exist constants $D_1, D_2 > 0$ such that

$$D_1 \le \frac{\mu(\Delta_{i_1 \cdots i_n})}{\|d_x f^n\|^{-1}} \le D_2$$

for every $n \in \mathbb{N}$ and $x \in \Delta_{i_1 \cdots i_n}$, where the sets $\Delta_{i_1 \cdots i_n}$ are obtained as in (4.5) from a given Markov partition $R_1, \ldots, R_p$ of J.

Now let $\Delta(\omega, r)$ be the sets constructed in the proof of Theorem 4.1.7 for each $\omega \in \Sigma_A^+$ and $r \in (0, 1)$, where A is the transition matrix obtained from the Markov partition. It follows from (4.8) that for every $\omega \in \Sigma_A^+$ and $r \in (0, 1)$ we have

$$C_1 \le \frac{\mu(\Delta(\omega, r))}{r^s} \le C_2,$$

for some constants $C_1, C_2 > 0$ (independent of ω and r). As in the proof of Theorem 4.1.7 each ball $B(x, r) \subset \mathbb{R}^m$ intersects at most a number C of the sets $\Delta(\omega, r)$, for some constant C independent of x and r. Therefore,

$$\mu(B(x, r)) \le C C_2 r^s. \tag{4.18}$$

On the other hand, by (4.12) we have $\operatorname{diam} \Delta(\omega, r) < r$. Hence, there exists $\omega \in \Sigma_A^+$ such that $B(x, r) \supset \Delta(\omega, r/2)$ (we note that the sets of the form $\Delta(\omega', r/2)$ that touch the boundary of $B(x, r)$ are not sufficient to cover the ball), and thus

$$\mu(B(x, r)) \ge \mu(\Delta(\omega, r/2)) \ge \frac{C}{2^s} r^s. \tag{4.19}$$

This shows that μ is equivalent to the s-dimensional Hausdorff measure in J, with Radon–Nikodym derivative bounded and bounded away from zero. Since $\mu(J) = 1$ the second statement follows immediately from the first.

For the last statement we note that by (4.18) and (4.19),

$$\lim_{r \to 0} \frac{\log \mu(B(x, r))}{\log r} = s$$

for every $x \in J$. Now we can apply Theorem 2.1.6 to obtain the desired statement. $\qquad \square$

Theorem 4.1.8 motivates the following definition.

Definition 4.1.9. The equilibrium measure μ of the function $s\varphi$ is called the *measure of maximal dimension* in J.

4.2 Hyperbolic sets and Markov partitions

We recall some basic notions in this section. These include stable and unstable invariant manifolds, product structure of hyperbolic sets, Markov partitions and some properties of their boundaries, as well as the associated symbolic dynamics. We also describe the product structure of Gibbs measures. These notions are used in the study of the dimension of hyperbolic sets of conformal maps, and in several other places in the book, and thus it is very convenient to have them collected in a unified manner in a single place.

4.2.1 Basic notions and product structure

Let $f: M \to M$ be a diffeomorphism of a smooth manifold, and let $\Lambda \subset M$ be a compact f-invariant set, that is, a compact set $\Lambda \subset M$ such that $f^{-1}\Lambda = \Lambda$.

Definition 4.2.1. We say that Λ is a *hyperbolic set* of f if for every point $x \in \Lambda$ there exists a decomposition of the tangent space

$$T_x M = E^s(x) \oplus E^u(x) \tag{4.20}$$

satisfying

$$d_x f E^s(x) = E^s(fx) \quad \text{and} \quad d_x f E^u(x) = E^u(fx),$$

and there exist constants $\lambda \in (0, 1)$ and $c > 0$ such that

$$\|d_x f^n | E^s(x)\| \le c\lambda^n \quad \text{and} \quad \|d_x f^{-n} | E^u(x)\| \le c\lambda^n$$

for every $x \in \Lambda$ and $n \in \mathbb{N}$.

Given $\rho > 0$, for each $x \in M$ we consider the sets

$$V_\rho^s(x) = \{y \in B(x, \rho) : d(f^n y, f^n x) < \rho \text{ for every } n > 0\} \tag{4.21}$$

and

$$V_\rho^u(x) = \{y \in B(x, \rho) : d(f^n y, f^n x) < \rho \text{ for every } n < 0\}, \tag{4.22}$$

where d is the distance in M and $B(x, \rho) \subset M$ is the open ball centered at x of radius ρ. We recall the following classical result.

Theorem 4.2.2 (Hadamard–Perron theorem). *If Λ is a hyperbolic set of a C^1 diffeomorphism, then there exists $\rho > 0$ such that for each $x \in \Lambda$ the sets $V_\rho^s(x)$ and $V_\rho^u(x)$ are C^1 manifolds containing x that satisfy*

$$T_x V_\rho^s(x) = E^s(x) \quad \text{and} \quad T_x V_\rho^u(x) = E^u(x). \tag{4.23}$$

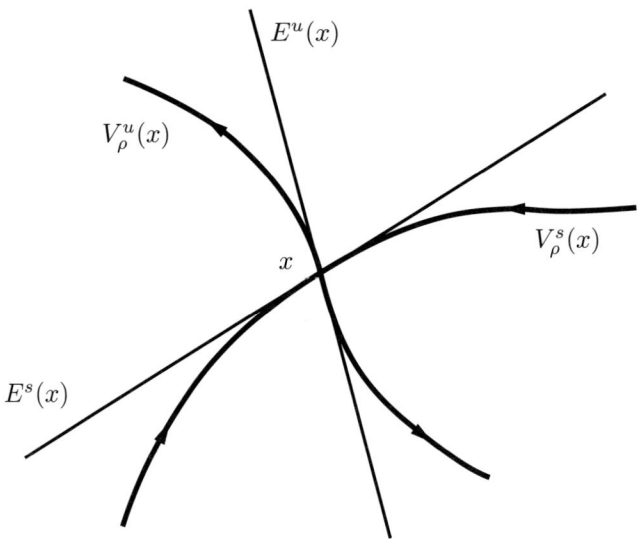

Figure 4.2: Local stable and unstable manifolds

We refer the reader to the book by Anosov [2, §4] for references and for a detailed account of the origins of the Hadamard–Perron theorem.

Definition 4.2.3. The manifolds $V_\rho^s(x)$ and $V_\rho^u(x)$ are called respectively a *local stable manifold* and a *local unstable manifold* at x (of size ρ).

It follows from (4.20) and (4.23) that these manifolds are transverse (see Figure 4.2). Furthermore, under the assumptions of Theorem 4.2.2 one can show that the sizes of $V_\rho^s(x)$ and $V_\rho^u(x)$ are uniformly bounded away from zero, i.e., there exists $\gamma = \gamma(\rho) > 0$ such that

$$V_\rho^s(x) \supset B^s(x, \gamma) \quad \text{and} \quad V_\rho^u(x) \supset B^u(x, \gamma)$$

for every $x \in \Lambda$, where $B^s(x, \gamma)$ and $B^u(x, \gamma)$ are the open balls centered at x of radius γ with respect to the distances induced by d respectively on $V_\rho^s(x)$ and $V_\rho^u(x)$. The continuous dependence of the spaces $E^s(x)$ and $E^u(x)$ in $x \in \Lambda$ and the smoothness of the local stable and unstable manifolds guarantee that there exists $\delta = \delta(\rho) > 0$ such that if $d(x, y) < \delta$ for two points x, $y \in \Lambda$, then the intersection $V_\rho^s(x) \cap V_\rho^u(y)$ contains exactly one point, although it may not be in Λ.

Definition 4.2.4. A hyperbolic set Λ of a diffeomorphism f is said to be *locally maximal* if there exists an open neighborhood U of Λ such that

$$\Lambda = \bigcap_{n \in \mathbb{Z}} f^n U.$$

One can show that for a locally maximal hyperbolic set, provided that δ is sufficiently small, the intersection $V_\rho^s(x) \cap V_\rho^u(y)$ consists of a single point in Λ. Thus, we can introduce the following definition.

Definition 4.2.5. Let Λ be a locally maximal hyperbolic set. For each $\delta > 0$ sufficiently small, the function

$$[\cdot, \cdot] \colon \{(x, y) \in \Lambda \times \Lambda : d(x, y) < \delta\} \to \Lambda$$

defined by

$$[x, y] = V_\rho^s(x) \cap V_\rho^u(y)$$

(see Figure 4.3) is called a *local product structure* of Λ.

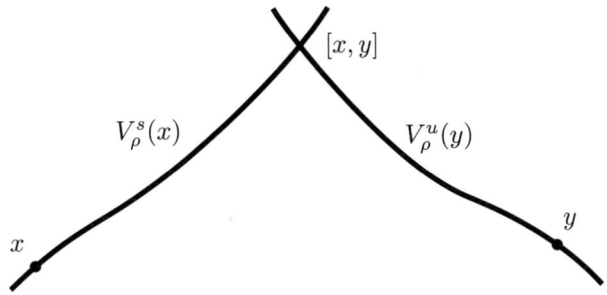

Figure 4.3: Local product structure

We also recall the notion of Markov partition of a locally maximal hyperbolic set Λ.

Definition 4.2.6. A nonempty closed set $R \subset \Lambda$ is called a *rectangle* if $\operatorname{diam} R < \delta$ (with δ as in Definition 4.2.5), $\overline{\operatorname{int} R} = R$, and $[x, y] \in R$ whenever x, $y \in R$. A finite cover of Λ by rectangles R_1, ..., R_p is called a *Markov partition* of Λ (with respect to f) if:

1. $\operatorname{int} R_i \cap \operatorname{int} R_j = \varnothing$ whenever $i \neq j$;

2. if $x \in f(\operatorname{int} R_i) \cap \operatorname{int} R_j$, then

$$f^{-1}(V_\rho^u(fx) \cap R_j) \subset V_\rho^u(x) \cap R_i \quad \text{and} \quad f(V_\rho^s(x) \cap R_i) \subset V_\rho^s(fx) \cap R_j.$$

We note that in Definition 4.2.6 the interior of each set R_i is computed with respect to the induced topology on Λ. Any hyperbolic set has Markov partitions with diameter as small as desired (we refer to [38] for details and references).

In a similar manner to that for repellers in Section 4.1, given the rectangles $R_1, \ldots, R_p$ of a Markov partition of the locally maximal hyperbolic set Λ, we define a $p \times p$ matrix $A = (a_{ij})$ with entries given by (4.2). We consider the space of sequences $\Sigma_p = \{1, \ldots, p\}^{\mathbb{Z}}$ and we equip it with the distance

$$d(\omega, \omega') = \sum_{k \in \mathbb{Z}} e^{-|k|} |i_k - i'_k|, \tag{4.24}$$

where $\omega = (\cdots i_{-1} i_0 i_1 \cdots)$ and $\omega' = (\cdots i'_{-1} i'_0 i'_1 \cdots)$. With this distance Σ_p becomes a compact metric space. We also consider the *shift map* $\sigma \colon \Sigma_p \to \Sigma_p$ defined by $(\sigma \omega)_n = i_{n+1}$ for each $\omega = (\cdots i_{-1} i_0 i_1 \cdots) \in \Sigma_p$ and $n \in \mathbb{Z}$.

Definition 4.2.7. The restriction of σ to the set

$$\Sigma_A = \{(\cdots i_{-1} i_0 i_1 \cdots) \in \Sigma_p : a_{i_{n+1} i_n} = 1 \text{ for every } n \in \mathbb{Z}\}$$

is called a *(two-sided) topological Markov chain* with *transition matrix* A.

It is easy to show that one can define a *coding map* $\chi \colon \Sigma_A \to \Lambda$ of the hyperbolic set Λ by

$$\chi(\cdots i_{-1} i_0 i_1 \cdots) = \bigcap_{n \in \mathbb{Z}} f^{-n} R_{i_n}. \tag{4.25}$$

The map χ is surjective and satisfies

$$\chi \circ \sigma = f \circ \chi$$

(i.e., the diagram in Figure 4.4 is commutative). In addition, χ is Hölder continuous, with the distance in Σ_A given by (4.24). In general the map χ is not invertible (although one can show that $\operatorname{card} \chi^{-1} x \le p^2$ for every $x \in \Lambda$).

$$
\begin{array}{ccc}
\Sigma_A & \xrightarrow{\ \sigma\ } & \Sigma_A \\
\chi \downarrow & & \downarrow \chi \\
\Lambda & \xrightarrow{\ f\ } & \Lambda
\end{array}
$$

Figure 4.4: Symbolic coding of a hyperbolic set

4.2.2　Boundaries of Markov partitions

Let Λ be a locally maximal hyperbolic set of a diffeomorphism f, and let $\mathcal{R} = \{R_1, \ldots, R_p\}$ be a Markov partition of Λ. The boundary of $\mathcal{R}$ is the union of the *stable boundary*

$$\partial^s \mathcal{R} = \bigcup_{i=1}^{p} \{x \in \partial R_i : V_\rho^u(x) \cap \operatorname{int} R_i \ne \varnothing\},$$

and of the *unstable boundary*

$$\partial^u \mathcal{R} = \bigcup_{i=1}^{p} \{x \in \partial R_i : V_\rho^s(x) \cap \operatorname{int} R_i \neq \varnothing\}.$$

That is, $\partial \mathcal{R} = \partial^s \mathcal{R} \cup \partial^u \mathcal{R}$. Moreover,

$$f(\partial^s \mathcal{R}) \subset \partial^s \mathcal{R} \quad \text{and} \quad f^{-1}(\partial^u \mathcal{R}) \subset \partial^u \mathcal{R}.$$

We note that the sets $\partial^s \mathcal{R}$, $\partial^u \mathcal{R}$, and $\partial \mathcal{R}$ are nowhere dense, and hence they have zero measure with respect to any ergodic f-invariant measure with full support.

The following statement shows that for a large class of measures the ρ-neighborhood of the set $\partial \mathcal{R}$ is at most polynomial in ρ. This was established by Barreira and Saussol in [17].

Theorem 4.2.8 (Boundaries of Markov partitions). *Let Λ be a locally maximal hyperbolic set of a $C^{1+\varepsilon}$ diffeomorphism f, for some $\varepsilon > 0$, such that f is topologically mixing on Λ, and let μ be the equilibrium measure of a Hölder continuous function in Λ. For any Markov partition $\mathcal{R}$ of Λ there exist constants $c > 0$ and $\nu > 0$ such that if $\rho > 0$, then*

$$\mu(\{x \in \Lambda : d(x, \partial \mathcal{R}) < \rho\}) \le c\rho^\nu.$$

Proof. For each $n \in \mathbb{N}$ we consider the distance d_n in (2.11) and we denote by $B_n(x, \delta)$ the open ball centered at x of radius δ with respect to the distance d_n.

Lemma 4.2.9. *For each sufficiently small $r > 0$, and each compact f-invariant set $K \subset \Lambda$ with $K \neq \Lambda$ there exist constants $c > 0$ and $\nu > 0$ such that for every $n \in \mathbb{N}$,*

$$\mu(\{x \in \Lambda : d_n(x, K) < r\}) \le ce^{-\nu n}.$$

Proof of the lemma. Under the assumptions in the theorem, there exists a unique equilibrium measure μ_K of the function $\varphi|K$ with respect to $f|K$. Clearly, we have $\operatorname{supp} \mu_K \subset K$. On the other hand, by the Gibbs property of μ we also have $\operatorname{supp} \mu = \Lambda$, and hence $\mu_K \neq \mu$. Furthermore, by the uniqueness of the equilibrium measure μ,

$$P_K(\varphi) = h_{\mu_K}(f) + \int_\Lambda \varphi \, d\mu_K < P_\Lambda(\varphi).$$

Now let $E_n \subset K$ be any set such that

$$K \subset \bigcup_{y \in E_n} B_n(y, r). \tag{4.26}$$

Clearly, setting

$$K_n = \{x \in \Lambda : d_n(x, K) < r\}$$

we have

$$K_n \subset \bigcup_{y \in E_n} B_n(y, 2r).$$

On the other hand, it is well known (see, for example, [84, Lemma 20.3.4]) that there exists $\zeta > 0$ such that

$$\mu(B_n(y, 2r)) \leq \zeta \exp\left(-nP_\Lambda(\varphi) + \sum_{k=0}^{n-1} \varphi(f^k y)\right)$$

for every $n \in \mathbb{N}$ and $y \in \Lambda$. Hence,

$$\mu(K_n) \leq \sum_{y \in E_n} \mu(B_n(y, 2r)) \leq \zeta \sum_{y \in E_n} \exp\left(-nP_\Lambda(\varphi) + \sum_{k=0}^{n-1} \varphi(f^k y)\right).$$

Moreover (see Definition 2.3.1), we have

$$P_K(\varphi) = \lim_{n \to \infty} \frac{1}{n} \log \inf_{E_n} \sum_{y \in E_n} \exp \sum_{k=0}^{n-1} \varphi(f^k y),$$

where the infimum is taken over all sets E_n for which (4.26) holds. Therefore,

$$\limsup_{n \to \infty} \frac{1}{n} \log \mu(K_n) \leq P_K(\varphi) - P_\Lambda(\varphi) < 0.$$

This completes the proof of the lemma.   □

Now we observe that

$$d_n(x, y) \leq \max\{\|d_x f\| : x \in \Lambda\}^n d(x, y) \tag{4.27}$$

for every $n \in \mathbb{N}$ and $x, y \in \Lambda$. For any compact set $K \subset \Lambda$ such that $f(K) \subset K \neq \Lambda$, it follows from Lemma 4.2.9 and (4.27) that there exist constants $c = c(f, K) > 0$ and $\nu = \nu(f, K) > 0$ such that

$$\mu(\{x \in \Lambda : d(x, K) < \rho\}) \leq c(f, K)\rho^{\nu(f, K)}. \tag{4.28}$$

Using (4.28) with $K = \partial^s \mathcal{R}$ we obtain

$$\mu(\{x \in \Lambda : d(x, \partial^s \mathcal{R}) < \rho\}) \leq c(f, \partial^s \mathcal{R})\rho^{\nu(f, \partial^s \mathcal{R})}. \tag{4.29}$$

Similarly, using (4.28) with $K = \partial^u \mathcal{R}$ now with respect to f^{-1} we obtain

$$\mu(\{x \in \Lambda : d(x, \partial^u \mathcal{R}) < \rho\}) \leq c(f^{-1}, \partial^u \mathcal{R})\rho^{\nu(f^{-1}, \partial^u \mathcal{R})}, \tag{4.30}$$

since the equilibrium measures of a function φ with respect to f are the same as the equilibrium measures of φ with respect to f^{-1}. Furthermore,

$$\{x \in \Lambda : d(x, \partial \mathcal{R}) < \rho\} \subset \{x \in \Lambda : d(x, \partial^s \mathcal{R}) < \rho\} \cup \{x \in \Lambda : d(x, \partial^u \mathcal{R}) < \rho\},$$

and the desired statement follows immediately from (4.29) and (4.30).   □

It is well known that $\partial \mathcal{R}$ has zero measure with respect to any equilibrium measure. This is a simple consequence of the fact that $\partial \mathcal{R}$ is a closed set with dense complement. On the other hand, to estimate the measure of a neighborhood of $\partial \mathcal{R}$ may be simpler when the boundary is piecewise regular (as in the case of hyperbolic automorphisms of the torus $\mathbb{T}^2$), but Markov partitions may have a very irregular boundary. In particular, it was proved by Bowen in [39] that $\partial \mathcal{R}$ is never piecewise regular in the case of hyperbolic automorphisms of the torus $\mathbb{T}^3$.

4.2.3 Product structure of Gibbs measures

Before describing the product structure of Gibbs measures, we explain how the symbolic dynamics can be decomposed into stable and unstable components. We also explain how this decomposition is related to the local product structure of a hyperbolic set (see Definition 4.2.5). We follow closely the appendix of [10].

Given a two-sided topological Markov chain $\sigma|\Sigma_A$ with $\Sigma_A \subset \{1, \dots, p\}^{\mathbb{Z}}$, we denote respectively by Σ_A^+ and Σ_A^- the sets of right-sided and left-sided infinite sequences on p symbols obtained from the sequences in Σ_A. More precisely, we denote by Σ_A^+ the set of sequences $(i_0 i_1 \cdots)$ such that

$$(i_0 i_1 \cdots) = (j_0 j_1 \cdots)$$

for some sequence $(\cdots j_{-1} j_0 j_1 \cdots) \in \Sigma_A$. Similarly, we denote by Σ_A^- the set of sequences $(\cdots i_{-1} i_0)$ such that

$$(\cdots i_{-1} i_0) = (\cdots j_{-1} j_0)$$

for some sequence $(\cdots j_{-1} j_0 j_1 \cdots) \in \Sigma_A$. We note that Σ_A^- is canonically identified with $\Sigma_{A^t}^+$, where A^t denotes the transpose of A, by the bijective map

$$\Sigma_A^- \ni (\cdots i_{-1} i_0) \mapsto (i_0 i_{-1} \cdots) \in \Sigma_{A^t}^+.$$

We also consider the one-sided topological Markov chains

$$\sigma^+ \colon \Sigma_A^+ \to \Sigma_A^+ \quad \text{and} \quad \sigma^- \colon \Sigma_A^- \to \Sigma_A^-$$

defined respectively by

$$\sigma^+(i_0 i_1 \cdots) = (i_1 i_2 \cdots) \quad \text{and} \quad \sigma^-(\cdots i_{-1} i_0) = (\cdots i_{-2} i_{-1}). \tag{4.31}$$

The following construction is described by Bowen in [38, Lemma 1.6]. We choose a number p of points $\omega_1, \dots, \omega_p \in \Sigma_A$ such that $\omega_i \in \Sigma_A \cap C_i$ for each i (here C_i is a cylinder set of length 1; see Definition 3.2.2). Set $\Omega = (\omega_1, \dots, \omega_p)$. We define the function $r_\Omega \colon \Sigma_A \to \Sigma_A$ by

$$r_\Omega(\cdots i_{-1} i_0 i_1 \cdots) = (\cdots j_{-2} j_{-1} i_0 i_1 i_2 \cdots),$$

where $(\cdots j_{-1}j_0j_1\cdots) = w_{i_0}$. Furthermore, given a function $\varphi\colon \Sigma_A \to \mathbb{R}$ we define a new function $\theta^u = \theta_\Omega^u\colon \Sigma_A \to \mathbb{R}$ by

$$\theta^u(\omega) = \varphi(r_\Omega(\omega)) + \sum_{j=0}^{\infty} \left[\varphi(\sigma^{j+1}r_\Omega(\omega)) - \varphi(\sigma^j r_\Omega(\sigma\omega))\right].$$

Now we introduce the concept of cohomology in dynamical systems.

Definition 4.2.10. Given a continuous transformation $f\colon X \to X$ of a topological space X, two functions $\varphi_1\colon X \to \mathbb{R}$ and $\varphi_2\colon X \to \mathbb{R}$ are said to be *cohomologous* on X with respect to f, or simply *cohomologous* if there exists a continuous function $\psi\colon X \to \mathbb{R}$ such that

$$\varphi_1 - \varphi_2 = \psi - \psi \circ f \quad \text{on} \quad X.$$

One can easily show that the functions θ^u and φ are cohomologous, and that they have the same topological pressure (see Lemma 1.6 in [38]). In particular, they have the same equilibrium measures.

Now let

$$\pi^+\colon \Sigma_A \to \Sigma_A^+ \quad \text{and} \quad \pi^-\colon \Sigma_A \to \Sigma_A^-$$

be the projections defined respectively by

$$\pi^+(\cdots i_{-1}i_0i_1\cdots) = (i_0i_1\cdots) \quad \text{and} \quad \pi^-(\cdots i_{-1}i_0i_1\cdots) = (\cdots i_{-1}i_0). \quad (4.32)$$

One can easily verify that

$$\theta^u(\cdots i_{-1}i_0i_1\cdots) = \theta^u(\cdots i'_{-1}i'_0i'_1\cdots)$$

whenever $i_j = i'_j$ for every $j \geq 0$. Similarly, one can define a function $\theta^s = \theta_\Omega^s$ in Σ_A such that

$$\theta^s(\cdots i_{-1}i_0i_1\cdots) = \theta^u(\cdots i'_{-1}i'_0i'_1\cdots)$$

whenever $i_j = i'_j$ for every $j \leq 0$. We thus have the following statement.

Proposition 4.2.11. *Given a continuous function $\varphi\colon \Sigma_A \to \mathbb{R}$, the following properties hold:*

1. *there exists a function $\varphi^u\colon \Sigma_A^+ \to \mathbb{R}$ such that $\theta^u = \varphi^u \circ \pi^+$ in Σ_A;*

2. *there exists a function $\varphi^s\colon \Sigma_A^- \to \mathbb{R}$ such that $\theta^s = \varphi^s \circ \pi^-$ in Σ_A;*

3. *the functions φ, $\varphi^u \circ \pi^+$, and $\varphi^s \circ \pi^-$ are cohomologous;*

4. *we have*

$$P_{\Sigma_A}(\varphi) = P_{\Sigma_A^+}(\varphi^u) = P_{\Sigma_A^-}(\varphi^s). \quad (4.33)$$

Now let $\varphi\colon \Sigma_A \to \mathbb{R}$ be a Hölder continuous function, and let ν be its unique equilibrium measure in Σ_A. One can easily show that the functions φ^u and φ^s are also Hölder continuous (see [38]). Let ν^u be the equilibrium measure of φ^u in Σ_A^+, and let ν^s be the equilibrium measure of φ^s in Σ_A^-. We also define a measure $\pi_*^+\nu$ in Σ_A^+ by

$$(\pi_*^+\nu)(B) = \nu((\pi^+)^{-1}B)$$

for every measurable subset $B \subset \Sigma_A^+$. Similarly, we define a measure $\pi_*^-\nu$ in Σ_A^- by

$$(\pi_*^-\nu)(B) = \nu((\pi^-)^{-1}B)$$

for every measurable subset $B \subset \Sigma_A^-$. We note that

$$\nu^u = \pi_*^+\nu \quad \text{in} \quad \Sigma_A^+,$$

since both are equilibrium measures for the function φ^u, and that

$$\nu^s = \pi_*^-\nu \quad \text{in} \quad \Sigma_A^-,$$

since both are equilibrium measures for the function φ^s. This implies that

$$\nu^u(C_{i_0\cdots i_n}^+) = \nu(C_{i_0\cdots i_n}) \quad \text{and} \quad \nu^s(C_{i_0\cdots i_n}^-) = \nu(C_{i_0\cdots i_n}), \tag{4.34}$$

where

$$C_{i_0\cdots i_n}^+ = \pi^+(C_{i_0\cdots i_n}) \quad \text{and} \quad C_{i_{-n}\cdots i_0}^- = \pi^-(C_{i_{-n}\cdots i_0}).$$

By Proposition 3.2 in [111] and (4.34) we have

$$\varphi^u(\omega^+) - P_{\Sigma_A^+}(\varphi^u) = \lim_{n\to\infty} \log \frac{\nu^u(C_{i_0\cdots i_n}^+)}{\nu^u(C_{i_1\cdots i_n}^+)} = \lim_{n\to\infty} \log \frac{\nu(C_{i_0\cdots i_n})}{\nu(C_{i_1\cdots i_n})}$$

for every $\omega^+ = (i_0 i_1 \cdots) \in \Sigma_A^+$ (with uniform convergence), and

$$\varphi^s(\omega^-) - P_{\Sigma_A^-}(\varphi^s) = \lim_{n\to\infty} \log \frac{\nu^s(C_{i_{-n}\cdots i_0}^-)}{\nu^s(C_{i_{-n}\cdots i_1}^-)} = \lim_{n\to\infty} \log \frac{\nu(C_{i_{-n}\cdots i_0})}{\nu(C_{i_{-n}\cdots i_1})}$$

for every $\omega^- = (\cdots i_{-1} i_0) \in \Sigma_A^-$ (with uniform convergence). In particular, this implies that the functions φ^u and φ^s are independent of Ω.

The following statement shows that any equilibrium measure of a Hölder continuous function in a topological Markov chain has a product structure. This follows immediately from the identities in (4.34).

Proposition 4.2.12. *There exist constants $K_1, K_2 > 0$ such that for every $n, m \in \mathbb{N}$ and $(\cdots i_{-1} i_0 i_1 \cdots) \in \Sigma_A$ we have*

$$K_1 \leq \frac{\nu(C_{i_{-m}\cdots i_n})}{\nu^u(C_{i_0\cdots i_n}^+)\nu^s(C_{i_{-m}\cdots i_0}^-)} \leq K_2.$$

Now let Λ be a locally maximal hyperbolic set of a diffeomorphism f, and let $R_1, \ldots, R_p$ be the elements of a Markov partition of Λ. For each $x \in \Lambda$, we set

$$A^u(x) = V_\rho^u(x) \cap R(x) \quad \text{and} \quad A^s(x) = V_\rho^s(x) \cap R(x), \qquad (4.35)$$

where $V_\rho^u(x)$ and $V_\rho^s(x)$ are the local unstable and stable manifolds at x (see (4.21) and (4.22)), and where $R(x)$ is a fixed element of the Markov partition that contains x. For each $\omega' \in \Sigma_A$ we have

$$\chi(\omega') \in V_\rho^u(x) \cap R(x) \quad \text{whenever} \quad \pi^- \omega' = \pi^- \omega$$

and

$$\chi(\omega') \in V_\rho^s(x) \cap R(x) \quad \text{whenever} \quad \pi^+ \omega' = \pi^+ \omega,$$

where $x = \chi(\omega)$ (here χ is the coding map in (4.25)). Therefore, writing $\omega = (\cdots i_{-1} i_0 i_1 \cdots)$, the set $A^u(x)$ can be identified with the cylinder set

$$C_{i_0}^+ = \{(j_0 j_1 \cdots) \in \Sigma_A^+ : j_0 = i_0\} \subset \Sigma_A^+, \qquad (4.36)$$

and the set $A^s(x)$ can be identified with the cylinder set

$$C_{i_0}^- = \{(\cdots j_{-1} j_0) \in \Sigma_A^- : j_0 = i_0\} \subset \Sigma_A^-. \qquad (4.37)$$

Now let $\psi \colon \Lambda \to \mathbb{R}$ be a Hölder continuous function, and let μ be its equilibrium measure in Λ. Then $\varphi = \psi \circ \chi$ is a Hölder continuous function in Σ_A. Furthermore, $\mu = \chi_* \nu$, where ν is the unique equilibrium measure of φ, that is,

$$\mu(B) = \nu(\chi^{-1} B) \quad \text{for every measurable subset} \quad B \subset \Lambda.$$

Moreover, for each $x \in \Lambda \cap \chi(C_{i_0})$, we define the measures

$$\mu_x^s = \chi_*(\nu^s | C_{i_0}) \quad \text{in} \quad A^s(x),$$

and

$$\mu_x^u = \chi_*(\nu^u | C_{i_0}) \quad \text{in} \quad A^u(x),$$

with ν^s and ν^u as above. The following statement is an immediate consequence of Proposition 4.2.12.

Proposition 4.2.13. *There exist constants $K_1, K_2 > 0$ such that for any $x \in \Lambda$ and any Borel sets $E \subset A^s(x)$ and $F \subset A^u(x)$ we have*

$$K_1 \mu_x^u(E) \mu_x^s(F) \le \mu([E, F]) \le K_2 \mu_x^u(E) \mu_x^s(F).$$

For μ-almost every $x \in \Lambda$, let ν_x^u and ν_x^s be respectively the conditional measures of μ in $A^u(x)$ and $A^s(x)$. We note that

$$\nu_x^u = \mu_x^u \quad \text{and} \quad \nu_x^s = \mu_x^s \qquad (4.38)$$

for μ-almost every $x \in \Lambda$.

4.3 Dimension of hyperbolic sets of conformal maps

We study in this section the dimension of hyperbolic sets of conformal diffeomorphisms. Let $f\colon M \to M$ be a diffeomorphism, and let $\Lambda \subset M$ be a hyperbolic set of f.

Definition 4.3.1. We say that f is *conformal* on Λ if the linear transformations $d_x f | E^s(x)$ and $d_x f | E^u(x)$ are multiples of isometries for every $x \in \Lambda$.

For example, if M is a surface and $\dim E^s(x) = \dim E^u(x) = 1$ for every $x \in \Lambda$, then f is conformal on Λ.

The following result is a version of Theorem 4.1.7 for hyperbolic sets. We consider the functions $\varphi_s \colon \Lambda \to \mathbb{R}$ and $\varphi_u \colon \Lambda \to \mathbb{R}$ defined by

$$\varphi_s(x) = \log \|d_x f | E^s(x)\| \quad \text{and} \quad \varphi_u(x) = -\log \|d_x f | E^u(x)\|. \tag{4.39}$$

Theorem 4.3.2 (Dimension of hyperbolic sets of conformal maps). *Let Λ be a locally maximal hyperbolic set of a $C^{1+\varepsilon}$ diffeomorphism, for some $\varepsilon > 0$, such that f is conformal and topologically mixing on Λ. Then*

$$\dim_H(V_\rho^s(x) \cap \Lambda) = \underline{\dim}_B(V_\rho^s(x) \cap \Lambda) = \overline{\dim}_B(V_\rho^s(x) \cap \Lambda) = t_s \tag{4.40}$$

and

$$\dim_H(V_\rho^u(x) \cap \Lambda) = \underline{\dim}_B(V_\rho^u(x) \cap \Lambda) = \overline{\dim}_B(V_\rho^u(x) \cap \Lambda) = t_u, \tag{4.41}$$

where t_s and t_u are the unique real numbers such that

$$P(t_s\varphi_s) = P(t_u\varphi_u) = 0. \tag{4.42}$$

Furthermore,

$$\dim_H \Lambda = \underline{\dim}_B \Lambda = \overline{\dim}_B \Lambda = t_s + t_u. \tag{4.43}$$

Proof. Let $R_1, \dots, R_p$ be the elements of a Markov partition of Λ with diameter at most ρ, where ρ is the size of the local stable and unstable manifolds (we recall that there exist Markov partitions with diameter as small as desired). Given $x, y \in \Lambda$ in the same rectangle R_i, we denote by $H_{x,y}^s \colon V_\rho^u(x) \to V_\rho^u(y)$ the *holonomy map along the stable manifolds*, given by

$$H_{x,y}^s(z) = [z, y].$$

Analogously, we denote by $H_{x,y}^u \colon V_\rho^s(x) \to V_\rho^s(y)$ the *holonomy map along the unstable manifolds*, given by

$$H_{x,y}^u(z) = [y, z].$$

We note that the maps $H_{x,y}^s$ and $H_{x,y}^u$ depend continuously on x and y. Furthermore, by results of Hasselblatt in [71], since f is conformal on Λ the stable and unstable distributions are smooth, and thus the local product structure is a Lipschitz homeomorphism with Lipschitz inverse. This implies that the holonomy maps $H^s_{x,y}$ and $H^u_{x,y}$ are Lipschitz.

Now we establish the identities in (4.40) and (4.41). We only consider $V^u_\rho(x)$ since the arguments for $V^s_\rho(x)$ are entirely analogous. Choose segments V_i of local unstable manifolds, for $i = 1, \ldots, p$, such that

$$V_i \cap \Lambda = V^u_\rho(x_i) \cap \Lambda \cap R_i \qquad (4.44)$$

for some point $x_i \in R_i$. We notice that $V_i \cap \operatorname{int} R_j = \varnothing$ whenever $j \neq i$. Since the holonomy maps are Lipschitz, the numbers $\dim_H(V_i \cap \Lambda)$, $\underline{\dim}_B(V_i \cap \Lambda)$, and $\overline{\dim}_B(V_i \cap \Lambda)$ are independent of x_i. Furthermore, since f is topologically mixing on Λ we have

$$\begin{aligned} \dim_H(V_i \cap \Lambda) &= \dim_H(V_j \cap \Lambda), \\ \underline{\dim}_B(V_i \cap \Lambda) &= \underline{\dim}_B(V_j \cap \Lambda), \\ \overline{\dim}_B(V_i \cap \Lambda) &= \overline{\dim}_B(V_j \cap \Lambda) \end{aligned} \qquad (4.45)$$

for every i and j. Set $V = \bigcup_{i=1}^p V_i$. We define

$$R_{i_0 \cdots i_n} = \bigcap_{k=0}^n f^{-k} R_{i_k} \quad \text{and} \quad V_{i_0 \cdots i_n} = V \cap R_{i_0 \cdots i_n}$$

for every $(\cdots i_{-1} i_0 i_1 \cdots) \in \Sigma_A$ and $n \in \mathbb{N}$, where A is the transition matrix obtained from the Markov partition as in (4.2).

We first obtain an upper estimate for the upper box dimension. Note that

$$f^n V_{i_0 \cdots i_n} \subset V^u_\rho(f^n x_{i_0}) \cap R_{i_n}$$

and

$$H^s_{f^n x_{i_0}, x_{i_n}} f^n (V_{i_0 \cdots i_n} \cap \Lambda) = V_{i_n} \cap \Lambda.$$

Furthermore, each point in $V_{i_n} \cap \Lambda$ has exactly one preimage under the holonomy map $H^s_{f^n x_{i_0}, x_{i_n}}$. Hence, if $\mathcal{U}$ is a cover of $V_{i_n} \cap \Lambda$, then the collection of sets $f^{-n}(H^s_{x_{i_n}, f^n x_{i_0}} \mathcal{U})$ is a cover of $V_{i_0 \cdots i_n} \cap \Lambda$, and thus,

$$N(V_{i_0 \cdots i_n} \cap \Lambda, r) \leq N(V_{i_n} \cap \Lambda, K^{-1} r \overline{\lambda}_{i_0 \cdots i_n})$$

for every $r > 0$, where $K \geq 1$ is a Lipschitz constant for the holonomy map, and

$$\overline{\lambda}_{i_0 \cdots i_n} = \max\{\|d_x f^{-n} | E^u(x)\| : x \in R_{i_0 \cdots i_n}\}.$$

Therefore,

$$N(V \cap \Lambda, r) \leq \sum_{i_0 \cdots i_n} N(V_{i_0 \cdots i_n} \cap \Lambda, r) \leq \sum_{i_0 \cdots i_n} N(V \cap \Lambda, K^{-1} r \overline{\lambda}_{i_0 \cdots i_n}).$$

Now take $s > \overline{\dim}_B(V \cap \Lambda)$. Then there exists $r_0 > 0$ such that $N(V \cap \Lambda, r) < r^{-s}$ for every $r \in (0, r_0)$. Setting

$$c_n(s) = \sum_{i_1 \cdots i_n} \overline{\lambda}_{i_0 \cdots i_n}^{\,s},$$

we obtain

$$N(V \cap \Lambda, r) \le r^{-s} K^s c_n(s)$$

for every $r < \lambda_n K r_0$, where

$$\lambda_n = \min_{i_0 \cdots i_n} \overline{\lambda}_{i_0 \cdots i_n}.$$

By induction we obtain

$$N(V \cap \Lambda, r) \le r^{-s} K^s c_n(s)^m$$

for every $r < (\lambda_n K)^m r_0$. Notice that $\lambda_n K < 1$ for all sufficiently large n. Therefore,

$$\frac{\log N(V \cap \Lambda, r)}{-\log r} \le s + \frac{m \log c_n(s)}{-\log r} \le s + \frac{m \log c_n(s)}{-\log[(\lambda_n K)^m r_0]},$$

and

$$\overline{\dim}_B(V \cap \Lambda) \le s + \limsup_{m \to +\infty} \frac{m \log c_n(s)}{-\log[(\lambda_n K)^m r_0]} = s - \frac{\log c_n(s)}{\log(\lambda_n K)}.$$

Letting $s \searrow \overline{\dim}_B(V \cap \Lambda)$ we obtain

$$c_n(\overline{\dim}_B(V \cap \Lambda)) \ge 1.$$

On the other hand,

$$c_n(s) = \sum_{i_0 \cdots i_n} \overline{\lambda}_{i_0 \cdots i_n}^{\,s} = \sum_{i_0 \cdots i_n} \exp \max_{x \in R_{i_0 \cdots i_n}} \left(s \sum_{k=0}^{n} \varphi_u(f^k x) \right),$$

and

$$P(s\varphi_u) = \lim_{n \to \infty} \frac{1}{n} \log c_n(s) \ge 0 = P(t_u \varphi_u).$$

Since the function $s \mapsto P(s\varphi_u)$ is strictly decreasing it follows that $s \le t_u$ for every $s > \overline{\dim}_B(V \cap \Lambda)$, and thus $\overline{\dim}_B(V \cap \Lambda) \le t_u$.

Now we consider the Hausdorff dimension. We proceed by contradiction. Assume on the contrary that $\dim_H(V \cap \Lambda) < t_u$, and let s be a positive number such that

$$\dim_H(V \cap \Lambda) < s < t_u. \tag{4.46}$$

Then $m(V \cap \Lambda, s) = 0$, and since $V \cap \Lambda$ is compact, for each $\delta > 0$ there is a finite cover $\mathcal{U}$ of $V \cap \Lambda$ by open balls such that

$$\sum_{U \in \mathcal{U}} (\operatorname{diam} U)^s < \delta^s. \tag{4.47}$$

For each $n \in \mathbb{N}$, let δ_n be a positive number such that

$$p_n(U) := \operatorname{card}\{(i_0 \cdots i_n) : U \cap R_{i_0 \cdots i_n} \neq \varnothing\} < p$$

whenever $\operatorname{diam} U < \delta_n$ (the existence of δ_n follows easily from the properties of the Markov partition). We note that $\delta_n \to 0$ as $n \to \infty$. It follows from (4.47) with $\delta = \delta_n$ that $\operatorname{diam} \mathcal{U} < \delta_n$, and hence that $p_n(U) < p$ for every $U \in \mathcal{U}$. Set $N = n + m - 1$, for some $m \in \mathbb{N}$ such that $A^m > 0$ (recall that f is topologically mixing on Λ). For each $(i_0 i_1 \cdots) \in \Sigma_A$ and $n \in \mathbb{N}$, we consider the cover $\mathcal{U}_{i_0 \cdots i_N}$ of V composed of the projections along the stable leaves onto V of the sets $f^N U$ with $U \in \mathcal{U}$ such that $U \cap R_{i_0 \cdots i_n} \neq \varnothing$. We have

$$\sum_{U \in \mathcal{U}_{i_0 \cdots i_N}} (\operatorname{diam} U)^s \leq \underline{\lambda}_{i_0 \cdots i_N}^{-s} \sum_{U \in \mathcal{U},\, U \cap R_{i_0 \cdots i_n} \neq \varnothing} (\operatorname{diam} U)^s,$$

where

$$\underline{\lambda}_{i_0 \cdots i_n} = \min\{\|d_x f^{-n}|E^u(x)\| : x \in R_{i_0 \cdots i_n}\}.$$

Now assume that for every $(i_0 i_1 \cdots) \in \Sigma_A$ and $n \in \mathbb{N}$ we have

$$\sum_{U \in \mathcal{U}_{i_0 \cdots i_N}} (\operatorname{diam} U)^s \geq \delta_n{}^s.$$

Then

$$
\begin{aligned}
p\delta_n{}^s &> p \sum_{U \in \mathcal{U}} (\operatorname{diam} U)^s \geq \sum_{U \in \mathcal{U}} p_n(U)(\operatorname{diam} U)^s \\
&= \sum_{i_0 \cdots i_n} \sum_{U \in \mathcal{U},\, U \cap R_{i_0 \cdots i_n} \neq \varnothing} (\operatorname{diam} U)^s \\
&\geq p^{-m+1} \sum_{i_0 \cdots i_N} \sum_{U \in \mathcal{U},\, U \cap R_{i_0 \cdots i_n} \neq \varnothing} (\operatorname{diam} U)^s \\
&\geq p^{-m+1} \sum_{i_0 \cdots i_N} \left(\underline{\lambda}_{i_0 \cdots i_N} \sum_{U \in \mathcal{U}_{i_0 \cdots i_N}} (\operatorname{diam} U)^s \right) \\
&\geq p^{-m+1} \delta_n{}^s \sum_{i_0 \cdots i_N} \underline{\lambda}_{i_0 \cdots i_N}{}^s.
\end{aligned}
\tag{4.48}
$$

We observe that by a property analogous to the bounded distortion in (4.11) there is a constant $C > 0$ (independent of $n \in \mathbb{N}$ and $(i_0 \cdots i_n)$) such that

$$C^{-1} \leq \frac{\overline{\lambda}_{i_0 \cdots i_n}}{\underline{\lambda}_{i_0 \cdots i_n}} \leq C.$$

Therefore, by (4.48),

$$P(s\varphi_u) = \lim_{N \to \infty} \frac{1}{N} \sum_{i_0 \cdots i_N} \overline{\lambda}_{i_0 \cdots i_n}{}^s$$

$$\leq \lim_{N \to \infty} \frac{1}{N} \sum_{i_0 \cdots i_N} \underline{\lambda}_{i_0 \cdots i_n}{}^s \leq 0. \tag{4.49}$$

Since the function $s \mapsto P(s\varphi_u)$ is strictly decreasing and $P(t_u\varphi_u) = 0$, we have that (4.49) contradicts (4.46). Hence, we must have

$$\sum_{U \in \mathcal{U}_{i_0 \cdots i_N}} (\operatorname{diam} U)^s < \delta_n{}^s \tag{4.50}$$

for some finite sequence $(i_0 \cdots i_N)$ and all sufficiently large n (recall that $N = n + m - 1$). Now we restart the process using the cover $\mathcal{V}_1 = \mathcal{U}_{i_0 \cdots i_N}$ to find inductively finite covers $\mathcal{V}_l$ of $V \cap \Lambda$ for each $l \in \mathbb{N}$. By (4.50) we have diam $\mathcal{V}_l < \delta_n$, and hence $p_n(U) < p$ for every $U \in \mathcal{V}_l$. This implies that card $\mathcal{V}_{l+1} < \operatorname{card} \mathcal{V}_l$, and hence card $\mathcal{V}_l = 1$ for some $l = l(n)$. Writing $\mathcal{V}_{l(n)} = \{U_n\}$ we obtain

$$\operatorname{diam}(V \cap \Lambda) \leq \operatorname{diam} U_n < \delta_n \to 0 \quad \text{as} \quad n \to \infty,$$

which is impossible. This contradiction shows that $\dim_H(V \cap \Lambda) \geq t_u$. We have thus shown that

$$\dim_H(V \cap \Lambda) = \underline{\dim}_B(V \cap \Lambda) = \overline{\dim}_B(V \cap \Lambda) = t_u.$$

By (4.44) and (4.45) this establishes (4.40). The identities in (4.41) follow from similar arguments.

It remains to establish (4.43). Since f is conformal on Λ the local product structure is a Lipschitz homeomorphism with Lipschitz inverse, and thus

$$\dim_H[V_\rho^s(x) \cap \Lambda, V_\rho^u(x) \cap \Lambda] = \dim_H((V_\rho^s(x) \cap \Lambda) \times (V_\rho^u(x) \cap \Lambda)),$$
$$\underline{\dim}_B[V_\rho^s(x) \cap \Lambda, V_\rho^u(x) \cap \Lambda] = \underline{\dim}_B((V_\rho^s(x) \cap \Lambda) \times (V_\rho^u(x) \cap \Lambda)), \tag{4.51}$$
$$\overline{\dim}_B[V_\rho^s(x) \cap \Lambda, V_\rho^u(x) \cap \Lambda] = \overline{\dim}_B((V_\rho^s(x) \cap \Lambda) \times (V_\rho^u(x) \cap \Lambda)).$$

Since the inequalities

$$\dim_H A + \dim_H B \leq \dim_H(A \times B)$$

and

$$\overline{\dim}_B(A \times B) \leq \overline{\dim}_B A + \overline{\dim}_B B$$

hold for any subsets A and B of $\mathbb{R}^m$, it follows from (4.40), (4.41), and (4.51) that

$$\dim_H[V_\rho^s(x) \cap \Lambda, V_\rho^u(x) \cap \Lambda] = \underline{\dim}_B[V_\rho^s(x) \cap \Lambda, V_\rho^u(x) \cap \Lambda]$$
$$= \overline{\dim}_B[V_\rho^s(x) \cap \Lambda, V_\rho^u(x) \cap \Lambda] \tag{4.52}$$
$$= t_s + t_u.$$

On the other hand, since Λ is locally maximal we have

$$[V_\rho^s(x) \cap \Lambda, V_\rho^u(x) \cap \Lambda] \subset \Lambda$$

for every $x \in \Lambda$, and we can choose points $x_1, \ldots, x_N \in \Lambda$ such that

$$\Lambda = \bigcup_{n=1}^{N} [V_\rho^s(x_n) \cap \Lambda, V_\rho^u(x_n) \cap \Lambda].$$

The identities in (4.43) follow now immediately from (4.52). $\square$

McCluskey and Manning showed in [100] that

$$\dim_H(V_\rho^s(x) \cap \Lambda) = t_s \quad \text{and} \quad \dim_H(V_\rho^u(x) \cap \Lambda) = t_u$$

for every $x \in \Lambda$. We note that [100] does not contain a proof of the identity $\dim_H \Lambda = t_s + t_u$. The equality between the Hausdorff dimension and the lower and upper box dimensions is due to Takens [152] in the case of C^2 diffeomorphisms (see also [109]), and to Palis and Viana [110] in the general C^1 case. Barreira [3] and Pesin [115] presented alternative proofs of Theorem 4.3.2 based on the thermodynamic formalism. Our proof of Theorem 4.3.2 follows closely [3], which in its turn is inspired in arguments in [152] (see also [109]). The proof of Pesin in [115] uses essentially the same arguments as the proof of Theorem 4.1.7. With the purpose of illustrating an alternative argument (which incidentally could also be used to establish Theorem 4.1.7) we follow instead [3]. We observe that the proof depends crucially on the fact that f is conformal on Λ. In fact, any local product structure is always a Hölder continuous homeomorphism with Hölder continuous inverse. But in general it is not more than Hölder continuous for a generic diffeomorphism in a certain open set, in view of work of Schmeling in [140] (see also [144]). On the other hand, when f is conformal on Λ, any local product structure is a Lipschitz homeomorphism with Lipschitz inverse. It is this property that allows us to add the dimensions in the stable and unstable directions.

4.4   Dimension for nonconformal maps: brief notes

We already observed that the dimension theory of invariant sets of nonconformal maps (both invertible and noninvertible) still lacks a satisfactory approach in its most general version. Indeed, in most related works the authors make additional assumptions essentially to avoid two main types of difficulties: the lack of a clear separation between different Lyapunov directions, connected with a possible small regularity of the associated distributions (which typically are only Hölder continuous); and the existence of number-theoretical properties that may cause a variation of the Hausdorff dimension with respect to a certain typical value (such as the one in [51], see Theorem 4.4.2). Other authors have obtained results not for a concrete invariant set, but instead for an invariant of a typical transformation, such as for example for Lebesgue almost all values in some parameter space (although usually without knowing what happens for any specific value of the parameter). Moreover, in the case of nonconformal transformations we are often only able to establish estimates instead of giving a formula for the dimension of the invariant set. Thus, sometimes the emphasis is on how to obtain sharp dimension estimates.

Nevertheless, there exist some partial results towards a nonconformal theory, for certain classes of repellers and hyperbolic sets, starting essentially with the seminal work of Douady and Oesterlé [48]. We formulate some of these results, without proofs. Given a linear map $L: \mathbb{R}^n \to \mathbb{R}^n$, let

$$\sigma_1(L) \geq \cdots \geq \sigma_n(L) \geq 0$$

be the *singular values* of L, that is, the eigenvalues of $(L^*L)^{1/2}$, counted with their multiplicities, where L^* is the adjoint of L. Following [48], for each $s \in [0, n]$ we set

$$\omega_s(L) = \sigma_1(L) \cdots \sigma_{\lfloor s \rfloor}(L)\sigma_{\lfloor s \rfloor+1}(L)^{s-\lfloor s \rfloor}$$

where $\lfloor s \rfloor$ is the integer part of s. Using these numbers, Falconer [54] computed the Hausdorff dimension of a class of repellers of nonconformal transformations (building on his former work [52]). His main result can be reformulated as follows.

Theorem 4.4.1. *Let J be a repeller of a C^2 transformation f which is topologically mixing on J. If*

$$\|(d_x f)^{-1}\|^2 \|d_x f\| < 1 \quad \text{for every} \quad x \in J,$$

then

$$\dim_B J = \overline{\dim}_B J \leq s,$$

where s is the unique real number such that $P(\Phi_s) = 0$ for the sequence Φ_s formed by the functions $\varphi_{n,s}: J \to \mathbb{R}$, $n \in \mathbb{N}$ given by

$$\varphi_{n,s}(x) = \log \omega_s((d_x f^n)^{-1}).$$

Under an additional geometric assumption, satisfied for example when J contains a nondifferentiable arc, the number s in Theorem 4.4.1 is equal to $\dim_H J$ (see [54]). In another direction, Hu [77] computed the box dimension of a class of repellers of nonconformal transformations that leave invariant a strong unstable foliation. His formula for the box dimension is also expressed in terms of the topological pressure. Related results were obtained earlier by Bedford in [26] (see also [27]), for a class of self-similar sets that are graphs of continuous functions. In many of the former works the maps are assumed to be piecewise affine, sometimes with constant rates of expansion.

In another direction, Falconer [51] studied a class of limit sets of geometric constructions obtained from the composition of affine transformations that are not necessarily conformal. Consider affine transformations $f_i: \mathbb{R}^n \to \mathbb{R}^n$, $i = 1, \ldots, p$ given by $f_i(x) = A_i x + b_i$ for some linear contraction A_i and some vector $b_i \in \mathbb{R}^n$. Then there is a unique nonempty compact set $J \subset \mathbb{R}^n$ such that $J = \bigcup_{i=1}^p f_i(J)$

(see [78]), and for every nonempty compact set $R \subset \mathbb{R}^n$ such that $f_i(R) \subset R$ for $i = 1, \ldots, p$ we have

$$J = \bigcap_{k=1}^{\infty} \bigcup_{i_1 \cdots i_k} (f_{i_1} \circ \cdots \circ f_{i_k})(R).$$

Set

$$s = \inf \left\{ d \in [0, n] : \sum_{k=1}^{\infty} \sum_{i_1 \cdots i_k} w_d(A_{i_1} \circ \cdots \circ A_{i_k}) < \infty \right\}.$$

We emphasize that the number s does not depend on the vectors $b_1, \ldots, b_p$.

Theorem 4.4.2. *We have* $\overline{\dim}_B J \leq s$. *In addition, if* $\|A_i\| < 1/2$ *for* $i = 1, \ldots, p$, *then for Lebesgue almost every* $(b_1, \ldots, b_p) \in (\mathbb{R}^n)^p$ *we have*

$$\dim_H J = \underline{\dim}_B J = \overline{\dim}_B J = s.$$

The statement in Theorem 4.4.2 is due to Falconer [51] when $\|A_i\| < 1/3$ for $i = 1, \ldots, p$, and to Solomyak [151] in the general case.

Related ideas were applied by Simon and Solomyak in [149] to compute the Hausdorff dimension of a class of solenoids in $\mathbb{R}^3$, obtained from $C^{1+\alpha}$ transformations of the form

$$(x, y, z) \mapsto (\varphi(x, z) + a, \psi(y, z) + b, \zeta(z)).$$

We recall that a *solenoid* is a hyperbolic set $\Lambda = \bigcap_{n=1}^{\infty} f^n T$, where $T \subset \mathbb{R}^3$ is diffeomorphic to a solid torus $S^1 \times D$ for some closed disk $D \subset \mathbb{R}^2$, and $f : T \to T$ is a diffeomorphism such that for each $x \in S^1$ the *section*

$$\Lambda_x = f(T) \cap (\{x\} \times D)$$

is a disjoint union of a fixed number of sets homeomorphic to a closed disk. Assuming that $|\partial\varphi/\partial x|, |\partial\psi/\partial y| < 1/2$ and $|\zeta'| > 1$, it is shown in [149] how to compute the Hausdorff dimension of the solenoid for Lebesgue almost every $(a, b) \in \mathbb{R}^2$. Again, the dimension is expressed implicitly in terms of the topological pressure.

Bothe [35] and then Simon [148] (also using his methods in [147] for noninvertible transformations) studied earlier the dimension of solenoids (see [115, 145] for a related discussion). In particular, it is shown in [35] that under certain conditions on the diffeomorphism the map $x \mapsto \dim_H \Lambda_x$ is constant (even though the holonomies are typically not Lipschitz). More recently, Hasselblatt and Schmeling conjectured in [74] (see also [73]) that, in spite of the difficulties due to the possible low regularity of the holonomies, the Hausdorff dimension of hyperbolic sets can be computed by adding the dimensions of the stable and unstable sections. They prove this conjecture for a class of solenoids, by showing that the Hausdorff dimension of the sections is in fact independent of the section.

Chapter 5

Measures of Maximal Dimension

We establish in this chapter the existence of ergodic measures of maximal dimension for hyperbolic sets of conformal diffeomorphisms. This is a dimensional version of the existence of ergodic measures of maximal entropy. A crucial difference is that while the entropy map is upper semicontinuous, the map $\nu \mapsto \dim_H \nu$ is neither upper semicontinuous nor lower semicontinuous. Our approach is based on the thermodynamic formalism. It turns out that for a generic diffeomorphism with a hyperbolic set, there exists an ergodic measure of maximal Hausdorff dimension in a particular two-parameter family of equilibrium measures. On the other hand, generically there exists no measure of full dimension, in strong contrast with what happens in the case of repellers (see Chapter 4).

5.1 Basic notions and basic properties

Let Λ be a locally maximal hyperbolic set of a $C^{1+\varepsilon}$ diffeomorphism f, for some $\varepsilon > 0$, such that f is conformal and topologically mixing on Λ.

In an analogous manner to the case of repellers (see (4.17)) one can ask whether there exists an invariant measure μ in Λ such that

$$\dim_H \Lambda = \dim_H \mu. \tag{5.1}$$

It happens that, in strong contrast with the case of repellers, in general the answer is negative. We show below (see the discussion after Definition 5.1.3) that there exists an invariant measure μ in Λ satisfying (5.1) if and only if the functions $t_s \varphi_s$ and $t_u \varphi_u$ (see (4.39) and (4.42)) are cohomologous (see Definition 4.2.10), that is, if there exists a continuous function $\psi \colon \Lambda \to \mathbb{R}$ such that

$$t_s \varphi_s - t_u \varphi_u = \psi - \psi \circ f. \tag{5.2}$$

By Livschitz's theorem (see, for example, [84, Theorem 19.2.1]), this happens if and only if

$$\|d_x f^n | E^s(x)\|^{t_s} \|d_x f^n | E^u(x)\|^{t_u} = 1$$

for every $x \in \Lambda$ and $n \in \mathbb{N}$ such that $f^n x = x$.

Instead, one can ask whether the supremum

$$\delta(f) := \sup\{\dim_H \nu : \nu \text{ is an } f\text{-invariant probability measure in } \Lambda\} \qquad (5.3)$$

is attained. A related quantity was introduced by Denker and Urbański in [46] (with the supremum in (5.3) replaced by the supremum over all ergodic measures with positive entropy). This quantity has been intensively studied in one-dimensional complex dynamics (see [155] for details).

Definition 5.1.1. Any invariant probability measure attaining the supremum in (5.3) is called a *measure of maximal Hausdorff dimension* or simply a *measure of maximal dimension*.

The measures of maximal dimension attain maximal complexity from the point of view of dimension theory. In particular, each of them is a dimensional counterpart of the measures of maximal entropy, which attain maximal complexity from the point of view of entropy theory. We recall that under the assumptions of Theorem 4.3.2 the entropy map $\mu \mapsto h_\mu(f)$ is upper semicontinuous, and thus there exist measures of maximal entropy. The same does not happen with the Hausdorff dimension since the dimension map $\mu \mapsto \dim_H \mu$ is *never* upper semicontinuous. To verify that this is the case it is sufficient to consider the sequence of measures $(\mu + (n-1)\delta)/n$ where $\dim_H \mu > 0$ and δ is an atomic measure. Nevertheless, it was proved by Barreira and Wolf in [23] that the supremum in (5.3) is always attained and that it is attained at an ergodic measure (see Theorem 5.2.1 below). Before presenting the result we introduce some auxiliary material.

Since f is of class $C^{1+\varepsilon}$, the stable and unstable distributions E^s and E^u are Hölder continuous (see, for example, [84, Section 19.1]). Hence, the functions φ_s and φ_u in (4.39) are also Hölder continuous. Now let $\mathcal{M}$ be the family of f-invariant probability Borel measures in Λ equipped with the weak* topology, and let $\mathcal{M}_E \subset \mathcal{M}$ be the subset of all ergodic measures. Then $\mathcal{M}$ is a compact metrizable space. For each $\nu \in \mathcal{M}$ we define

$$\lambda_s(\nu) = \int_\Lambda \varphi_s \, d\nu \quad \text{and} \quad \lambda_u(\nu) = -\int_\Lambda \varphi_u \, d\nu. \qquad (5.4)$$

It was shown by Young in [165] (see Theorem 13.2.3 below) that

$$\text{if} \quad \nu \in \mathcal{M}_E, \quad \text{then} \quad \dim_H \nu = d(\nu), \qquad (5.5)$$

where

$$d(\nu) := h_\nu(f) \left(\frac{1}{\lambda_u(\nu)} - \frac{1}{\lambda_s(\nu)} \right).$$

We also need several additional properties of the topological pressure which we formulate without proof (we refer to [132] for details). Given $\varepsilon \in (0, 1]$, let $C^\varepsilon(\Lambda)$ be the space of Hölder continuous functions $\varphi \colon \Lambda \to \mathbb{R}$ with Hölder exponent ε.

Theorem 5.1.2. *The following properties hold:*

1. *the map $\varphi \mapsto P(\varphi)$ is analytic in $C^\varepsilon(\Lambda)$;*

2. *each function $\varphi \in C^\varepsilon(\Lambda)$ has a unique equilibrium measure $\nu_\varphi \in \mathfrak{M}$; furthermore ν_φ is ergodic, and given $\psi \in C^\varepsilon(\Lambda)$ we have*

$$\frac{d}{dt}P(\varphi + t\psi)\Big|_{t=0} = \int_\Lambda \psi \, d\nu_\varphi; \tag{5.6}$$

3. *for each $\varphi, \psi \in C^\varepsilon(\Lambda)$ we have $\nu_\varphi = \nu_\psi$ if and only if $\varphi - \psi$ is cohomologous to a constant;*

4. *for each $\varphi, \psi \in C^\varepsilon(\Lambda)$ and $t \in \mathbb{R}$ we have*

$$\frac{d^2}{dt^2}P(\varphi + t\psi) \geq 0, \tag{5.7}$$

with equality if and only if ψ is cohomologous to a constant.

The approach in [23] to establish the existence of (ergodic) measures of maximal dimension is based on the study of the function $Q \colon \mathbb{R}^2 \to \mathbb{R}$ defined by

$$Q(p, q) = P(p\varphi_u + q\varphi_s).$$

Since φ_s and φ_u are Hölder continuous, by property 1 in Theorem 5.1.2 the function Q is analytic. Furthermore, by property 2, for each $(p, q) \in \mathbb{R}^2$ the function $p\varphi_u + q\varphi_s$ has a unique equilibrium measure, say $\nu_{p,q} \in \mathfrak{M}_E$. We use the notation

$$\lambda_u(p, q) = \lambda_u(\nu_{p,q}), \quad \lambda_s(p, q) = \lambda_s(\nu_{p,q}), \quad h(p, q) = h_{\nu_{p,q}}(f).$$

Accordingly, we also think of λ_u, λ_s, and h as functions in $\mathbb{R}^2$. Note that by Theorem 2.3.3 we have

$$Q(p, q) = h(p, q) - p\lambda_u(p, q) + q\lambda_s(p, q). \tag{5.8}$$

We now briefly describe how these functions relate to dimension theory. Note that

$$0 = h(t_u, 0) - t_u\lambda_u(t_u, 0) \geq h_\nu(f) - t_u\lambda_u(\nu)$$

with strict inequality if and only if $\nu \neq \nu_{t_u,0}$, and that

$$0 = h(0, t_s) + t_s\lambda_s(0, t_s) \geq h_\nu(f) + t_s\lambda_s(\nu)$$

with strict inequality if and only if $\nu \neq \nu_{0,t_s}$. Therefore,

$$t_u = \max_{\nu \in \mathfrak{M}} \frac{h_\nu(f)}{\lambda_u(\nu)} = \frac{h(t_u, 0)}{\lambda_u(t_u, 0)} \quad \text{and} \quad t_s = \max_{\nu \in \mathfrak{M}} \frac{h_\nu(f)}{-\lambda_s(\nu)} = -\frac{h(0, t_s)}{\lambda_s(0, t_s)}. \tag{5.9}$$

Furthermore, the maxima in (5.9) are respectively uniquely attained at the measures $\nu_{t_u,0}$ and ν_{0,t_s}.

Definition 5.1.3. A measure μ in Λ is called a *measure of full dimension* if $\dim_H \Lambda = \dim_H \mu$.

Together with (5.5) and (4.43), the uniqueness of the maxima in (5.9) implies that there exists a measure $\mu \in \mathcal{M}_E$ of full dimension if and only if $\nu_{t_u,0} = \nu_{0,t_s}$, in which case $\mu = \nu_{t_u,0} = \nu_{0,t_s}$. In view of (4.42) and Theorem 5.1.2 this occurs if and only if identity (5.2) holds for some continuous function $\psi \colon \Lambda \to \mathbb{R}$. We notice that if there exists a measure of full dimension in $\mathcal{M}_E$, then it is unique.

Example 5.1.4. We show that if f preserves volume, then there exists an ergodic invariant measure of full dimension, and that in this case we have $t_u = t_s$. A short argument is the following. If $f^n x = x \in \Lambda$ and $k \in \mathbb{Z}$, then

$$1 = |\det d_x f^{kn}| = \exp[k\varphi_s(f^n x) - k\varphi_u(f^n x)] \sin \angle(E^u(x), E^s(x)).$$

Letting $k \to \pm\infty$ we obtain

$$\varphi_s(f^n x) - \varphi_u(f^n x) = 0$$

whenever $f^n x = x \in \Lambda$. Therefore, by Livschitz's theorem (see, for example, [84, Theorem 19.2.1]), the function $\varphi_s - \varphi_u$ is cohomologous to zero. This implies that $Q(t,0) = Q(0,t)$ for every $t \in \mathbb{R}$, and hence $t_u = t_s$. Therefore $t_u \varphi_u$ is cohomologous to $t_s \varphi_s$, and $\nu_{t_u,0} = \nu_{0,t_s}$ is the ergodic invariant measure of full dimension.

In the case of hyperbolic polynomial automorphisms of $\mathbb{C}^2$ it was shown by Wolf in [164] that if there exists an ergodic invariant measure of full dimension, then either the map is volume preserving, or φ_s and φ_u are both cohomologous to a constant (in which case the ergodic invariant measure of full dimension coincides with the measure of maximal entropy).

5.2   Existence of measures of maximal dimension

The existence of ergodic invariant measures of maximal dimension was established by Barreira and Wolf in [23] for diffeomorphisms on surfaces with one-dimensional stable and unstable distributions. Their approach extends without change to the more general case of hyperbolic sets of conformal maps.

Theorem 5.2.1 (Measures of maximal dimension). *Let Λ be a locally maximal hyperbolic set of a $C^{1+\varepsilon}$ diffeomorphism f, for some $\varepsilon > 0$, such that f is conformal and topologically mixing on Λ. Then*

$$\delta(f) = \max\{\dim_H \mu : \mu \in \mathcal{M}_E\}.$$

Proof. Since the maps $\nu \mapsto \lambda_u(\nu)$ and $\nu \mapsto \lambda_s(\nu)$ defined by (5.4) are continuous in $\mathcal{M}$, and $\mathcal{M}$ is compact, we can set

$$\lambda_u^{\min} = \min \lambda_u(\mathcal{M}), \quad \lambda_u^{\max} = \max \lambda_u(\mathcal{M}),$$

and

$$\lambda_s^{\min} = \min \lambda_s(\mathcal{M}), \quad \lambda_s^{\max} = \max \lambda_s(\mathcal{M}).$$

We consider the intervals

$$I_u = (\lambda_u^{\min}, \lambda_u^{\max}) \quad \text{and} \quad I_s = (\lambda_s^{\min}, \lambda_s^{\max}).$$

Note that $I_u \neq \varnothing$ if and only if φ_u is not cohomologous to a constant, and that $I_s \neq \varnothing$ if and only if φ_s is not cohomologous to a constant. We also consider the functions

$$d_u(p, q) = h(p, q)/\lambda_u(p, q) \quad \text{and} \quad d_s(p, q) = -h(p, q)/\lambda_s(p, q). \tag{5.10}$$

It follows from (5.6) that

$$\frac{\partial Q}{\partial p} = -\lambda_u \quad \text{and} \quad \frac{\partial Q}{\partial q} = \lambda_s. \tag{5.11}$$

Since Q is analytic, the functions λ_u and λ_s are also analytic. We conclude from (5.8) that h is analytic, and it follows from (5.10) that the functions d_u and d_s are also analytic. Now we establish a few additional properties of these functions.

Lemma 5.2.2. *The following properties hold:*

1. *if φ_u is not cohomologous to a constant and $q \in \mathbb{R}$, then:*

   (a) *$\lambda_u(\cdot, q)$ is strictly decreasing and $\{\lambda_u(p, q) : p \in \mathbb{R}\} = I_u$;*

   (b) *$h(\cdot, 0)$ is strictly decreasing in $[0, \infty)$;*

   (c) *$d_u(\cdot, 0)$ is strictly increasing in $(-\infty, t_u]$ and is strictly decreasing in $[t_u, \infty)$.*

2. *if φ_s is not cohomologous to a constant and $p \in \mathbb{R}$, then:*

   (a) *$\lambda_s(p, \cdot)$ is strictly decreasing and $\{\lambda_s(p, q) : q \in \mathbb{R}\} = I_s$;*

   (b) *$h(0, \cdot)$ is strictly decreasing in $[0, \infty)$;*

   (c) *$d_s(0, \cdot)$ is strictly increasing in $(-\infty, t_s]$ and is strictly decreasing in $[t_s, \infty)$.*

Proof of the lemma. Assume that φ_u is not cohomologous to a constant and fix $q \in \mathbb{R}$. By (5.7) and (5.11) we have

$$\frac{\partial \lambda_u}{\partial p} = -\frac{\partial^2 Q}{\partial p^2} < 0, \tag{5.12}$$

and thus $\lambda_u(\cdot, q)$ is strictly decreasing. The continuity of the function $\lambda_u(\cdot, q)$ implies that $\{\lambda_u(p, q) : p \in \mathbb{R}\}$ is an open interval. We claim that

$$\lim_{p \to \infty} \lambda_u(p, q) = \lambda_u^{\min} \quad \text{and} \quad \lim_{p \to -\infty} \lambda_u(p, q) = \lambda_u^{\max}. \tag{5.13}$$

If the first identity did not hold, then there would exist $\nu \in \mathcal{M}$ and $\delta > 0$ such that $\lambda_u(\nu) + \delta < \lambda_u(p, q)$ for every $p \in \mathbb{R}$. Now we take $p > 0$ satisfying

$$p\delta > h(f) - q\lambda_s(\nu) + q\lambda_s(p, q)$$

(such a p always exists since the function $\lambda_s(\cdot, q)$ is bounded). We obtain

$$
\begin{aligned}
Q(p, q) &= h(p, q) - p\lambda_u(p, q) + q\lambda_s(p, q) \\
&< h(f) - p(\lambda_u(\nu) + \delta) + q\lambda_s(p, q) \\
&< h_\nu(f) - p\lambda_u(\nu) + q\lambda_s(\nu),
\end{aligned}
$$

which contradicts Theorem 2.3.3. This establishes the first identity in (5.13). A similar argument establishes the second identity and property 1a follows.

It follows from (5.8) that

$$h(p, 0) = Q(p, 0) + p\lambda_u(p, 0).$$

Using (5.11) and (5.12) it is straightforward to verify that

$$\frac{\partial h}{\partial p}(p, 0) = p\frac{\partial \lambda_u}{\partial p}(p, 0). \tag{5.14}$$

This establishes property 1b.

Finally, using (5.8), (5.12), and (5.14) we obtain

$$
\begin{aligned}
\frac{\partial d_u}{\partial p}(p, 0) &= \frac{p\partial \lambda_u/\partial p(p, 0)\lambda_u(p, 0) - h(p, 0)\partial \lambda_u/\partial p(p, 0)}{\lambda_u(p, 0)^2} \\
&= -\frac{\partial \lambda_u}{\partial p}(p, 0)\frac{Q(p, 0)}{\lambda_u(p, 0)^2} \\
&= \frac{\partial^2 Q}{\partial p^2}(p, 0)\frac{Q(p, 0)}{\lambda_u(p, 0)^2}.
\end{aligned}
\tag{5.15}
$$

On the other hand, it follows from Theorem 2.3.3 that the function $Q(\cdot, q)$ is strictly decreasing. This implies that

$$Q(p, 0) > Q(t_u, 0) = 0 \quad \text{for} \quad p < t_u,$$

and that

$$Q(p, 0) < Q(t_u, 0) = 0 \quad \text{for} \quad p > t_u.$$

Property 1c follows now immediately from (5.12) and (5.15).

The proofs of the statements for the stable part are entirely analogous.   □

Using Lemma 5.2.2 we can introduce two curves that are crucial for our approach.

Lemma 5.2.3. *The following properties hold:*

1. *for each $a \in I_u$ there exists a unique function $\gamma_u \colon \mathbb{R} \to \mathbb{R}$ satisfying*

$$\lambda_u(\gamma_u(q), q) = a \quad \text{for every} \quad q \in \mathbb{R},$$

and γ_u is analytic;

2. *for each $b \in I_s$ there exists a unique function $\gamma_s \colon \mathbb{R} \to \mathbb{R}$ satisfying*

$$\lambda_s(p, \gamma_s(p)) = b \quad \text{for every} \quad p \in \mathbb{R},$$

and γ_s is analytic.

Proof of the lemma. We only prove the second statement. The proof of the first statement is entirely analogous. Let $b \in I_s$. In particular $I_s \neq \varnothing$, and φ_s is not cohomologous to a constant. By statement 2a in Lemma 5.2.2 and (5.11), for each $p \in \mathbb{R}$ there exists a unique number $\gamma_s(p) \in \mathbb{R}$ such that

$$\frac{\partial Q}{\partial q}(p, \gamma_s(p)) = \lambda_s(p, \gamma_s(p)) = b.$$

Since φ_s is not cohomologous to a constant, we have $\partial^2 Q/\partial q^2(p, q) > 0$ for every $(p, q) \in \mathbb{R}^2$. It follows from the implicit function theorem that the map $p \mapsto \gamma_s(p)$ is analytic. $\square$

We proceed with the proof of the theorem. Let $(\nu_n)_{n \in \mathbb{N}}$ be a sequence of measures in $\mathcal{M}_E$ such that

$$\lim_{n \to \infty} \dim_H \nu_n = \sup\{\dim_H \nu : \nu \in \mathcal{M}_E\}. \tag{5.16}$$

Since $\mathcal{M}$ is compact in the weak$*$ topology, we can also assume that $(\nu_n)_{n \in \mathbb{N}}$ converges to some measure $m \in \mathcal{M}$. Since the entropy map $\mathcal{M} \ni \nu \mapsto h_\nu(f)$ is upper semicontinuous, it follows from (5.5) and from the continuity of the maps $\nu \mapsto \lambda_u(\nu)$ and $\nu \mapsto \lambda_s(\nu)$ that

$$\lim_{n \to \infty} \dim_H \nu_n \leq d(m). \tag{5.17}$$

By (5.16) and (5.17) we obtain

$$\sup\{\dim_H \nu : \nu \in \mathcal{M}_E\} \leq d(m). \tag{5.18}$$

Therefore, in order to establish the existence of a measure $\mu \in \mathcal{M}_E$ satisfying

$$\dim_H \mu = \sup\{\dim_H \nu : \nu \in \mathcal{M}_E\}, \tag{5.19}$$

it is sufficient to show that there exists $\mu \in \mathcal{M}_E$ such that

$$\dim_H \mu = d(m). \tag{5.20}$$

Clearly, any measure $\mu \in \mathcal{M}_E$ satisfing (5.20) also satisfies (5.19). We note that when m is ergodic, it follows from (5.5) that $\dim_H m = d(m)$, and hence (5.19) holds for $\mu = m$. However, m may not be ergodic.

Set $a = \lambda_u(m)$ and $b = \lambda_s(m)$. By Lemma 5.2.3, when $a \in I_u$ (respectively $b \in I_s$) we can consider the curve γ_u (respectively γ_s) associated to the number a (respectively b). We first prove some auxiliary statements.

Lemma 5.2.4. *If $\lambda_s(m) \in I_s$, then there exists $p \in [0, h_m(f)/\lambda_u(m)]$ such that $\lambda_u(p, \gamma_s(p)) = \lambda_u(m)$.*

Proof of the lemma. The assumption $\lambda_s(m) \in I_s$ guarantees that the function γ_s is well-defined. Since $\nu_{p,\gamma_s(p)}$ is the equilibrium measure of $p\varphi_u + \gamma_s(p)\varphi_s$ we have

$$h(p, \gamma_s(p)) - p\lambda_u(p, \gamma_s(p)) + \gamma_s(p)\lambda_s(p, \gamma_s(p))$$
$$\geq h_m(f) - p\lambda_u(m) + \gamma_s(p)\lambda_s(m) \quad (5.21)$$

for every $p \in \mathbb{R}$. Note that $\lambda_u(p, \gamma_s(p)) > 0$. It is straightforward to verify that

$$\frac{h(p, \gamma_s(p))}{\lambda_u(p, \gamma_s(p))} - \frac{h_m(f)}{\lambda_u(m)} \geq \left(1 - \frac{\lambda_u(m)}{\lambda_u(p, \gamma_s(p))}\right)\left(p - \frac{h_m(f)}{\lambda_u(m)}\right). \quad (5.22)$$

Let $\kappa = h_m(f)/\lambda_u(m)$. Setting $p = \kappa$, it follows from (5.22) that

$$h(\kappa, \gamma_s(\kappa))/\lambda_u(\kappa, \gamma_s(\kappa)) \geq h_m(f)/\lambda_u(m). \quad (5.23)$$

Now assume that

$$\lambda_u(\kappa, \gamma_s(\kappa)) > \lambda_u(m).$$

By (5.23) we obtain $h(\kappa, \gamma_s(\kappa)) > h_m(f)$. It follows from (5.5) and (5.23) that $\dim_H \nu_{\kappa,\gamma_s(\kappa)} > d(m)$. This contradicts (5.18), and thus we must have

$$\lambda_u(\kappa, \gamma_s(\kappa)) \leq \lambda_u(m). \quad (5.24)$$

On the other hand, it follows from (5.5) and (5.18) that

$$\frac{h(0, \gamma_s(0))}{\lambda_u(0, \gamma_s(0))} - \frac{h(0, \gamma_s(0))}{\lambda_s(m)} \leq \frac{h_m(f)}{\lambda_u(m)} - \frac{h_m(f)}{\lambda_s(m)}. \quad (5.25)$$

Setting $p = 0$ in (5.21) we obtain $h(0, \gamma_s(0)) \geq h_m(f)$, and it follows from (5.25) that

$$\lambda_u(0, \gamma_s(0)) \geq \lambda_u(m). \quad (5.26)$$

By the continuity of the function $p \mapsto \lambda_u(p, \gamma_u(p))$ together with (5.24) and (5.26), there exists $p \in [0, \kappa]$ such that $\lambda_u(p, \gamma_s(p)) = \lambda_u(m)$. This completes the proof of the lemma. $\qquad\square$

Lemma 5.2.5. *Assume that neither φ_u nor φ_s are cohomologous to a constant. Then $\lambda_u(m) \in I_u$ if and only if $\lambda_s(m) \in I_s$.*

Proof of the lemma. Assume that $\lambda_s(m) \in I_s$. By Lemma 5.2.4, there exists p such that $\lambda_u(p, \gamma_s(p)) = \lambda_u(m)$. Lemma 5.2.2 implies that $\lambda_u(p, \gamma_s(p)) \in I_u$, and hence $\lambda_u(m) \in I_u$. A similar argument together with the corresponding version of Lemma 5.2.4 show that $\lambda_s(m) \in I_s$ whenever $\lambda_u(m) \in I_u$. □

We note that by Lemma 5.2.5 it is sufficient to consider the following four cases:

1. $\lambda_s(m) \in I_s$ and $\lambda_u(m) \in I_u$;

2. $\lambda_s(m) \in I_s$ and φ_u is cohomologous to a constant;

3. $\lambda_u(m) \in I_u$ and φ_s is cohomologous to a constant;

4. $\lambda_s(m) \notin I_s$ and $\lambda_u(m) \notin I_u$.

We continue with an auxiliary statement.

Lemma 5.2.6. *If $p, q \in \mathbb{R}$ are such that $\lambda_u(p, q) = \lambda_u(m)$ and $\lambda_s(p, q) = \lambda_s(m)$, then $m = \nu_{p,q}$.*

Proof of the lemma. We have

$$h(p, q) + \int_\Lambda (p\varphi_u + q\varphi_s) \, d\nu_{p,q} = h(p, q) - p\lambda_u(m) + q\lambda_s(m)$$

$$\geq h_m(f) + \int_\Lambda (p\varphi_u + q\varphi_s) \, dm.$$

Hence $h(p, q) \geq h_m(f)$, with equality if and only if $\nu_{p,q} = m$. On the other hand, combining (5.5) with (5.18) yields that $h(p, q) \leq h_m(f)$. Therefore $h(p, q) = h_m(f)$ and $m = \nu_{p,q}$. □

Now we consider each of the above four cases.

Lemma 5.2.7. *If $\lambda_u(m) \in I_u$ and $\lambda_s(m) \in I_s$, then there exist $p, q \in \mathbb{R}$ such that $(p, \gamma_s(p)) = (\gamma_u(q), q)$ and $m = \nu_{p,q}$.*

Proof of the lemma. The hypotheses guarantee that the curves γ_u and γ_s are well-defined. Since $\lambda_s(p, \gamma_s(p)) = \lambda_s(m)$, it follows from Lemma 5.2.4 and the uniqueness of γ_u that $(p, \gamma_s(p)) = (\gamma_u(q), q)$ for some $p, q \in \mathbb{R}$. In particular,

$$\lambda_u(p, q) = \lambda_u(m) \quad \text{and} \quad \lambda_s(p, q) = \lambda_s(m).$$

It follows from Lemma 5.2.6 that $m = \nu_{p,q}$. □

Lemma 5.2.8. *If $\lambda_s(m) \in I_s$ and φ_u is cohomologous to a constant, then there exist $p, q \in \mathbb{R}$ such that $m = \nu_{p,q}$.*

Proof of the lemma. Since $\lambda_s(m) \in I_s$, the curve γ_s is well-defined, and we have $\lambda_s(p, \gamma_s(p)) = \lambda_s(m)$ for every p. On the other hand, the cohomological assumption ensures that $\lambda_u(p, \gamma_s(p)) = \lambda_u(m)$. Setting $q = \gamma_s(p)$ we obtain

$$\lambda_u(p,q) = \lambda_u(m) \quad \text{and} \quad \lambda_s(p,q) = \lambda_s(m).$$

It follows from Lemma 5.2.6 that $m = \nu_{p,q}$.   □

An analogous argument establishes the following.

Lemma 5.2.9. *If $\lambda_u(m) \in I_u$ and φ_s is cohomologous to a constant, then there exist $p, q \in \mathbb{R}$ such that $m = \nu_{p,q}$.*

Finally we consider the fourth case.

Lemma 5.2.10. *If $\lambda_u(m) \notin I_u$ and $\lambda_s(m) \notin I_s$, then:*

1. $\lambda_u(m) = \lambda_u^{\min}$ *and* $\lambda_s(m) = \lambda_s^{\max}$;

2. *there exists a measure $\nu \in \mathcal{M}_E$ such that*

$$\lambda_u(\nu) = \lambda_u(m), \quad \lambda_s(\nu) = \lambda_s(m), \quad \text{and} \quad h_\nu(f) = h_m(f).$$

Proof of the lemma. We first establish property 1. When $I_u = I_s = \varnothing$ (that is, when φ_u and φ_s are both cohomologous to constants) there is nothing to prove. Now assume that

$$I_u = \varnothing, \quad I_s \neq \varnothing, \quad \text{and} \quad \lambda_s(m) = \lambda_s^{\min}. \tag{5.27}$$

Since $\nu_{0,0}$ is the measure of maximal entropy, we have $h(0,0) \geq h_m(f)$. Hence, it follows from $\lambda_u(0,0) = \lambda_u^{\min}$, statement 2a in Lemma 5.2.2, and (5.5) that $\dim_H \nu_{0,0} > d(m)$. But this contradicts (5.18), and hence (5.27) cannot occur. Analogously we can show that it is impossible to have $I_s = \varnothing$, $I_u \neq \varnothing$, and $\lambda_u(m) = \lambda_u^{\max}$.

In order to complete the proof of property 1 it remains to consider the case when $I_u \neq \varnothing$ and $I_s \neq \varnothing$. We then have

$$\lambda_u(m) \in \partial I_u = \{\lambda_u^{\min}, \lambda_u^{\max}\} \quad \text{and} \quad \lambda_s(m) \in \partial I_s = \{\lambda_s^{\min}, \lambda_s^{\max}\}.$$

Assume first that

$$\lambda_u(m) = \lambda_u^{\max} \quad \text{and} \quad \lambda_s(m) = \lambda_s^{\min}. \tag{5.28}$$

Since $\nu_{0,0}$ is the measure of maximal entropy, we have $h(0,0) \geq h_m(f)$. On the other hand, it follows from Lemma 5.2.2 that

$$\lambda_u(0,0) < \lambda_u(m) \quad \text{and} \quad \lambda_s(0,0) > \lambda_s(m).$$

By (5.5) we obtain $\dim_H \nu_{0,0} > d(m)$. But this contradicts (5.18), and hence (5.28) cannot occur. Assume now that

$$\lambda_u(m) = \lambda_u^{\min} \quad \text{and} \quad \lambda_s(m) = \lambda_s^{\min}. \tag{5.29}$$

We claim that
$$h(p, 0) > h_m(f) \tag{5.30}$$
for every $p > 0$. Otherwise, if $h(p, 0) \le h_m(f)$ for some $p > 0$, then it would follow from Lemma 5.2.2 that
$$h(p, 0) - p\lambda_u(p, 0) < h_m(f) - p\lambda_u(m).$$
But this is impossible since $\nu_{p,0}$ is the equilibrium measure of $p\varphi_u$. We also claim that
$$d_u(p, 0) \ge h_m(f)/\lambda_u(m) \tag{5.31}$$
for all sufficiently large p (see (5.10) for the definition of the function d_u). Otherwise, Lemma 5.2.2 would guarantee the existence of $p_0 \in \mathbb{R}$ and $\delta > 0$ such that
$$d_u(p, 0) + \delta < h_m(f)/\lambda_u(m)$$
for every $p \ge p_0$. It would then follow from (5.13) that $h_m(f) > h(p, 0)$ for all sufficiently large p. But this contradicts (5.30), and hence (5.31) holds for all sufficiently large p. It follows from (5.29)–(5.31) that
$$\dim_H \nu_{p,0} = d_u(p, 0) + d_s(p, 0) \ge \frac{h_m(f)}{\lambda_u(m)} - \frac{h(p, 0)}{\lambda_s(p, 0)} > d(m)$$
for all sufficiently large p. This contradicts (5.18), and hence (5.29) cannot occur. Analogously one can show that it is impossible to have
$$\lambda_u(m) = \lambda_u^{\max} \quad \text{and} \quad \lambda_s(m) = \lambda_s^{\max}.$$
Therefore, property 1 holds.

To establish property 2 we consider an ergodic decomposition τ of the measure m, i.e., a probability Borel measure in the metrizable space $\mathcal{M}$ with $\tau(\mathcal{M}_E) = 1$ such that
$$\int_{\mathcal{M}} \int_{\Lambda} \varphi \, d\nu \, d\tau(\nu) = \int_{\Lambda} \varphi \, dm \tag{5.32}$$
for every continuous function $\varphi \colon \Lambda \to \mathbb{R}$ (see also Definition 13.1.2 below). Setting $\varphi = \varphi_u$ in (5.32) yields
$$\lambda_u^{\min} = \lambda_u(m) = \int_{\mathcal{M}} \lambda_u(\nu) \, d\tau(\nu).$$
Since $\lambda_u(\nu) \ge \lambda_u^{\min}$ for every $\nu \in \mathcal{M}$, there exists $A_1 \subset \mathcal{M}_E$ with $\tau(A_1) = 1$ such that $\lambda_u(\nu) = \lambda_u^{\min}$ for every $\nu \in A_1$. Analogously, there exists $A_2 \subset \mathcal{M}_E$ with $\tau(A_2) = 1$ such that $\lambda_s(\nu) = \lambda_s^{\max}$ for every $\nu \in A_2$. We conclude from (5.5) and (5.18) that $h_\nu(f) \le h_m(f)$ for every $\nu \in A_1 \cap A_2$. On the other hand, since
$$\tau(A_1 \cap A_2) = 1 \quad \text{and} \quad h_m(f) = \int_{\mathcal{M}} h_\nu(f) \, d\tau(\nu)$$
(see, for example, [45]), there exists $A \subset A_1 \cap A_2$ with $\tau(A) = 1$ such that $h_\nu(f) = h_m(f)$ for every $\nu \in A$. This completes the proof of the lemma. $\qquad \square$

By Lemmas 5.2.7–5.2.10, in each of the four cases there exists a measure $\mu \in \mathcal{M}_E$ satisfying (5.20), namely each measure $\nu_{p,q}$ in Lemmas 5.2.7–5.2.9, and each measure ν in statement 2 in Lemma 5.2.10. This completes the proof of the theorem.   □

See [164] for a related result of Wolf in the particular case of polynomial automorphisms of $\mathbb{C}^2$. It was shown by Rams in [127] that in general there exists no unique ergodic invariant measure of maximal dimension, even in the case of linear horseshoes (more precisely, it was shown in [127] that there exists a one-parameter family of Bernoulli measures of maximal dimension). We refer to [23] for further properties of the measures of maximal dimension.

Part II

Multifractal Analysis: Core Theory

Chapter 6

Multifractal Analysis of Equilibrium Measures

The objective of this chapter is to present the multifractal analysis of repellers and hyperbolic sets of conformal maps. Multifractal analysis is a subarea of the dimension theory of dynamical systems. Briefly speaking, it studies the complexity of the level sets of invariant local quantities obtained from a dynamical system. For example, we can consider Birkhoff averages, Lyapunov exponents, pointwise dimensions, and local entropies. These functions are usually only measurable and thus their level sets are rarely manifolds. Therefore, to measure the complexity of these sets it is appropriate to use quantities such as the topological entropy or the Hausdorff dimension.

6.1 Dimension spectrum for repellers

Let ν be a finite Borel measure in a metric space X, and let $Y \subset X$ be the set of points $x \in X$ for which the pointwise dimension in (2.9) is well-defined. We consider the level sets of the pointwise dimension, that is, the sets

$$K_\alpha = \left\{ x \in Y : \lim_{r \to 0} \frac{\log \nu(B(x,r))}{\log r} = \alpha \right\} \tag{6.1}$$

for each $\alpha \in [-\infty, +\infty]$, where $B(x,r) \subset X$ is the ball of radius r centered at x. These sets are pairwise disjoint and we obtain a *multifractal decomposition* of X given by

$$X = (X \setminus Y) \cup \bigcup_{\alpha \in [-\infty, +\infty]} K_\alpha. \tag{6.2}$$

One way to measure the complexity of the sets K_α is to compute their Hausdorff dimension.

Definition 6.1.1. The *dimension spectrum* of the measure ν is the function

$$\mathcal{D} = \mathcal{D}_\nu \colon \{\alpha \in [-\infty, +\infty] : K_\alpha \neq \varnothing\} \to \mathbb{R}$$

defined by

$$\mathcal{D}(\alpha) = \dim_H K_\alpha.$$

Now let J be a repeller of a $C^{1+\varepsilon}$ transformation f, for some $\varepsilon > 0$ (see Section 4.1 for the definition). We always assume in this section that f is conformal and topologically mixing on J. Consider a Hölder continuous function $\psi \colon J \to \mathbb{R}$ with $P(\psi) = 0$, where P denotes the topological pressure on J. We define a function $T \colon \mathbb{R} \to \mathbb{R}$ by requiring that

$$P(-T(q)\log\|df\| + q\psi) = 0 \tag{6.3}$$

for every $q \in \mathbb{R}$. To verify that the function T is well-defined, we first observe that the statement in Theorem 5.1.2 also holds for repellers, simply with Λ replaced by J (we refer to [132] for details). In particular, the map $Q \colon \mathbb{R}^2 \to \mathbb{R}$ defined by

$$Q(t, q) = P(-t\log\|df\| + q\psi) \tag{6.4}$$

is analytic. Furthermore, by (5.6),

$$\frac{\partial Q}{\partial t}(t, q) = -\int_J \log\|df\|\, d\nu_{t,q},$$

where $\nu_{t,q}$ is the unique equilibrium measure of $-t\log\|df\| + q\psi$. Proceeding as in (4.7) and using (4.1) we obtain

$$\frac{\partial Q}{\partial t}(t, q) \leq -\log\beta < 0 \quad \text{for every} \quad t, q \in \mathbb{R}.$$

It follows from the implicit function theorem that T is well-defined and analytic. See Figure 6.1 for a typical graph of the function T.

The following result of Pesin and Weiss in [119] describes the dimension spectrum $\mathcal{D}$ of a Gibbs measure in a repeller of a conformal map. See Figure 6.2 for a typical graph of the function $\mathcal{D}$. Set

$$\alpha(q) = -T'(q)$$

and let $(\underline{\alpha}, \overline{\alpha})$ be the range of the function $\alpha(q)$. We also denote by μ the equilibrium measure of $-(\dim_H J)\log\|df\|$ (this is the measure of maximal dimension; see Theorem 4.1.8), and by ν_q the equilibrium measure of the function

$$-T(q)\log\|df\| + q\psi.$$

Notice that $\mu = \nu_0$.

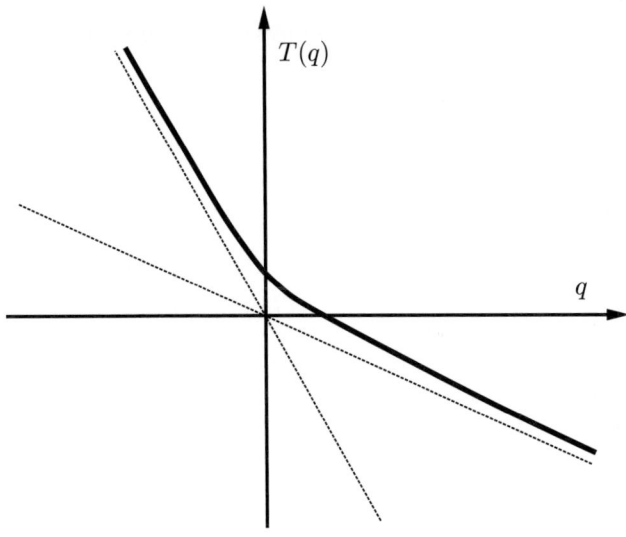

Figure 6.1: Typical graph of the function T

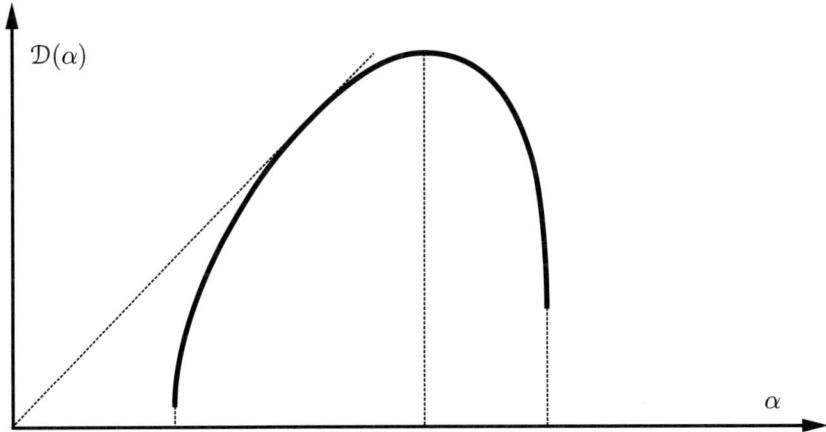

Figure 6.2: Typical graph of the function $\mathcal{D}$ (in fact, for a generic potential, the spectrum $\mathcal{D}$ is zero at the endpoints of its domain; see Theorem 7.6.1)

Theorem 6.1.2 (Multifractal analysis of equilibrium measures). *Let J be a repeller of a $C^{1+\varepsilon}$ transformation f, for some $\varepsilon > 0$, such that f is conformal and topologically mixing on J. If ν is the equilibrium measure of a Hölder continuous function $\psi\colon J \to \mathbb{R}$ with $P(\psi) = 0$, then:*

1. *the set $K_{\alpha(q)}$ is f-invariant and dense for every $q \in \mathbb{R}$;*

2. *if $\nu = \mu$, then $\underline{\alpha} = \overline{\alpha} = \dim_H J$ and $\mathcal{D}$ is a delta function;*

3. *if $\nu \neq \mu$, then $\mathcal{D}\colon (\underline{\alpha}, \overline{\alpha}) \to \mathbb{R}$ is analytic and strictly convex;*

4. *$\mathcal{D}$ is the Legendre transform of T, that is, for each $q \in \mathbb{R}$ we have*

$$\mathcal{D}(\alpha(q)) = T(q) + q\alpha(q); \tag{6.5}$$

5. *for each $q \in \mathbb{R}$ we have $\nu_q(K_{\alpha(q)}) = 1$ and*

$$\lim_{r \to 0} \frac{\log \nu_q(B(x,r))}{\log r} = T(q) + q\alpha(q)$$

for ν_q-almost every $x \in K_{\alpha(q)}$.

Proof. Let $R_1, \dots, R_p$ be the elements of a Markov partition of J. Since ν_q is an equilibrium measure of a Hölder continuous function it is a Gibbs measure and there exist constants $D_1, D_2 > 0$ such that

$$D_1 \leq \frac{\nu_q(\Delta_{i_1 \cdots i_n})}{\|d_x f^n\|^{-T(q)} \exp\left(q \sum_{k=0}^{n-1} \psi(f^k x)\right)} \leq D_2 \tag{6.6}$$

for every $n \in \mathbb{N}$ and $x \in \Delta_{i_1 \cdots i_n}$. On the other hand, taking derivatives with respect to q in (6.3) we obtain

$$0 = T'(q)\frac{\partial Q}{\partial t}(T(q), q) + \frac{\partial Q}{\partial q}(T(q), q),$$

and using (5.6),

$$0 = -T'(q) \int_J \log \|df\| \, d\nu_{T(q),q} + \int_J \psi \, d\nu_{T(q),q}.$$

Since $\nu_{T(q),q} = \nu_q$ we obtain

$$\alpha(q) = -T'(q) = -\frac{\int_J \psi \, d\nu_q}{\int_J \log \|df\| \, d\nu_q}. \tag{6.7}$$

For each $q \in \mathbb{R}$ set

$$J_q = \left\{ \omega \in \Sigma_A^+ : -\lim_{n \to \infty} \frac{\sum_{k=0}^{n-1} \psi(\chi(\sigma^k \omega))}{\log \|d_{\chi(\omega)} f^n\|} = \alpha(q) \right\}, \tag{6.8}$$

where the transition matrix A is obtained from the Markov partition as in (4.2). Given $\delta > 0$, for each $\omega \in J_q$ there exists $r(\omega) > 0$ such that for $r \in (0, r(\omega))$ we have

$$\alpha(q) - \delta < \frac{\sum_{k=0}^{n(\omega,r)-1} \psi(\chi(\sigma^k \omega))}{\log \|d_{\chi(\omega)} f^{n(\omega,r)}\|} < \alpha(q) + \delta. \tag{6.9}$$

Now we consider appropriate Pesin sets. Namely, given $l > 0$ we define

$$Q_l = \{\omega \in J_q : r(\omega) \geq 1/l\}. \tag{6.10}$$

Clearly,

$$Q_l \subset Q_{l+1} \quad \text{and} \quad J_q = \bigcup_{l>0} Q_l. \tag{6.11}$$

Since ν_q is ergodic, it follows from (6.7) that $\nu_q(\chi(J_q)) = 1$, and thus there exists $l_0 > 0$ such that $\nu_q(\chi(Q_l)) > 0$ for every $l > l_0$. Now we construct a Moran cover of $\chi(Q_l)$ for $l > l_0$. Given $\omega = (i_1 i_2 \cdots) \in Q_l$ and $r \in (0,1)$, we consider the unique integer $n = n(\omega, r)$ satisfying (4.8). Then the sets $\Delta(\omega, r)$ in (4.9) are pairwise disjoint and form a cover of $\chi(Q_l)$. Proceeding as in the proof of Theorem 4.1.7 we show that there exists a constant $C > 0$ (independent of r) such that each ball $B(x, r) \subset \mathbb{R}^m$ intersects at most a number C of the sets $\Delta(\omega, r)$.

Lemma 6.1.3. *For ν_q-almost every $x \in \chi(J_q)$ we have $\underline{d}_{\nu_q}(x) \geq T(q) + q\alpha(q)$.*

Proof of the lemma. Let $\tilde{\Delta}_j = \Delta(\omega_j, r)$, for $j = 1, \ldots, N(r)$, be the sets in the Moran cover of $\chi(Q_l)$, where $\omega_j \in Q_l$ for each j. Given $r < 1/l$ it follows from (6.6) and (6.9) that

$$\nu_q(B(x,r) \cap \chi(Q_l)) \leq \sum_{\tilde{\Delta}_j \cap B(x,r) \neq \varnothing} \nu_q(\Delta(\omega_j, r))$$

$$\leq D_2 \sum_{\tilde{\Delta}_j \cap B(x,r) \neq \varnothing} \|d_{\chi(\omega_j)} f^{n(\omega_j,r)}\|^{-T(q)}$$

$$\times \exp\left(q \sum_{k=0}^{n(\omega_j,r)-1} \psi(f^k(\chi(\omega_j)))\right)$$

$$\leq D_2 \sum_{\tilde{\Delta}_j \cap B(x,r) \neq \varnothing} \|d_{\chi(\omega_j)} f^{n(\omega_j,r)}\|^{-T(q)-q(\alpha(q)-\delta)}.$$

Using (4.8) we conclude that there exists $C' > 0$ such that

$$\nu_q(B(x,r) \cap \chi(Q_l)) \leq C' r^{T(q)+q(\alpha(q)-\delta)} \tag{6.12}$$

for every $x \in J$ and $r \in (0, 1/l)$. By the Borel density lemma (see, for example, [59, Theorem 2.9.11]), for ν_q-almost every $x \in \chi(Q_l)$ we have

$$\lim_{r \to 0} \frac{\nu_q(B(x,r) \cap \chi(Q_l))}{\nu_q(B(x,r))} = 1,$$

and thus there exists $\rho(x) > 0$ such that for every $r \in (0, \rho(x))$,

$$\nu_q(B(x,r)) \leq 2\nu_q(B(x,r) \cap \chi(Q_l)).$$

Together with (6.12) this implies that for ν_q-almost every $x \in \chi(Q_l)$,

$$
\begin{aligned}
\underline{d}_{\nu_q}(x) &= \liminf_{r \to 0} \frac{\log \nu_q(B(x,r))}{\log r} \\
&\geq \liminf_{r \to 0} \frac{\log \nu_q(B(x,r) \cap \chi(Q_l))}{\log r} \\
&\geq T(q) + q(\alpha(q) - \delta).
\end{aligned}
$$

By (6.11) we conclude that

$$\underline{d}_{\nu_q}(x) \geq T(q) + q(\alpha(q) - \delta)$$

for ν_q-almost every $x \in \chi(J_q)$. Since δ is arbitrary this implies the desired result.
□

By Lemma 6.1.3 it follows from Theorem 2.1.5 that

$$\dim_H \chi(J_q) \geq \dim_H \nu_q \geq T(q) + q\alpha(q). \qquad (6.13)$$

Lemma 6.1.4. *For every $x \in \chi(J_q)$ we have $\overline{d}_{\nu_q}(x) \leq T(q) + q\alpha(q)$.*

Proof of the lemma. We continue to consider the Moran cover constructed above. It follows from (4.12) that $B(x, 2r) \supset \Delta(\omega, r)$ for every $x = \chi(\omega)$ with $\omega \in Q_l$ and $r \in (0, 1)$, provided that the diameter of the Markov partition is sufficiently small. Hence, using (4.8), (6.6), and (6.9), for each $x \in \chi(Q_l)$ and $r < 1/l$ we obtain

$$
\begin{aligned}
\nu_q(B(x, 2r)) &\geq \nu_q(\Delta(\omega, r)) \\
&\geq D_1 \|d_x f^{n(\omega,r)}\|^{-T(q)} \exp\left(q \sum_{k=0}^{n(\omega,r)-1} \psi(f^k x) \right) \\
&\geq D_1 \|d_x f^{n(\omega,r)}\|^{-T(q)-q(\alpha(q)+\delta)} \\
&\geq D r^{T(q)+q(\alpha(q)+\delta)},
\end{aligned}
$$

for some constant $D > 0$ (independent of x and r). Therefore, for each $x \in \chi(Q_l)$ we have

$$\overline{d}_{\nu_q}(x) = \limsup_{r \to 0} \frac{\log \nu_q(B(x,r))}{\log r} \leq T(q) + q(\alpha(q) + \delta).$$

By (6.11) and the arbitrariness of δ we obtain the desired result.
□

By Lemma 6.1.4 it follows from Theorem 2.1.5 that

$$\dim_H \chi(J_q) \leq T(q) + q\alpha(q).$$

Together with (6.13) this implies that

$$\dim_H \chi(J_q) = T(q) + q\alpha(q). \tag{6.14}$$

Furthermore, by Lemmas 6.1.3 and 6.1.4, for ν_q-almost every $x \in \chi(J_q)$ we have

$$\underline{d}_{\nu_q}(x) = \overline{d}_{\nu_q}(x) = T(q) + q\alpha(q). \tag{6.15}$$

Now we establish a crucial property of the measure ν that allows us to transfer the results at the level of symbolic dynamics to results for the repeller. Since ν is the equilibrium measure of ψ it is a Gibbs measure and there exist constants $D_1, D_2 > 0$ such that

$$D_1 \le \frac{\nu(\Delta_{i_1 \cdots i_n})}{\exp \sum_{k=0}^{n-1} \psi(f^k x)} \le D_2 \tag{6.16}$$

for every $n \in \mathbb{N}$ and $x \in \Delta_{i_1 \cdots i_n}$.

Lemma 6.1.5. *There exists $C > 0$ such that for every $x \in J$ and $r > 0$ we have*

$$\nu(B(x, 2r)) \le C\nu(B(x, r)).$$

Proof of the lemma. Given $r > 0$, we consider a Moran cover $\tilde{\Delta}_1, \ldots, \tilde{\Delta}_{N(x,r)}$ of the ball $B(x, 2r)$ by sets $\tilde{\Delta}_j = \Delta(\omega_j, r/2)$ as in (4.9) with $\omega_j \in \Sigma_A^+$ for each j. Proceeding as in the proof of Theorem 4.1.7 we can show that there exists $N > 0$ such that $N(x, r) \le N$ for every $x \in J$ and $r > 0$. It follows from (6.16) that

$$
\begin{aligned}
\nu(B(x, 2r)) &\le \sum_{j=1}^{N(x,r)} \nu(\tilde{\Delta}_j) \\
&\le D_2 \sum_{j=1}^{N(x,r)} \exp \sum_{k=0}^{n(\omega_j, r/2)-1} \psi(f^k \chi(\omega_j)).
\end{aligned}
\tag{6.17}
$$

Since ψ is Hölder continuous, proceeding in a similar manner to that in (4.10) and (4.11) we show that there exist constants $D_1, D_2 > 0$ such that

$$D_1 \le \frac{\prod_{k=0}^{n-1} \exp \psi(f^k x)}{\prod_{k=0}^{n-1} \exp \psi(f^k y)} \le D_2 \tag{6.18}$$

for every $n \in \mathbb{N}$ and $x, y \in \Delta_{i_1 \cdots i_n}$. Furthermore, it follows easily from (4.11) and the choice of $n(\omega, r)$ in (4.8) that there exists $\kappa > 0$ such that

$$|n(\omega, r) - n(\omega', r)| \le \kappa$$

for every $\omega, \omega' \in \Sigma_A^+$ with $\chi(\omega') \in \Delta(\omega, r)$ and $r > 0$. Together with (6.18) this implies that there exist constants $D_1', D_2' > 0$ (independent of x and r) such that

$$D_1' \le \frac{\prod_{k=0}^{n(\omega_j, r/2)-1} \exp \psi(f^k \chi(\omega_j))}{\prod_{k=0}^{n(\omega_l, r/2)-1} \exp \psi(f^k \chi(\omega_l))} \le D_2' \tag{6.19}$$

for every $j, l \in \{1, \ldots, N(x,r)\}$ and $r > 0$. Now we observe that by (4.12) there exists l such that $\tilde{\Delta}_l \subset B(x, r)$. It follows from (6.17) and (6.19) that

$$\nu(B(x, 2r)) \le D_2 N(x, r) D_2' \exp \sum_{k=0}^{n(\omega_l, r/2)-1} \psi(f^k \chi(\omega_l))$$

$$\le D_2 N D_2' D_1^{-1} \nu(\tilde{\Delta}_l)$$

$$\le D_2 N D_2' D_1^{-1} \nu(B(x, r)).$$

This completes the proof of the lemma. $\qquad\square$

Lemma 6.1.6. *We have $\chi(J_q) = K_{\alpha(q)}$.*

Proof of the lemma. Let $r \in (0, 1)$. Taking $n = n(\omega, r)$ as in (4.8) and proceeding as in the proof of Theorem 4.1.7 (see (4.12) and (4.13)), we show that there exists $\kappa > 0$ (independent of r) such that for each $x = \chi(\omega) \in J$ there exists $y \in \Delta(\omega, r)$ for which

$$B(y, \kappa r) \subset \Delta(\omega, r) \subset B(x, 2r). \tag{6.20}$$

On the other hand, since $\operatorname{diam} \Delta(\omega, r) < r$ we have $B(x, r) \subset B(y, 2r)$. Therefore, by Lemma 6.1.5 and (6.20) we have

$$\nu(\Delta(\omega, r)) \le \nu(B(x, 2r)) \le C\nu(B(x, r))$$

$$\le C\nu(B(y, 2r)) \le C^{n+2} \nu(B(y, r2^{-n}))$$

$$\le C^{n+2} \nu(B(y, \kappa r)) \le C^{n+2} \nu(\Delta(\omega, r)),$$

provided that $n \in \mathbb{N}$ is chosen so that $2^{-n} < \kappa$. This implies that if either of the two limits

$$\lim_{r \to 0} \frac{\log \nu(\Delta(\omega, r))}{\log r} \quad \text{and} \quad \lim_{r \to 0} \frac{\log \nu(B(x, r))}{\log r} \tag{6.21}$$

exists, then the other one also exists and has the same value. On the other hand, by (6.16) and (4.8), if the first limit in (6.21) exists, then

$$\lim_{r \to 0} \frac{\log \nu(\Delta(\omega, r))}{\log r} = \lim_{r \to 0} \frac{\sum_{k=0}^{n(\omega, r)-1} \psi(f^k x)}{\log r}$$

$$= \lim_{r \to 0} \frac{\sum_{k=0}^{n(\omega, r)-1} \psi(\chi(\sigma^k \omega))}{-\log \|d_{\chi(\omega)} f^{n(\omega, r)}\|}.$$

Therefore, by (6.8), $\omega \in J_q$ if and only if

$$\lim_{r \to 0} \frac{\log \nu(\Delta(\omega, r))}{\log r} = \alpha(q),$$

and thus, by (6.21), $\omega \in J_q$ if and only if $x \in K_{\alpha(q)}$. In other words we have $K_{\alpha(q)} = \chi(J_q)$. $\qquad\square$

By Lemma 6.1.6, the last two statements in Theorem 6.1.2 follow immediately from (6.14) and (6.15).

For the third statement in the theorem recall that we have already shown that $T(q)$ and thus also that $\alpha(q)$ are analytic functions. We continue with an auxiliary statement.

Lemma 6.1.7. *We have $T'(q) \leq 0$ and $T''(q) \geq 0$ for every $q \in \mathbb{R}$. Furthermore, $T''(q) > 0$ for every $q \in \mathbb{R}$ if and only if ν is not the equilibrium measure of $-(\dim_H J) \log \|df\|$.*

Proof of the lemma. By (6.7) we have

$$T'(q) = \frac{\int_J \psi \, d\nu_q}{\int_J \log \|df\| \, d\nu_q}. \qquad (6.22)$$

In view of (4.1) we obtain

$$\int_J \log \|df\| \, d\nu_q \geq \log \beta > 0.$$

Furthermore,

$$0 = P(\psi) \geq h_{\nu_q}(f) + \int_J \psi \, d\nu_q,$$

and

$$\int_J \psi \, d\nu_q \leq -h_{\nu_q}(f) < 0.$$

It follows from (6.22) that $T'(q) < 0$ for every $q \in \mathbb{R}$. On the other hand, using the function Q in (6.4) we obtain

$$T'(q) = -\frac{\partial Q/\partial q}{\partial Q/\partial t}$$

and thus,

$$T''(q) = -\frac{T'(q)^2 \partial^2 Q/\partial t^2 + 2T'(q)\partial^2 Q/\partial q\partial t + \partial^2 Q/\partial q^2}{\partial Q/\partial t}, \qquad (6.23)$$

where all partial derivatives are computed at the point $(T(q), q)$. Now we use the formula for the second derivative of the topological pressure given by Ruelle in [132]. Namely,

$$\left.\frac{\partial^2 P(\varphi + t_1\varphi_1 + t_2\varphi_2)}{\partial t_1 \partial t_2}\right|_{t_1=t_2=0} = B_\varphi(\varphi_1, \varphi_2), \qquad (6.24)$$

where B_φ is the bilinear form in $C^\varepsilon(J)$ given by

$$B_\varphi(\varphi_1, \varphi_2) = \sum_{k=0}^{\infty} \left(\int_J \varphi_1(\varphi_2 \circ f^k) \, d\nu_\varphi - \int_J \varphi_1 \, d\nu_\varphi \int_J \varphi_2 \, d\nu_\varphi \right), \qquad (6.25)$$

where ν_φ is the equilibrium measure of φ. It is also shown in [132] that $B_\varphi(\psi, \psi) \geq 0$, and that $B_\varphi(\psi, \psi) > 0$ if and only if ψ is not cohomologous to a constant. It follows from (6.25) that

$$\frac{\partial^2 Q}{\partial t^2}(T(q), q) = B_{\varphi_q}(\log \|df\|, \log \|df\|),$$

$$\frac{\partial^2 Q}{\partial q^2}(T(q), q) = B_{\varphi_q}(\psi, \psi),$$

$$\frac{\partial^2 Q}{\partial q \partial t}(T(q), q) = -B_{\varphi_q}(\psi, \log \|df\|),$$

where

$$\varphi_q = -T(q)\log\|df\| + q\psi.$$

Therefore, by (6.23) we obtain

$$T''(q) = \frac{B_{\varphi_q}(\psi - T'(q)\log\|df\|, \psi - T'(q)\log\|df\|)}{\int_J \log\|df\| \, d\nu_q}.$$

This shows that $T''(q) \geq 0$. Furthermore, $T''(q) > 0$ if and only if the function $\psi - T'(q)\log\|df\|$ is not cohomologous to a constant.

When $u = \log\|df\|$ is cohomologous to a constant we have $T''(q) > 0$ for every $q \in \mathbb{R}$ if and only if ψ is not cohomologous to a constant, and thus if and only if ψ is not cohomologous to $-su$, where $s = \dim_H J$.

Now we assume that u is not cohomologous to a constant. When ψ is cohomologous to $-su$ we have $\nu = \mu = \nu_0$, and by (6.22),

$$T'(0) = \frac{\int_J \psi \, d\nu_0}{\int_J u \, d\nu_0} = \frac{\int_J \psi \, d\nu}{\int_J u \, d\nu}.$$

Since

$$0 = P(\psi) = h_\nu(f) + \int_J \psi \, d\nu,$$

we obtain

$$T'(0) = -\frac{h_\nu(f)}{\int_J u \, d\nu} = -\frac{h_\mu(f)}{\int_J u \, d\mu},$$

and it follows from

$$0 = P(-su) = h_\mu(f) - s\int_J u \, d\mu$$

that $T'(0) = -s$. Therefore, $\psi - T'(0)u = \psi + su$ is cohomologous to a constant, and $T''(0) = 0$. Conversely, assume that $T''(q) = 0$ for some $q \in \mathbb{R}$. Then $\psi - T'(q)u$ is cohomologous to a constant, say $c \in \mathbb{R}$. Integrating with respect to the measure ν_q we obtain

$$\int_J \psi \, d\nu_q - T'(q)\int_J u \, d\nu_q = c,$$

and it follows from (6.22) that $c = 0$. That is, ψ and $T'(q)u$ are cohomologous, and hence

$$P(T'(q)u) = P(\psi) = 0.$$

This implies that $T'(q) = -s$, and ψ is cohomologous to $-su$.

We have thus shown that $T''(q) > 0$ for every $q \in \mathbb{R}$ if and only if ψ is not cohomologous to $-su$, that is, if and only if $\nu \neq \mu$. □

It follows from (6.5) that

$$\mathcal{D}'(\alpha(q))\alpha'(q) = T'(q) + \alpha(q) + q\alpha'(q) = q\alpha'(q),$$

and hence, $\mathcal{D}'(\alpha(q)) = q$. Taking derivatives we obtain

$$\mathcal{D}''(\alpha(q))\alpha'(q) = 1,$$

and by Lemma 6.1.7, we conclude that

$$\mathcal{D}''(\alpha(q)) = -\frac{1}{T''(q)} < 0 \quad \text{for every} \quad q \in \mathbb{R} \tag{6.26}$$

if and only if ν is not the equilibrium measure of $-(\dim_H J)\log \|df\|$. This establishes the strict convexity of the function $\mathcal{D}$ in statement 3 in the theorem. The analyticity of $\mathcal{D}$ can be obtained directly from the formula $\mathcal{D}'(\alpha(q)) = q$ by inverting the analytic function $q \mapsto \alpha(q)$ (recall that $\alpha'(q) = -T''(q) < 0$ and thus α is strictly decreasing). This establishes the analyticity of $\mathcal{D}'$ and thus also of $\mathcal{D}$.

For the second statement in the theorem we note that if $\nu = \mu$, then by Theorem 4.1.8 there exist constants $D_1, D_2 > 0$ such that

$$D_1 < \frac{\nu(B(x,r))}{r^{\dim_H J}} < D_2$$

for every $x \in J$ and every sufficiently small $r > 0$. Therefore,

$$\lim_{r \to 0} \frac{\log \nu(B(x,r))}{\log r} = \dim_H J$$

for every $x \in J$. In particular, $\underline{\alpha} = \overline{\alpha} = \dim_H J$ and $\mathcal{D}$ is a delta function.

Finally, the first statement in the theorem follows easily from the fact that at the level of symbolic dynamics the limit in (6.8) only depends on the future. More precisely, it follows from (4.4) that

$$\chi \circ \sigma^k = f^k \circ \chi$$

for every $k \in \mathbb{N}$, where χ is the coding map in (4.3). We thus obtain

$$\frac{\sum_{k=0}^{n-1} \psi(\chi(\sigma^k \omega))}{-\log \|d_{\chi(\omega)} f^n\|} = \frac{\sum_{k=0}^{n-1} \psi(f^k x)}{-\log \|d_x f^n\|},$$

where $x = \chi(\omega)$. Therefore, if $x \in K_\alpha$, then $y = \chi(\omega') \in K_\alpha$ for every ω' in the set

$$K = \{\omega' \in \Sigma_A^+ : \sigma^n \omega' = \omega \text{ for some } n \in \mathbb{N}\}.$$

We note that K is dense in Σ_A^+ (recall that $A^m > 0$ for some $m \in \mathbb{N}$, since f is topologically mixing on J), and the same happens with $\chi(K)$. Since $\chi(K) \subset K_\alpha$ we conclude that K_α is dense. This completes the proof of the theorem. $\qquad\square$

Statement 1 in Theorem 6.1.2 was observed by Barreira and Schmeling in [21]. In [142], Schmeling showed that the domain of $\mathcal{D}$ coincides with $[\underline{\alpha}, \overline{\alpha}]$, i.e., that $K_\alpha \neq \varnothing$ if and only if $\alpha \in [\underline{\alpha}, \overline{\alpha}]$ (see Theorem 7.4.1). One can also show that

$$\underline{\alpha} = \inf_{\tau \in \mathcal{M}_E} -\frac{\int_J \psi \, d\tau}{\int_J \log \|df\| \, d\tau} \quad \text{and} \quad \overline{\alpha} = \sup_{\tau \in \mathcal{M}_E} -\frac{\int_J \psi \, d\tau}{\int_J \log \|df\| \, d\tau},$$

where $\mathcal{M}_E$ is the family of all ergodic f-invariant probability Borel measures in J (see Theorem 7.4.1). Furthermore, the maximum of the function $\mathcal{D}$ is equal to $\dim_H J$ (see Figure 6.2).

Theorem 6.1.2 reveals an enormous complexity of multifractal decompositions. In particular, it shows that the decomposition in (6.2) is composed of an *uncountable* number of pairwise disjoint invariant sets, each of them dense and with positive Hausdorff dimension. We will see in Section 8.1 that the set $X \setminus Y$ in (6.2) is also very complex: even though it has zero measure with respect to *any* finite invariant measure (this is a simple consequence of Birkhoff's ergodic theorem), we will see that it has full Hausdorff dimension (see Section 8.5).

In [4], Barreira and Gelfert considered repellers of nonconformal transformations satisfying a certain cone condition, and obtained a version of multifractal analysis for the topological entropy of the level sets of the Lyapunov exponents. We note that due to the nonconformality, one cannot use Birkhoff's ergodic theorem neither Gibbs measures. Related works are due to Feng and Lau [62] and Feng [60, 61], with a study of products of nonnegative matrices and their thermodynamic properties, and to Barreira and Radu [12], with lower bounds for the dimension spectra for a class of repellers of nonconformal transformations.

We mention briefly a few directions of research concerning nonuniformly hyperbolic systems and countable topological Markov chains. For *finite* topological Markov chains the dimension and entropy spectra of an equilibrium measure of a Hölder continuous function (see Section 7.1) has bounded domain and is analytic. In strong contrast, in the case of nonuniformly hyperbolic systems and countable topological Markov chains the spectrum may have unbounded domain and need not be analytic. In [123], Pollicott and Weiss presented a multifractal analysis of the Lyapunov exponent for the Gauss map and the Manneville–Pomeau transformation. Related results were obtained by Yuri in [166]. In [97, 98, 99], Mauldin and Urbański developed the theory of infinite conformal iterated function systems, studying in particular the Hausdorff dimension of the limit set (see also [69]). Related results were obtained by Nakaishi in [103]. In [88], Kesseböhmer and Stratmann established a detailed multifractal analysis for Stern–Brocot intervals, continued fractions, and certain Diophantine growth rates, building on their former work [87]. In [79], Iommi obtained a detailed multifractal analysis for countable topological Markov chains, using the so-called Gurevich pressure introduced by Sarig in [135] (building on former work of Gurevich [66] on the notion of topological entropy for countable Markov chains). In [5], Barreira and Iommi considered the case of suspension flows over a countable topological Markov chain, building also on work of Savchenko [139] on the notion of topological entropy. In [80], Iommi and Skorulski studied the multifractal analysis of conformal measures for the exponential family $z \mapsto \lambda e^z$ with $\lambda \in (0, 1/e)$ (we note that in this setting the Julia set is not compact and that the dynamics is not Markov on the Julia set). They use a construction described by Urbański and Zdunik in [157].

6.2   Dimension spectrum for hyperbolic sets

Now let f be a $C^{1+\varepsilon}$ diffeomorphism, for some $\varepsilon > 0$, and let Λ be a locally maximal hyperbolic of f. We always assume in this section that f is conformal and topologically mixing on Λ (see Definition 4.3.1).

Consider a Hölder continuous function $\psi \colon \Lambda \to \mathbb{R}$ with $P(\psi) = 0$, where P denotes the topological pressure on Λ. We define functions $T_s \colon \Lambda \to \mathbb{R}$ and $T_u \colon \Lambda \to \mathbb{R}$ by requiring that

$$P(T_s(q) \log \|df|E^s\| + q\psi) = 0$$

and

$$P(-T_u(q) \log \|df|E^u\| + q\psi) = 0$$

for every $q \in \mathbb{R}$. Proceeding in a similar manner to that in Section 6.1 we can easily verify that the functions T_s and T_u are well-defined and analytic. We set

$$T(q) = T_s(q) + T_u(q) \quad \text{and} \quad \alpha(q) = -T'(q).$$

We also denote by $(\underline{\alpha}, \overline{\alpha})$ the range of $\alpha(q)$.

Consider a Markov partition of Λ and the associated transition matrix A with entries given by (4.2). Given $q \in \mathbb{R}$ we construct a measure ν_q in each rectangle of the Markov partition as follows. By Proposition 4.2.11 there exist functions

$$\varphi_q^s \colon \Sigma_A^- \to \mathbb{R} \quad \text{and} \quad \varphi_q^u \colon \Sigma_A^+ \to \mathbb{R}$$

such that $\varphi_q^s \circ \pi^-$ and $\varphi_q^u \circ \pi^+$ are cohomologous respectively to the functions

$$\left(T_s(q) \log \|df|E^s\| + q\psi\right) \circ \chi \quad \text{and} \quad \left(-T_u(q) \log \|df|E^u\| + q\psi\right) \circ \chi,$$

where χ is the coding map in (4.25), and where π^- and π^+ are the projections defined by (4.32). Moreover, let μ_q^s be the equilibrium measure of φ_q^s in Σ_A^- (with respect to σ^-), and let μ_q^u be the equilibrium measure of φ_q^u in Σ_A^+ (with respect to σ^+). Given $x \in \Lambda$ we consider a rectangle $R(x)$ of the Markov partition containing x. We have $R(x) = \chi(C_{i_0})$ where $x = \chi(\cdots i_0 \cdots)$. We also consider the measures

$$\nu_q^s = \chi_*(\mu_q^s | C_{i_0}^-) \quad \text{in} \quad A^s(x),$$

and

$$\nu_q^u = \chi_*(\mu_q^u | C_{i_0}^+) \quad \text{in} \quad A^u(x),$$

with $A^s(x)$ and $A^u(x)$ as in (4.35). Finally, we define a measure ν_q in the rectangle $R(x) = [A^s(x), A^u(x)]$ by

$$\nu_q = \nu_q^s \times \nu_q^u.$$

The following result of Pesin and Weiss in [120] (see also [115] for a related discussion) describes the dimension spectrum $\mathcal{D}$ of the equilibrium measure of the function ψ.

Theorem 6.2.1 (Multifractal analysis of equilibrium measures). *Let Λ be a locally maximal hyperbolic set of a $C^{1+\varepsilon}$ diffeomorphism f, for some $\varepsilon > 0$, such that f is conformal and topologically mixing on Λ. If ν is the equilibrium measure of a Hölder continuous function $\psi \colon \Lambda \to \mathbb{R}$ with $P(\psi) = 0$, then:*

1. *if $\dim_H \nu = \dim_H \Lambda$, then $\underline{\alpha} = \overline{\alpha} = \dim_H \Lambda$ and $\mathcal{D}$ is a delta function;*

2. *if $\dim_H \nu \neq \dim_H \Lambda$, then $\mathcal{D} \colon (\underline{\alpha}, \overline{\alpha}) \to \mathbb{R}$ is analytic and strictly convex;*

3. *$\mathcal{D}$ is the Legendre transform of T, that is, for each $q \in \mathbb{R}$ we have*

$$\mathcal{D}(\alpha(q)) = T(q) + q\alpha(q);$$

4. *for each $q \in \mathbb{R}$ we have $\nu_q(K_{\alpha(q)}) = 1$ and*

$$\lim_{r \to 0} \frac{\log \nu_q(B(x, r))}{\log r} = T(q) + q\alpha(q)$$

for ν_q-almost every $x \in K_{\alpha(q)}$.

Proof. With some appropriate modifications the proof follows closely the arguments in the proof of Theorem 6.1.2 in the case of repellers. Let $R_1, \ldots, R_p$ be the elements of a Markov partition of Λ. By Proposition 4.2.11 there exist functions $\psi^s, d^s \colon \Sigma_A^- \to \mathbb{R}$ and $\psi^u, s^u \colon \Sigma_A^+ \to \mathbb{R}$ such that:

1. $\psi \circ \chi$, $\psi^s \circ \pi^-$, and $\psi^u \circ \pi^+$ are cohomologous;

2. $\log \|df^{-1}|E^s\| \circ \chi$ and $d^s \circ \pi^-$ are cohomologous;

3. $\log \|df|E^u\| \circ \chi$ and $d^u \circ \pi^+$ are cohomologous.

Given $\omega \in \Sigma_A$ and $r \in (0,1)$, we consider the unique integers $n = n(\omega, r)$ and $m = m(\omega, r)$ such that

$$\|d_x f^{-n} | E^s(x)\|^{-1} < r \le \|d_x f^{-(n-1)} | E^s(x)\|^{-1} \tag{6.27}$$

and

$$\|d_x f^m | E^u(x)\|^{-1} < r \le \|d_x f^{m-1} | E^u(x)\|^{-1}, \tag{6.28}$$

where $x = \chi(\omega)$. For each $q \in \mathbb{R}$, let J_q be the set of points $\omega \in \Sigma_A$ such that

$$- \lim_{r \to 0} \left(\frac{\sum_{k=0}^{n(\omega,r)-1} \psi^s(\sigma^k \omega^-)}{\sum_{k=0}^{n(\omega,r)-1} d^s(\sigma^k \omega^-)} + \frac{\sum_{k=0}^{m(\omega,r)-1} \psi^u(\sigma^k \omega^+)}{\sum_{k=0}^{m(\omega,r)-1} d^u(\sigma^k \omega^+)} \right) = \alpha(q),$$

where $\omega^- = \pi^- \omega$ and $\omega^+ = \pi^+ \omega$ (see (4.32) for the definition of π^- and π^+). By (4.33), we have

$$P_{\Sigma_A^-}(-T_s(q)d^s + q\psi^s) = P_{\Sigma_A^+}(-T_u(q)d^u + q\psi^u) = 0.$$

Therefore,

$$0 = -T_s'(q) \int_{\Sigma_A^-} d^s \, d\mu_q^s + \int_{\Sigma_A^-} \psi^s \, d\mu_q^s,$$

and

$$0 = -T_u'(q) \int_{\Sigma_A^+} d^u \, d\mu_q^u + \int_{\Sigma_A^+} \psi^u \, d\mu_q^u.$$

We thus obtain

$$\alpha_s(q) := -T_s'(q) = -\frac{\int_{\Sigma_A^-} \psi^s \, d\mu_q^s}{\int_{\Sigma_A^-} d^s \, d\mu_q^s}$$

and

$$\alpha_u(q) := -T_u'(q) = -\frac{\int_{\Sigma_A^+} \psi^u \, d\mu_q^u}{\int_{\Sigma_A^+} d^u \, d\mu_q^u}.$$

Since the measures μ_q^s and μ_q^u are ergodic, by Birkhoff's ergodic theorem for μ_q^s-almost every $\omega^- \in \Sigma_A^-$ we have

$$\lim_{n \to \infty} -\frac{\sum_{k=0}^{n-1} \psi^s(\sigma^k \omega^-)}{\sum_{k=0}^{n-1} d^s(\sigma^k \omega^-)} = \alpha_s(q),$$

and for μ_q^u-almost every $\omega^+ \in \Sigma_A^+$ we have

$$\lim_{m \to \infty} -\frac{\sum_{k=0}^{m-1} \psi^u(\sigma^k \omega^+)}{\sum_{k=0}^{m-1} d^u(\sigma^k \omega^+)} = \alpha_u(q).$$

Therefore, given $\delta > 0$ and $\omega \in \Sigma_A$ there exists $r(\omega) > 0$ such that for $r \in (0, r(\omega))$ we have

$$\alpha_s(q) - \delta < -\frac{\sum_{k=0}^{n(\omega,r)-1} \psi^s(\sigma^k \omega^-)}{\sum_{k=0}^{n(\omega,r)-1} d^s(\sigma^k \omega^-)} < \alpha_s(q) + \delta, \tag{6.29}$$

and

$$\alpha_u(q) - \delta < -\frac{\sum_{k=0}^{m(\omega,r)-1} \psi^u(\sigma^k \omega^+)}{\sum_{k=0}^{m(\omega,r)-1} d^u(\sigma^k \omega^+)} < \alpha_u(q) + \delta, \tag{6.30}$$

where again $\omega^- = \pi^- \omega$ and $\omega^+ = \pi^+ \omega$. Now we consider appropriate Pesin sets. Namely, given $l > 0$ we consider the sets Q_l in (6.10). Clearly, (6.11) holds.

Now we observe that since μ_q^s and μ_q^u are equilibrium measures of Hölder continuous functions they are Gibbs measures. That is, there exist constants $D_1, D_2 > 0$ such that for every $n, m \in \mathbb{N}$ and $\omega = (\cdots i_{-1} i_0 i_1 \cdots) \in \Sigma_A$ we have

$$D_1 \leq \frac{\mu_q^s(C_{i_{-n} \cdots i_0}^-)}{\exp\left(-T_s(q) \sum_{k=0}^{n-1} d^s(\sigma^k \omega^-) + q \sum_{k=0}^{n-1} \psi^s(\sigma^k \omega^-)\right)} \leq D_2 \tag{6.31}$$

and

$$D_1 \leq \frac{\mu_q^u(C_{i_0 \cdots i_m}^+)}{\exp\left(-T_u(q) \sum_{k=0}^{m-1} d^u(\sigma^k \omega^+) + q \sum_{k=0}^{m-1} \psi^u(\sigma^k \omega^+)\right)} \leq D_2. \tag{6.32}$$

Repeating arguments in the proof of Lemma 6.1.3 we show that

$$\underline{d}_{\nu_q^s}(y) \geq T_s(q) + q(\alpha_s(q) - \delta)$$

for ν_q^s-almost every $y \in A^s(x) \cap \chi(Q_l)$, and

$$\underline{d}_{\nu_q^u}(z) \geq T_u(q) + q(\alpha_u(q) - \delta)$$

for ν_q^u-almost every $z \in A^u(x) \cap \chi(Q_l)$ (see (4.35) for the definition of the sets $A^s(x)$ and $A^u(x)$). It follows from (6.11) and the arbitrariness of δ that

$$\underline{d}_{\nu_q^s}(y) \geq T_s(q) + q\alpha_s(q)$$

for ν_q^s-almost every $y \in A^s(x) \cap \chi(J_q)$, and

$$\underline{d}_{\nu_q^u}(z) \geq T_u(q) + q\alpha_u(q)$$

for ν_q^u-almost every $z \in A^u(x) \cap \chi(J_q)$. Since $\nu_q = \nu_q^s \times \nu_q^u$ it follows from Proposition 4.2.13 that

$$\begin{aligned}
\underline{d}_{\nu_q}(x) &= \liminf_{r \to 0} \frac{\log \nu_q(B(x,r))}{\log r} \\
&\geq \underline{d}_{\nu_q^s}(x) + \underline{d}_{\nu_q^u}(x) \geq T(q) + q\alpha(q)
\end{aligned} \tag{6.33}$$

for ν_q-almost every $x \in \chi(J_q)$. By Theorem 2.1.5 and (6.33) we obtain

$$\dim_H \chi(J_q) \geq T(q) + q\alpha(q). \tag{6.34}$$

On the other hand, it follows from the choice of n and m in (6.27) and (6.28) that there exists a constant $K > 0$ such that for every $x = \chi(\omega) \in \Lambda$ and $r \in (0,1)$ we have

$$\Delta(\omega, r) := \bigcup_{j=-n(\omega,r)}^{m(\omega,r)} f^{-k} R_{i_j} \subset B(x, Kr). \tag{6.35}$$

It follows from (6.31) and (6.32) that for every $x = \chi(\omega)$ with $\omega \in Q_l$, and $r < 1/l$, setting $n = n(\omega, r)$ and $m = m(\omega, r)$ we have

$$\nu_q(B(x, Kr)) \geq \nu_q(\Delta(\omega, r)) = \mu_q^s(C_{i_0 \cdots i_n}^-) \mu_q^u(C_{i_0 \cdots i_m}^+)$$

$$\geq D_1^2 \exp\left(-T_s(q) \sum_{k=0}^{n-1} d^s(\sigma^k \omega^-) + q \sum_{k=0}^{n-1} \psi^s(\sigma^k \omega^-)\right)$$

$$\times \exp\left(-T_u(q) \sum_{k=0}^{m-1} d^u(\sigma^k \omega^+) + q \sum_{k=0}^{m-1} \psi^u(\sigma^k \omega^+)\right)$$

$$= D_1^2 \exp\left(-T_s(q) \log \|d_x f^{-n}|E^s(x)\| + q \sum_{k=0}^{n-1} \psi(f^{-k}x)\right)$$

$$\times \exp\left(-T_u(q) \log \|d_x f^m|E^u(x)\| + q \sum_{k=0}^{m-1} \psi(f^k x)\right).$$

Therefore, by (6.27), (6.28), (6.29), and (6.30) we obtain

$$\bar{d}_{\nu_q}(x) = \limsup_{r \to 0} \frac{\log \nu_q(B(x, r))}{\log r}$$

$$\leq T_s(q) \limsup_{r \to 0} \frac{-\log \|d_x f^{-n(\omega, r)}|E^s(x)\|}{\log r}$$

$$+ T_u(q) \limsup_{r \to 0} \frac{-\log \|d_x f^{m(\omega, r)}|E^u(x)\|}{\log r}$$

$$+ \limsup_{r \to 0} \frac{q \sum_{k=-(n(\omega,r)-1)}^{m(\omega,r)-1} \psi(f^k x)}{\log r}$$

$$\leq T_s(q) + T_u(q) + q(\alpha_s(q) + 2\delta)$$

for every $x \in \chi(Q_l)$. It follows from (6.11) and the arbitrariness of δ that

$$\bar{d}_{\nu_q}(x) \leq T(q) + q\alpha(q). \tag{6.36}$$

By Theorem 2.1.5 it follows from (6.36) that

$$\dim_H \chi(J_q) \leq T(q) + q\alpha(q).$$

Together with (6.34) this implies that

$$\dim_H \chi(J_q) = T(q) + q\alpha(q). \tag{6.37}$$

We want to show that $\chi(J_q) = K_{\alpha(q)}$. For this we first establish a version of Lemma 6.1.5 for the measure ν.

Lemma 6.2.2. *Given $\gamma > 1$, there exists $K > 0$ such that for every $y \in R(x)$ and every sufficiently small $r > 0$ we have*

$$\nu(B(y, \gamma r)) \le K\nu(B(y, r)).$$

Proof of the lemma. For ν-almost every $y \in \Lambda$, let ν_y^s and ν_y^u be respectively the conditional measures of ν in $A^s(x)$ and $A^u(x)$ (see Section 4.2.3). Modifying in a straightforward manner the arguments in the proof of Lemma 6.1.5, we show that there exists $C > 0$ such that for every $y \in \Lambda$ and every sufficiently small $r > 0$ we have

$$\nu_y^u(B^u(y, 2r)) \le C\nu_y^u(B^u(y, r)) \quad \text{and} \quad \nu_y^s(B^s(y, 2r)) \le C\nu_y^s(B^s(y, r)), \tag{6.38}$$

where $B^u(y, r)$ and $B^s(y, r)$ are the open balls centered at y of radius r with respect to the distances induced respectively on the local unstable and stable manifolds $V^u(y)$ and $V^s(y)$. Now we observe that given $\gamma > 1$ there exists $\kappa > 1$ such that

$$\Lambda \cap B(y, \gamma r) \subset \left[\Lambda \cap B^s(y, \kappa r), \Lambda \cap B^u(y, \kappa r)\right] \tag{6.39}$$

and

$$\left[\Lambda \cap B^s(y, r/\kappa), \Lambda \cap B^u(y, r/\kappa)\right] \subset \Lambda \cap B(y, r) \tag{6.40}$$

for every $y \in \Lambda$ and every sufficiently small $r > 0$. It follows from (6.39) and Proposition 4.2.13 that for some constant $c > 0$ (independent of y and r) we have

$$\nu(B(y, \gamma r)) \le c\nu_y^u(B^u(y, \kappa r))\nu_y^s(B^s(y, \kappa r)).$$

Applying (6.38) a number n of times such that $\kappa 2^{-n} < 1/\kappa$, we obtain

$$\nu(B(y, \gamma r)) \le C^{2n}\nu_y^u(B^u(y, r/\kappa))\nu_y^s(B^s(y, r/\kappa)).$$

It follows from (6.40) and Proposition 4.2.13 that for some constant $\bar{c} > 0$ (independent of y and r) we have

$$\nu(B(y, \gamma r)) \le \bar{c}C^{2n}\nu(B(y, r)),$$

and the desired inequality holds with $K = \bar{c}C^{2n}$. $\qquad \square$

Lemma 6.2.3. *We have $\chi(J_q) = K_{\alpha(q)}$.*

Proof of the lemma. Let $r \in (0,1)$. Taking $n = n(\omega, r)$ and $m = m(\omega, r)$ as in (6.27) and (6.28), and proceeding in a similar manner to that in the proof of Theorem 4.1.7, we show that there exists $\kappa > 0$ (independent of r) such that for each $x = \chi(\omega) \in \Lambda$ there exists $y \in \Delta(\omega, r)$ (see (6.35) for the definition) for which

$$B(y, \kappa r) \subset \Delta(\omega, r) \subset B(x, Kr).$$

On the other hand, since diam $\Delta(\omega, r) < dr$ for some constant $d > 0$ (independent of ω and r), we have $B(x, r) \subset B(y, \bar{d}r)$ for some constant $\bar{d} > 0$ (independent of x and r). Using Lemma 6.2.2, we can repeat arguments in the proof of Lemma 6.1.6 to show that if either of the two limits

$$\lim_{r \to 0} \frac{\log \nu(\Delta(\omega, r))}{\log r} \quad \text{and} \quad \lim_{r \to 0} \frac{\log \nu(B(x, r))}{\log r} \tag{6.41}$$

exists, then the other one also exists and has the same value. On the other hand, since ν is the equilibrium measure of ψ and $P(\psi) = 0$, if the first limit exists then

$$a := \lim_{r \to 0} \frac{\log \nu(\Delta(\omega, r))}{\log r} = \lim_{r \to 0} \frac{\sum_{k=-(n(\omega,r)-1)}^{m(\omega,r)-1} \psi(f^k x)}{\log r}.$$

It follows from (6.27) and (6.28) that $x = \chi(\omega) \in \chi(J_q)$ if and only if

$$a = \lim_{r \to 0} \left(\frac{\sum_{k=0}^{n(\omega,r)-1} \psi(f^{-k}x)}{\log r} + \frac{\sum_{k=0}^{m(\omega,r)-1} \psi(f^k x)}{\log r} \right)$$

$$= \lim_{r \to 0} \left(\frac{\sum_{k=0}^{n(\omega,r)-1} \psi(f^{-k}x)}{-\log \|d_x f^{-n(\omega,r)} | E^s(x)\|} + \frac{\sum_{k=0}^{m(\omega,r)-1} \psi(f^k x)}{-\log \|d_x f^{m(\omega,r)} | E^u(x)\|} \right) = \alpha(q).$$

That is, $x \in \chi(J_q)$ if and only if the second limit in (6.41) is equal to $K_{\alpha(q)}$. □

The two last statements in Theorem 6.2.1 follow readily from Lemma 6.2.3 together with (6.33), (6.36), and (6.37). The two first statements in the theorem can be obtained by modifying arguments in the proof of Theorem 6.1.2. In particular, we can show that

$$T''(q) = T_s''(q) + T_u''(q) \geq 0.$$

If $\dim_H \nu \neq \dim_H \Lambda$, then by the discussion after Definition 5.1.3 the functions

$$\psi_s = t_s \log \|df|E^s\| \quad \text{and} \quad \psi_u = -t_u \log \|df|E^u\|$$

are not cohomologous. In particular, either ψ is not cohomologous to ψ_s or ψ is not cohomologous to ψ_u (or to both). Repeating arguments in the proof of Lemma 6.1.7 we show that either $T_s''(q) > 0$ for every $q \in \mathbb{R}$ or $T_u''(q) > 0$ for every $q \in \mathbb{R}$. Therefore, if $\dim_H \nu \neq \dim_H \Lambda$ then $T''(q) > 0$ for every $q \in \mathbb{R}$, and it follows from (6.26) that the function $\mathcal{D}$ is strictly convex. The analyticity of $\mathcal{D}$ can be obtained as in the proof of Theorem 6.1.2.

Finally, when $\dim_H \nu = \dim_H \Lambda$, again by the discussion after Definition 5.1.3, ν is simultaneously the equilibrium measure of ψ_s and ψ_u. Therefore, there exist constants $D_1, D_2 > 0$ such that

$$D_1 \le \frac{\nu(\Delta(\omega, r))}{\|d_x f^{-n(\omega,r)}|E^s(x)\|^{-t_s}\|d_x f^{m(\omega,r)}|E^u(x)\|^{-t_u}} \le D_2$$

for every $x = \chi(\omega) \in \Lambda$ and $r \in (0, 1)$. Hence, by (6.27) and (6.28) we obtain

$$\lim_{r \to 0} \frac{\log \nu(\Delta(\omega, r))}{\log r} = t_s + t_u$$

for every $\omega \in \Sigma_A$. It follows from Theorem 4.3.2 and the equality between the limits in (6.41) that

$$d_\nu(x) = t_s + t_u = \dim_H \Lambda$$

for every $x \in \Lambda$, and thus $\underline{\alpha} = \overline{\alpha} = \dim_H \Lambda$. $\square$

Statement 3 in Theorem 6.2.1 was first established by Simpelaere in [150]. In the case of hyperbolic flows appropriate versions of Theorem 6.2.1 were obtained by Pesin and Sadovskaya in [117] and by Barreira and Saussol in [14] (in the case of entropy spectra; see Section 7.1), using the symbolic dynamics developed by Bowen [36] and Ratner [129].

Chapter 7

General Concept of Multifractal Analysis

The concept of multifractal analysis, that was studied for repellers and hyperbolic sets in the former chapter, can be extended to other classes of dynamical systems and other invariant local quantities, besides the pointwise dimension considered in (6.1). With the purpose of unifying the theory, in [9] Barreira, Pesin and Schmeling proposed a general concept of multifractal analysis that we describe in this chapter. In particular, this provides many spectra that can be seen as potential multifractal moduli, in the sense that they may contain nontrivial information about the dynamical system. In particular, we describe in detail the multifractal analysis of the so-called u-dimension, which allows us to unify and generalize the results in Chapter 6.

7.1 General concept and basic notions

Consider a function $g\colon Y \to [-\infty, +\infty]$ in a subset Y of the space X. The level sets
$$K_\alpha^g = \{x \in Y : g(x) = \alpha\}$$
are pairwise disjoint, and in a similar manner to that in (6.2) we obtain a *multifractal decomposition* of X given by
$$X = (X \setminus Y) \cup \bigcup_{\alpha \in [-\infty, +\infty]} K_\alpha^g. \tag{7.1}$$

Now let G be a function defined in the set of subsets of X.

Definition 7.1.1. The *multifractal spectrum* $\mathcal{F}\colon [-\infty, +\infty] \to \mathbb{R}$ of the pair (g, G) is defined by
$$\mathcal{F}(\alpha) = G(K_\alpha^g).$$

When X is a smooth manifold and g is differentiable, each level set K_α^g is a hypersurface for all values of α that are not critical values of g. But in the theory of multifractal analysis we are mostly interested in studying the level sets of functions that are not differentiable, which naturally occur in ergodic theory. In fact, functions such as Birkhoff averages, Lyapunov exponents, pointwise dimensions, and local entropies are typically only measurable.

Now we describe some of the functions g and G that naturally occur in the theory of dynamical systems. Let X be a compact metric space and let $f\colon X \to X$ be a continuous function. We define functions G_D and G_E by

$$G_D(Z) = \dim_H Z \quad \text{and} \quad G_E(Z) = h(f|Z),$$

obtained from the Hausdorff dimension and the topological entropy (here we are using the notion of topological entropy for arbitrary subsets of a compact space, introduced by Bowen in [37]; see Section 7.2).

Definition 7.1.2. We call the multifractal spectra generated by G_D and G_E respectively *dimension spectra* and *entropy spectra*.

Let μ be a finite Borel measure in X, and let $Y \subset X$ be the set of points $x \in X$ for which the pointwise dimension

$$g_D(x) = g_{D,\mu}(x) = \lim_{r \to 0} \frac{\log \mu(B(x,r))}{\log r}$$

is well-defined. We obtain two multifractal spectra

$$\mathcal{D}_D = \mathcal{D}_{D,\mu} \quad \text{and} \quad \mathcal{D}_E = \mathcal{D}_{E,\mu}$$

generated respectively by the pairs (g_D, G_D) and (g_D, G_E). The spectrum $\mathcal{D}_D = \mathcal{D}$ was already considered in Sections 6.1 and 6.2. We observe that for $C^{1+\varepsilon}$ diffeomorphisms and hyperbolic invariant measures, Theorem 14.3.4 below ensures that $\mu(X \setminus Y) = 0$.

Now let $f\colon X \to X$ be a measurable transformation preserving a probability measure μ in X. Given a partition ξ of X into measurable subsets, for each $n \in \mathbb{N}$ we define a new partition of X by $\xi_n = \bigvee_{k=0}^{n-1} f^{-k}\xi$. Let $Y \subset X$ be the set of points $x \in X$ for which the *local entropy*

$$g_E(x) = g_{E,\mu}(x) = \lim_{n \to \infty} -\frac{1}{n} \log \mu(\xi_n(x))$$

is well-defined, where $\xi_n(x)$ denotes the element of ξ_n containing x (which is well-defined for μ-almost every $x \in X$). We obtain two multifractal spectra

$$\mathcal{E}_D = \mathcal{E}_{D,\mu} \quad \text{and} \quad \mathcal{E}_E = \mathcal{E}_{E,\mu}$$

generated respectively by the pairs (g_E, G_D) and (g_E, G_E). It follows readily from Shannon–McMillan–Breiman's theorem (see, for example, [84]) that $\mu(X \setminus Y) = 0$.

Moreover, if ξ is a generating partition and μ is ergodic, then $g_E(x) = h_\mu(f)$ for μ-almost every $x \in X$.

For repellers and hyperbolic sets of $C^{1+\varepsilon}$ conformal maps, Pesin and Weiss obtained in [119, 120] a multifractal analysis of the spectrum $\mathcal{D}_D$ (see Theorems 6.1.2 and 6.2.1). Barreira, Pesin and Schmeling, building on results in [119], obtained in [9] a multifractal analysis of the spectrum $\mathcal{E}_E$ for repellers of $C^{1+\varepsilon}$ expanding maps that are not necessarily conformal. In [153], Takens and Verbitski obtained a multifractal analysis of the spectrum $\mathcal{E}_E$ for expansive homeomorphisms with specification and equilibrium measures of a certain class of continuous functions (we note that these systems need not have Markov partitions).

We note that the spectra $\mathcal{D}_D$ and $\mathcal{E}_E$ are of different nature from the spectra $\mathcal{D}_E$ and $\mathcal{E}_D$. Namely, the first two relate local quantities—the pointwise dimension and the local entropy—with global quantities that are naturally associated to them—respectively the Hausdorff dimension and the topological entropy. This is not the case of the spectra $\mathcal{D}_E$ and $\mathcal{E}_D$ that put together quite different local and global quantities. Because of this we refer to them as *mixed spectra*. We describe their multifractal properties in Section 9.4.

We can also consider multifractal spectra defined by the Lyapunov exponents. Here we mention only a particular case, that essentially corresponds to considering only the top Lyapunov exponent. This particular case is well-adapted to the study of conformal dynamics. Let X be a smooth manifold and let $f\colon X \to X$ be a C^1 map. Consider the set $Y \subset X$ of points $x \in X$ for which the limit

$$g_L(x) = \lim_{n \to +\infty} \frac{1}{n} \log \|d_x f^n\|$$

exists. We obtain two multifractal spectra $\mathcal{L}_D$ and $\mathcal{L}_E$ generated respectively by the pairs (g_L, G_D) and (g_L, G_E). By Theorem 14.2.3 below (or by Kingman's subadditive ergodic theorem), if μ is an f-invariant probability Borel measure in X, then $\mu(X \setminus Y) = 0$. When the Lyapunov exponent takes more than one value at each point the study of the associated multifractal spectra becomes much more complicated.

7.2 The notion of u-dimension

We first recall the notion of topological pressure for *arbitrary* subsets of a compact metric space. It was introduced by Pesin and Pitskel in [116] as a Carathéodory characteristic. We note that the level sets of the pointwise dimension and of the local entropy are often noncompact. In fact, in some appropriate sense this is the generic situation. For example, under the hypotheses of Theorem 6.1.2 the level sets K_α are all noncompact if and only if ν is not the measure of maximal dimension μ in Theorem 4.1.8, that is, if and only if the function ψ is not cohomologous to $-(\dim_H J) \log \|df\|$ (see also Theorem 8.4.2).

Let $f\colon X \to X$ be a continuous transformation of a compact metric space, and let $\mathcal{U}$ be a finite open cover of X. Using the notation in Section 3.3.1, we say that the collection $\Gamma \subset \bigcup_{n\geq 1} W_n(\mathcal{U})$ *covers* the set $Z \subset X$ provided that

$$\bigcup_{\mathbf{U}\in\Gamma} X(\mathbf{U}) \supset Z.$$

Consider a continuous function $\varphi\colon X \to \mathbb{R}$. Given $\mathbf{U} \in W_n(\mathcal{U})$ we define

$$\varphi(\mathbf{U}) = \begin{cases} \sup_{X(\mathbf{U})} \sum_{k=0}^{n-1} \varphi \circ f^k & \text{if } X(\mathbf{U}) \neq \varnothing \\ -\infty & \text{if } X(\mathbf{U}) = \varnothing \end{cases}. \tag{7.2}$$

For each $Z \subset X$ and $\alpha \in \mathbb{R}$ we set

$$M(Z,\alpha,\varphi,\mathcal{U}) = \lim_{n\to\infty} \inf_{\Gamma} \sum_{\mathbf{U}\in\Gamma} \exp(-\alpha m(\mathbf{U}) + \varphi(\mathbf{U})),$$

where the infimum is taken over all finite or countable collections $\Gamma \subset \bigcup_{k\geq n} W_k(\mathcal{U})$ covering Z. Denoting by $\operatorname{diam}\mathcal{U}$ the diameter of the cover $\mathcal{U}$, one can easily show that

$$P_Z(\varphi) := \lim_{\operatorname{diam}\mathcal{U}\to 0} \inf\{\alpha \in \mathbb{R} : M(Z,\alpha,\varphi,\mathcal{U}) = 0\}$$

is well-defined.

Definition 7.2.1. The number $P_Z(\varphi)$ is called the *topological pressure* of φ in the set $Z \subset X$ (with respect to f).

We emphasize that Z need not be compact nor f-invariant. When $Z = X$ the number $P_X(\varphi)$ coincides with the topological pressure for compact sets introduced by Ruelle in [131] in the case of expansive maps, and by Walters in [161] in the general case. When $\varphi = 0$, the number

$$h(f|Z) := P_Z(0) = \lim_{\operatorname{diam}\mathcal{U}\to 0} \inf\left\{\alpha \in \mathbb{R} : \lim_{n\to\infty} \inf_{\Gamma} \sum_{\mathbf{U}\in\Gamma} \exp(-\alpha m(\mathbf{U})) = 0\right\}, \tag{7.3}$$

where the infimum in Γ is taken over all finite or countable collections $\Gamma \subset \bigcup_{k\geq n} W_k(\mathcal{U})$ covering Z, is called the *topological entropy* of f in the set Z (with respect to f). It was introduced by Pesin and Pitskel in [116], and coincides with the notion of topological entropy for noncompact sets introduced earlier by Bowen in [37]. Again we emphasize that the set Z need not be compact nor f-invariant. When $Z = X$ we recover the notion of topological entropy in Definition 2.3.2.

Now we recall a related notion introduced by Barreira and Schmeling in [21]. Let f be a continuous transformation of a compact metric space X, and let $\mathcal{U}$ be a finite open cover of X. Let also $u\colon X \to \mathbb{R}$ be a positive continuous function. Given $\mathbf{U} \in W_n(\mathcal{U})$ we define $u(\mathbf{U})$ as in (7.2). For each $Z \subset X$ and $\alpha \in \mathbb{R}$ we set

$$N(Z,\alpha,u,\mathcal{U}) = \lim_{n\to\infty} \inf_{\Gamma} \sum_{\mathbf{U}\in\Gamma} \exp(-\alpha u(\mathbf{U})), \tag{7.4}$$

where the infimum is taken over all finite or countable collections $\Gamma \subset \bigcup_{k \geq n} \mathcal{W}_k(\mathcal{U})$ covering Z. Set

$$\dim_{u,\mathcal{U}} Z = \inf\{\alpha \in \mathbb{R}: N(Z, \alpha, u, \mathcal{U}) = 0\}. \tag{7.5}$$

We can easily show that

$$\dim_u Z := \lim_{\operatorname{diam} \mathcal{U} \to 0} \dim_{u,\mathcal{U}} Z$$

is well-defined.

Definition 7.2.2. The number $\dim_u Z$ is called the *u-dimension* of the set Z (with respect to f).

The following result gives a relation between the u-dimension and the topological pressure. It follows easily from the definitions.

Proposition 7.2.3. *We have* $\dim_u Z = \alpha$, *where* α *is the unique real number such that* $P_Z(-\alpha u) = 0$.

Furthermore, given a probability Borel measure μ in X, we set

$$\dim_{u,\mathcal{U}} \mu = \inf\{\dim_{u,\mathcal{U}} Z: \mu(Z) = 1\}.$$

One can easily show that the number

$$\dim_u \mu := \lim_{\operatorname{diam} \mathcal{U} \to 0} \dim_{u,\mathcal{U}} \mu$$

is well-defined, and we call it the *u-dimension* of μ.

We are particularly interested in the following examples.

Example 7.2.4. If $u = 1$, then $\dim_u Z = h(f|Z)$ for every set $Z \subset X$, and $\dim_u \mu = h_\mu(f)$ for every probability measure μ in X.

Example 7.2.5. Let X be a repeller of a $C^{1+\varepsilon}$ conformal expanding map f, for some $\varepsilon > 0$ (see Section 4.1). If $u = \log \|df\|$, then $\dim_u Z = \dim_H Z$ for every set $Z \subset X$, and $\dim_u \mu = \dim_H \mu$ for every probability measure μ in X. Indeed, since f is of class $C^{1+\varepsilon}$, it follows from the bounded distortion property that there exist constants c_1, $c_2 > 0$ such that

$$c_1(\operatorname{diam} X(\mathbf{U}))^\alpha \leq \exp(-\alpha u(\mathbf{U})) \leq c_2(\operatorname{diam} X(\mathbf{U}))^\alpha \tag{7.6}$$

for every $\mathbf{U} \in \bigcup_{n \geq 1} \mathcal{W}_n(\mathcal{U})$ and $\alpha \in \mathbb{R}$. The two identities follow immediately from (7.6) and Theorem 4.1.7.

We also introduce local quantities that are generalizations of the lower and upper pointwise dimensions.

Definition 7.2.6. The *lower* and *upper u-pointwise dimensions* of μ at the point $x \in X$ are defined by

$$\underline{d}_{\mu,u}(x) = \lim_{\operatorname{diam}\mathcal{U}\to 0} \liminf_{n\to\infty} \inf_{\mathbf{U}} -\frac{\log \mu(X(\mathbf{U}))}{u(\mathbf{U})}$$

and

$$\overline{d}_{\mu,u}(x) = \lim_{\operatorname{diam}\mathcal{U}\to 0} \limsup_{n\to\infty} \sup_{\mathbf{U}} -\frac{\log \mu(X(\mathbf{U}))}{u(\mathbf{U})},$$

where the infimum and supremum are taken over all vectors $\mathbf{U} \in \mathcal{W}_n(\mathcal{U})$ such that $x \in X(\mathbf{U})$.

The following statement was formulated by Barreira and Schmeling in [21].

Proposition 7.2.7. *If μ is an ergodic f-invariant probability Borel measure in X, then*

$$\dim_u \mu = \lim_{\operatorname{diam}\mathcal{U}\to 0} \underline{d}_{\mu,u}(x,\mathcal{U}) = \lim_{\operatorname{diam}\mathcal{U}\to 0} \overline{d}_{\mu,u}(x,\mathcal{U}) = \frac{h_\mu(f)}{\int_X u\, d\mu} \qquad (7.7)$$

for μ-almost every $x \in X$.

Proof. The proof follows similar arguments to those of Pesin in [115, Proposition 11.1]. Given $x \in X$, $n \in \mathbb{N}$, and $\delta > 0$ we consider the sets

$$B(x,n,\delta) = \big\{y \in X : d(f^i x, f^i y) < \delta \text{ for } i = 0,\dots,n-1\big\},$$

where d is the distance in X. Now let $\mathcal{U}$ be a finite open cover of X and let $\delta(\mathcal{U})$ be its Lebesgue number. One can easily verify that for every $x \in X$ and $\mathbf{U} \in \bigcup_{n\geq 1} \mathcal{W}_n(\mathcal{U})$ such that $x \in X(\mathbf{U})$ we have

$$B\big(x, m(\mathbf{U}), \delta(\mathcal{U})/2\big) \subset X(\mathbf{U}) \subset B\big(x, m(\mathbf{U}), 2\operatorname{diam}\mathcal{U}\big). \qquad (7.8)$$

We recall the *local entropy formula* established by Brin and Katok in [43].

Lemma 7.2.8. *Let μ be an f-invariant probability Borel measure in X. For μ-almost every $x \in X$ we have*

$$\begin{aligned}
h_\mu(x) : &= \lim_{\delta\to 0} \limsup_{n\to\infty} -\frac{1}{n} \log \mu(B(x,n,\delta)) \\
&= \lim_{\delta\to 0} \liminf_{n\to\infty} -\frac{1}{n} \log \mu(B(x,n,\delta)).
\end{aligned} \qquad (7.9)$$

Furthermore, the function $x \mapsto h_\mu(x)$ is μ-integrable, is f-invariant μ-almost everywhere, and satisfies

$$h_\mu(f) = \int_X h_\mu(x)\, d\mu(x). \qquad (7.10)$$

When μ is ergodic, it follows from Lemma 7.2.8 and (7.8) that

$$
\begin{aligned}
h_\mu(f) &= \lim_{\operatorname{diam}\mathcal{U}\to 0}\ \liminf_{n\to\infty}\inf_{\mathbf{U}}\ -\frac{1}{n}\log\mu(X(\mathbf{U})) \\
&= \lim_{\operatorname{diam}\mathcal{U}\to 0}\ \limsup_{n\to\infty}\sup_{\mathbf{U}}\ -\frac{1}{n}\log\mu(X(\mathbf{U})),
\end{aligned}
\tag{7.11}
$$

where the infimum and supremum are taken over all vectors $\mathbf{U}\in\mathcal{W}_n(\mathcal{U})$ such that $x\in X(\mathbf{U})$.

On the other hand, since u is continuous and X is compact, given $\varepsilon>0$ there exists $\delta>0$ such that

$$
|u(x)-u(y)|<\varepsilon \quad\text{whenever}\quad d(x,y)<\delta.
$$

This implies that if $\operatorname{diam}\mathcal{U}<\delta$, then for μ-almost every $x\in X$ we have

$$
\left|\liminf_{n\to\infty}\inf_{\mathbf{U}}\ \sup_{y\in X(\mathbf{U})}\ \frac{1}{n}\sum_{k=0}^{n-1}u(f^k y)-\int_X u\,d\mu\right|\le\varepsilon
$$

and

$$
\left|\limsup_{n\to\infty}\sup_{\mathbf{U}}\ \sup_{y\in X(\mathbf{U})}\ \frac{1}{n}\sum_{k=0}^{n-1}u(f^k y)-\int_X u\,d\mu\right|\le\varepsilon,
$$

where the infimum and supremum in $\mathbf{U}$ are taken over all $\mathbf{U}\in\mathcal{W}_n(\mathcal{U})$ such that $x\in X(\mathbf{U})$. Since ε is arbitrary we obtain

$$
\begin{aligned}
\int_X u\,d\mu &= \lim_{\operatorname{diam}\mathcal{U}\to 0}\ \liminf_{n\to\infty}\inf_{\mathbf{U}}\ \sup_{y\in X(\mathbf{U})}\ \frac{1}{n}\sum_{k=0}^{n-1}u(f^k y) \\
&= \lim_{\operatorname{diam}\mathcal{U}\to 0}\ \limsup_{n\to\infty}\sup_{\mathbf{U}}\ \sup_{y\in X(\mathbf{U})}\ \frac{1}{n}\sum_{k=0}^{n-1}u(f^k y),
\end{aligned}
$$

that is,

$$
\begin{aligned}
\int_X u\,d\mu &= \lim_{\operatorname{diam}\mathcal{U}\to 0}\ \liminf_{n\to\infty}\inf_{\mathbf{U}}\ \frac{u(\mathbf{U})}{n} \\
&= \lim_{\operatorname{diam}\mathcal{U}\to 0}\ \limsup_{n\to\infty}\sup_{\mathbf{U}}\ \frac{u(\mathbf{U})}{n}.
\end{aligned}
\tag{7.12}
$$

It follows from (7.11) and (7.12) that

$$
\lim_{\operatorname{diam}\mathcal{U}\to 0}\underline{d}_{\mu,u}(x,\mathcal{U})=\lim_{\operatorname{diam}\mathcal{U}\to 0}\overline{d}_{\mu,u}(x,\mathcal{U})=\frac{h_\mu(f)}{\int_X u\,d\mu}.
$$

The first identity in (7.7) can be obtained with a simple modification of the arguments in the proofs of Theorems 2.1.5 and 2.1.6. $\qquad\square$

7.3   Multifractal analysis of u-dimension

We present in this section a complete multifractal analysis of the spectra generated by the u-dimension, for a one-sided or two-sided topological Markov chain $\sigma|\Sigma$ (see Definitions 4.1.3 and 4.2.7), that is, such that either $\Sigma \subset \Sigma_p^+$ or $\Sigma \subset \Sigma_p$. Let μ be a probability Borel measure in Σ and let $u\colon \Sigma \to \mathbb{R}^+$ be a continuous function. For each $x \in \Sigma$ we write

$$\underline{d}_{\mu,u}(x) = \liminf_{n\to\infty} -\frac{\log \mu(C_n(x))}{(S_n u)(x)}$$

and

$$\overline{d}_{\mu,u}(x) = \limsup_{n\to\infty} -\frac{\log \mu(C_n(x))}{(S_n u)(x)},$$

where $C_n(x)$ is the cylinder set of length n that contains x, and where

$$(S_n u)(x) = \sum_{k=0}^{n-1} u(\sigma^k x). \tag{7.13}$$

We can easily show that if u is Hölder continuous, then

$$\underline{d}_{\mu,u}(x) = \underline{d}_{\mu,u}(x,\mathcal{U}) \quad \text{and} \quad \overline{d}_{\mu,u}(x) = \overline{d}_{\mu,u}(x,\mathcal{U})$$

for every $x \in \Sigma$ and every open cover $\mathcal{U}$ of Σ by cylinder sets (not necessarily all with the same length). For every $\alpha \in \mathbb{R}$, we consider the set

$$K_\alpha = \{x \in \Sigma\colon \underline{d}_{\mu,u}(x) = \overline{d}_{\mu,u}(x) = \alpha\}.$$

Whenever $K_\alpha \neq \varnothing$ and $x \in K_\alpha$, we denote the common value α of $\underline{d}_{\mu,u}(x)$ and $\overline{d}_{\mu,u}(x)$ by $d_{\mu,u}(x)$, and we call it the u-*pointwise dimension* of μ at x.

Definition 7.3.1. The u-*dimension spectrum* (for the u-pointwise dimensions) of the measure μ is the function

$$\mathcal{D}_u = \mathcal{D}_{u,\mu}\colon \{\alpha \in \mathbb{R} : K_\alpha \neq \varnothing\} \to \mathbb{R}$$

defined by

$$\mathcal{D}_u(\alpha) = \dim_u K_\alpha.$$

Now let φ be a continuous function in Σ. For each $q \in \mathbb{R}$, we consider the function

$$\varphi_q = -T_u(q)u + q\varphi, \tag{7.14}$$

where $T_u(q)$ is the unique real number such that $P(\varphi_q) = 0$ (here P denotes the topological pressure in Σ; see Proposition 3.2.1). Proceeding as in Section 6.1 we can show that if the functions u and φ are Hölder continuous, then $T_u\colon \mathbb{R} \to \mathbb{R}$ is analytic. We then set

$$\alpha_u(q) = -T_u'(q)$$

and we denote by $(\underline{\alpha}_u, \overline{\alpha}_u)$ the range of the function α_u. We also denote by ν_q and m_u respectively the equilibrium measures of φ_q and $-(\dim_u \Sigma)u$ with respect to σ (see Definition 3.2.3).

The following statement contains a complete multifractal analysis of the spectrum $\mathcal{D}_u$ for topological Markov chains. It was formulated by Barreira and Schmeling in [21].

Theorem 7.3.2. *Given a one-sided or two-sided topologically mixing topological Markov chain $\sigma|\Sigma$, let u and φ be Hölder continuous functions in Σ, such that u is positive and $P(\varphi) = 0$, and let μ be the equilibrium measure of φ with respect to σ. Then the following properties hold:*

1. *the set $K_{\alpha_u(q)}$ is σ-invariant and dense for every $q \in \mathbb{R}$;*

2. *for μ-almost every $x \in \Sigma$ we have*

$$\underline{d}_{\mu,u}(x) = \overline{d}_{\mu,u}(x) = -\frac{\int_\Sigma \varphi \, d\mu}{\int_\Sigma u \, d\mu} = \frac{h_\mu(\sigma)}{\int_\Sigma u \, d\mu};$$

3. *the function $T_u \colon \mathbb{R} \to \mathbb{R}$ is analytic, and satisfies $T_u'(q) \le 0$ and $T_u''(q) \ge 0$ for every $q \in \mathbb{R}$; moreover, $T_u(0) = \dim_u \Sigma$ and $T_u(1) = 0$;*

4. *if $\mu = m_u$, then $\underline{\alpha}_u = \overline{\alpha}_u = \dim_u \Sigma$ and $\mathcal{D}_u$ is a delta function;*

5. *if $\mu \ne m_u$, then the functions $\mathcal{D}_u \colon (\underline{\alpha}_u, \overline{\alpha}_u) \to \mathbb{R}$ and T_u are analytic and strictly convex;*

6. *$\mathcal{D}_u$ is the Legendre transform of T_u, that is, for each $q \in \mathbb{R}$ we have*

$$\mathcal{D}_u(\alpha_u(q)) = T_u(q) + q\alpha_u(q);$$

moreover,

$$\alpha_u(q) = -\frac{\int_\Sigma \varphi \, d\nu_q}{\int_\Sigma u \, d\nu_q}; \tag{7.15}$$

7. *for each $q \in \mathbb{R}$ we have $\nu_q(K_{\alpha_u(q)}) = 1$ and*

$$\lim_{n\to\infty} -\frac{\log \nu_q(C_n(x))}{(S_n u)(x)} = T_u(q) + q\alpha_u(q)$$

for ν_q-almost every $x \in K_{\alpha_u(q)}$; moreover,

$$\overline{d}_{\nu_q,u}(x) \le T_u(q) + q\alpha_u(q)$$

for every $x \in K_{\alpha_u(q)}$, and $\mathcal{D}_u(\alpha_u(q)) = \dim_u \nu_q$ for each $q \in \mathbb{R}$.

Proof. For one-sided shifts statement 1 follows immediately from the σ-invariance of $K_{\alpha_u(q)}$. For two-sided shifts we note that $\sigma^{-1}|\Sigma$ is a topological Markov chain with transition matrix equal to the transpose of the transition matrix of $\sigma|\Sigma$, and thus $\sigma^{-1}|\Sigma$ is also topologically mixing. Since

$$K_{\alpha_u(q)} = \Sigma \cap \pi^{-1}(\pi K_{\alpha_u(q)}),$$

where $\pi \colon \{1,\ldots,p\}^{\mathbb{Z}} \to \{1,\ldots,p\}^{\mathbb{N}}$ is the canonical projection, we conclude that the set $K_{\alpha_u(q)}$ is dense.

Since μ is ergodic, statement 2 follows immediately from Proposition 7.2.7.

The proof of statements 3–7 can essentially be obtained by repeating arguments in the proof of Theorem 6.1.2, simply replacing J and Σ by $\log\|df\|$ and u, and thus the details are omitted. For statement 7 note that by Proposition 7.2.3 we have $P_{\Sigma}(-su) = 0$, where $s = \dim_u\Sigma$, and since m_u is the equilibrium measure of the function $-su$ there exist constants c_1, $c_2 > 0$ such that

$$c_1 \exp[-s(S_n u)(x)] \le m_u(C_n(x)) \le c_2 \exp[-s(S_n u)(x)]$$

for every $n \in \mathbb{N}$ and $x \in \Sigma$. Therefore, if $\mu = m_u$ then

$$d_{\mu,u}(x) = \lim_{n\to\infty} -\frac{\log m_u(C_n(x))}{S_n u(x)} = \dim_u\Sigma$$

for every $x \in \Sigma$. This implies that

$$K_\alpha = \begin{cases} \Sigma & \text{if } \alpha = \dim_u\Sigma, \\ \varnothing & \text{if } \alpha \neq \dim_u\Sigma, \end{cases}$$

and thus,

$$\mathcal{D}_u(\alpha) = \begin{cases} \dim_u\Sigma & \text{if } \alpha = \dim_u\Sigma, \\ 0 & \text{if } \alpha \neq \dim_u\Sigma. \end{cases}$$

This completes the proof of the theorem.   □

We note that the measure m_u is in fact the unique σ-invariant probability Borel measure in Σ with

$$\dim_u m_u = \dim_u\Sigma. \tag{7.16}$$

It is also the unique ergodic σ-invariant probability Borel measure in Σ satisfying (7.16).

Definition 7.3.3. The measure m_u is called the *measure of maximal u-dimension*. For each $q \in \mathbb{R}$, the measure ν_q is called the *full measure* for the spectrum $\mathcal{D}_u$ at the point $\alpha_u(q)$.

7.4 Domain of the spectra

The statements in the remaining sections of this chapter were established by
Schmeling in [142] for the dimension and entropy spectra. Following [21] we re-
formulate them for the u-dimension. We continue to denote by $(\underline{\alpha}_u, \overline{\alpha}_u)$ the range
of α_u, that is,

$$\underline{\alpha}_u = \lim_{q \to +\infty} \alpha_u(q) \quad \text{and} \quad \overline{\alpha}_u = \lim_{q \to -\infty} \alpha_u(q).$$

Theorem 7.4.1. *The following properties hold:*

1. *we have*

$$\underline{\alpha}_u = \inf_{x \in \Sigma} \underline{d}_{\mu,u}(x) = \inf_{\nu \in \mathcal{M}_E} -\frac{\int_\Sigma \varphi \, d\nu}{\int_\Sigma u \, d\nu}$$

and

$$\overline{\alpha}_u = \sup_{x \in \Sigma} \overline{d}_{\mu,u}(x) = \sup_{\nu \in \mathcal{M}_E} -\frac{\int_\Sigma \varphi \, d\nu}{\int_\Sigma u \, d\nu};$$

2. *for each $\alpha \in \mathbb{R}$, we have $K_\alpha = \varnothing$ if and only if $\alpha \notin [\underline{\alpha}_u, \overline{\alpha}_u]$.*

Proof. We first prove an auxiliary statement.

Lemma 7.4.2. *We have*

$$\inf_{\nu \in \mathcal{M}_E} -\frac{\int_\Sigma \varphi \, d\nu}{\int_\Sigma u \, d\nu} = \underline{\alpha}_u \quad \text{and} \quad \sup_{\nu \in \mathcal{M}_E} -\frac{\int_\Sigma \varphi \, d\nu}{\int_\Sigma u \, d\nu} = \overline{\alpha}_u. \tag{7.17}$$

Proof of the lemma. We only prove the first identity. The proof of the second one
is entirely analogous. In view of Theorem 7.3.2 it remains to prove that

$$\inf_{\nu \in \mathcal{M}_E} -\frac{\int_\Sigma \varphi \, d\nu}{\int_\Sigma u \, d\nu} \le \underline{\alpha}_u.$$

We first observe that

$$T_u(q) = \inf_{\nu \in \mathcal{M}_E} \frac{h_\nu(\sigma) + q \int_\Sigma \varphi \, d\nu}{\int_\Sigma u \, d\nu} = \frac{h_{\nu_q}(\sigma) + q \int_\Sigma \varphi \, d\nu_q}{\int_\Sigma u \, d\nu_q},$$

that is,

$$T_u(q) = \inf_{\nu \in \mathcal{M}_E} \left(\dim_H \nu + q \frac{\int_\Sigma \varphi \, d\nu}{\int_\Sigma u \, d\nu} \right) = \dim_H \nu_q + q \frac{\int_\Sigma \varphi \, d\nu_q}{\int_\Sigma u \, d\nu_q}. \tag{7.18}$$

Indeed, since $P(\varphi_q) = 0$ it follows from Theorem 2.3.3 that

$$0 \ge h_\nu(\sigma) - T_u(q) \int_\Sigma u \, d\nu + q \int_\Sigma \varphi \, d\nu$$

for every $\nu \in \mathcal{M}_E$, with equality if and only if $\nu = \nu_q$. Therefore,

$$T_u(q) \geq \frac{h_\nu(\sigma) + q \int_\Sigma \varphi \, d\nu}{\int_\Sigma u \, d\nu},$$

with equality if and only if $\nu = \nu_q$. This establishes (7.18).

Given $\varepsilon > 0$ and $q < 0$ such that $\alpha_u(q) - \underline{\alpha}_u < \varepsilon$, it follows from (7.15) and (7.18) that

$$(1 - q)\underline{\alpha}_u \geq (q - 1)\frac{\int_\Sigma \varphi \, d\nu_q}{\int_\Sigma u \, d\nu_q} - (1 - q)\varepsilon$$

$$= \dim_H \nu_q + q\frac{\int_\Sigma \varphi \, d\nu_q}{\int_\Sigma u \, d\nu_q} - (1 - q)\varepsilon$$

$$= T_u(q) - (1 - q)\varepsilon$$

$$= \inf_{\nu \in \mathcal{M}_E} \left(\dim_H \nu + q\frac{\int_\Sigma \varphi \, d\nu}{\int_\Sigma u \, d\nu}\right) - (1 - q)\varepsilon$$

$$\geq -q \inf_{\nu \in \mathcal{M}_E} -\frac{\int_\Sigma \varphi \, d\nu}{\int_\Sigma u \, d\nu} - (1 - q)\varepsilon.$$

Dividing by $-q$ and letting $q \to -\infty$ we obtain the first identity in (7.17). $\qquad \square$

We proceed with the proof of the theorem. In view of Theorem 7.3.2, to obtain the first identity in statement 1 we prove that

$$\underline{\alpha}_u \leq \inf_{x \in \Sigma} \underline{d}_{\mu,u}(x).$$

Let us assume that this was not the case. Then there would exist $\delta > 0$, $x \in \Sigma$, and an increasing sequence $n_k = n_k(x)$ of natural numbers such that

$$-\frac{\sum_{j=0}^{n_k} \varphi(\sigma^j x)}{\sum_{j=0}^{n_k} u(\sigma^j x)} < \underline{\alpha}_u - \delta.$$

Now let ν be an accumulation point of the sequence of measures

$$\nu_k = \frac{1}{n_k} \sum_{j=0}^{n_k - 1} \delta_{\sigma^j x},$$

where δ_p is the probability delta measure at p, that is, $\delta_p(\{p\}) = 1$. We have

$$-\frac{\int_\Sigma \varphi \, d\nu_k}{\int_\Sigma u \, d\nu_k} = -\frac{\sum_{j=0}^{n_k} \varphi(\sigma^j x)}{\sum_{j=0}^{n_k} u(\sigma^j x)} < \underline{\alpha}_u - \delta,$$

and hence,

$$-\frac{\int_\Sigma \varphi \, d\nu}{\int_\Sigma u \, d\nu} \leq \underline{\alpha}_u - \delta,$$

which contradicts Lemma 7.4.2. This establishes the first identity.

Statement 2 is an immediate consequence of the first statement. $\qquad \square$

7.5 Existence of spectra with prescribed data

Now we establish the existence of u-dimension spectra for any prescribed data at the endpoints of the domain of the spectrum. We note that

$$(\underline{\alpha}_u, \mathcal{D}_u(\underline{\alpha}_u), \overline{\alpha}_u, \mathcal{D}_u(\overline{\alpha}_u)) \in \mathbb{B},$$

where

$$\mathbb{B} = \left\{ (x_1, y_1, x_2, y_2) \in \mathbb{R}^4 \colon y_1 \le x_1 \le \dim_u \Sigma \text{ and } y_2 \le \dim_u \Sigma \le x_2 \right\}.$$

Given $\theta \in (0, 1]$, we denote by $C^\theta(\Sigma)$ the space of Hölder continuous functions $\varphi \colon \Sigma \to \mathbb{R}$ with Hölder exponent θ, equipped with the norm

$$\|\varphi\|_\theta = \sup\{|\varphi(x)| : x \in \Sigma\} + \sup \left\{ \frac{|\varphi(x) - \varphi(y)|}{d(x, y)^\theta} : x, y \in \Sigma \text{ and } x \ne y \right\}. \quad (7.19)$$

We note that $C^\theta(\Sigma)$ is a Baire space with the induced topology.

The following statement was obtained by Schmeling in [142].

Theorem 7.5.1. *Given a one-sided or two-sided topologically mixing topological Markov chain $\sigma|\Sigma$, let u be a positive Hölder continuous function in Σ. For each $(x_1, y_1, x_2, y_2) \in \text{int}\,\mathbb{B}$ there exists a Hölder continuous function $\varphi \colon \Sigma \to \mathbb{R}$ such that the spectrum $\mathcal{D}_u$ of the equilibrium measure of φ satisfies*

$$(\underline{\alpha}_u, \mathcal{D}_u(\underline{\alpha}_u), \overline{\alpha}_u, \mathcal{D}_u(\overline{\alpha}_u)) = (x_1, y_1, x_2, y_2).$$

Proof. We start with an auxiliary result.

Lemma 7.5.2. *Given numbers $d_i < \dim_u \Sigma$, $i = 1, 2$, there exist disjoint closed σ-invariant sets $S_i \subset \Sigma$ with $\dim_u S_i = d_i$, and numbers λ_i, $i = 1, 2$ such that for every $x \in S_i$ we have*

$$\lim_{n \to \infty} \frac{1}{n} \sum_{j=0}^{n-1} u(\sigma^j x) = \lambda_i.$$

Proof of the lemma. We construct inductively topological Markov chains $\sigma|S_{i,n}$ that approximate the sets $\sigma|S_i$. Set

$$\lambda_u = \int_\Sigma u \, dm_u, \quad h_u = h_{m_u}(\sigma), \quad \underline{\lambda} = \min u, \quad \text{and} \quad \overline{\lambda} = \max\{u, 1\}.$$

Fix $\varepsilon > 0$ such that

$$\varepsilon < \min(\dim_u \Sigma - d_i)\lambda_u \quad \text{for} \quad i = 1, 2.$$

For each $m \in \mathbb{N}$ we consider the sets

$$\Lambda_\varepsilon^m = \left\{ x \in \Sigma \colon \left| \frac{1}{n} \sum_{j=0}^{n-1} u(\sigma^j x) - \lambda_u \right| \le L\varepsilon \text{ for every } n \ge m \right\},$$

where
$$L = 1/\min\{4(d_i + 1) : i = 1, 2\},$$

and
$$H_\varepsilon^m = \left\{x \in \Sigma : \left|h_u + \frac{1}{n}\log m_u(C_n(x))\right| \leq \varepsilon \text{ for every } n \geq m\right\}.$$

We also write $\Gamma_\varepsilon^m = \Lambda_\varepsilon^m \cap H_\varepsilon^m$, and we define
$$\kappa = \min\{m \in \mathbb{N} : m_u(\Gamma_\varepsilon^m) > 1/2\}.$$

By Birkhoff's ergodic theorem and Shannon–McMillan–Breiman's theorem, the number κ is finite. Set
$$n = \max\left\{\left\lfloor\frac{4(K + M)\overline{\lambda}(d_i + 1)}{\varepsilon}\right\rfloor + 1, \left\lfloor\frac{1 + \log 4}{(\dim_u\Sigma - d_i)\lambda_u - \varepsilon}\right\rfloor, \kappa\right\},$$

where $\lfloor\cdot\rfloor$ denotes the integer part, $K = \|u\|_\varrho e^{\overline{\lambda}}/(1 - e^{\underline{\lambda}})$, and M is the smallest positive integer such that $A^M > 0$ (recall that $\sigma|\Sigma$ is topologically mixing).

Given $m > n$ we set
$$\mathcal{C}_m = \{C_m(x) : x \in \Gamma_\varepsilon^n\}.$$

For each $C \in \mathcal{C}_m$ we have $m_u(C) \leq e^{-mh_u + m\varepsilon}$, and
$$\text{card}\, \mathcal{C}_m \geq \frac{m_u(\Gamma_\varepsilon^m)}{\max m_u(C)} \geq \frac{1}{2}e^{mh_u - m\varepsilon}.$$

Since $\varepsilon < (\dim_u\Sigma - d_1)\lambda_u$, by the choice of n there exists a subset $\mathcal{C}_{1,1} \subset \mathcal{C}_n$ with
$$\text{card}\, \mathcal{C}_{1,1} = \lfloor\exp[n(\lambda_u d_1 + d_1\varepsilon/4)]\rfloor + 1.$$

Now we consider the set $S_{1,1}$ of points $x \in \Sigma$ with the property that there exists $l < n + M$ such that
$$C_n(\sigma^{l+j(n+M)}x) \in \mathcal{C}_{1,1} \quad \text{for every} \quad j \in \mathbb{N}.$$

We note that $\sigma|S_{1,1}$ is a topological Markov chain. Since $A^M > 0$ we have $S_{1,1} \neq \varnothing$,
$$\liminf_{k\to\infty}\frac{1}{k}\log\text{card}\{C_k(x) : x \in S_{1,1}\} \geq \liminf_{k\to\infty}\frac{1}{k}\log\left[(\text{card}\, \mathcal{C}_{1,1})^{\lfloor k/(n+M)\rfloor - 1}\right]$$
$$\geq \lambda_u d_1 + \varepsilon/2,$$

and
$$\limsup_{k\to\infty}\frac{1}{k}\log\text{card}\{C_k(x) : x \in S_{1,1}\}$$
$$\leq \limsup_{k\to\infty}\frac{1}{k}\log\left[(\text{card}\, \mathcal{C}_{1,1})^{\lfloor k/(n+M)\rfloor}p^{\lfloor k/(n+M)\rfloor + l}\right]$$
$$\leq \lambda_u d_1 + 3\varepsilon/2,$$

where A is assumed to be a $p \times p$ matrix. Hence,

$$\lambda_u d_1 + \varepsilon/2 \le h(\sigma|S_{1,1}) \le \lambda_u d_1 + 3\varepsilon/2. \tag{7.20}$$

On the other hand, for each $x \in S_{1,1}$ we have

$$\limsup_{k \to \infty} \frac{1}{k} \sum_{j=0}^{k-1} u(\sigma^j x)$$

$$\le \limsup_{k \to \infty} \sum_{l=0}^{\lfloor (n+M)/k \rfloor + 1} \frac{1}{n+M} \sum_{j=0}^{n-1} \left[u(\sigma^{j+l(n+M)} x) + M\bar{\lambda} \right]$$

$$\le \lambda_u + L\varepsilon + \frac{K}{n+M} + \frac{M}{n+M} \le \lambda_u + (L+1)\varepsilon,$$

and

$$\liminf_{k \to \infty} \frac{1}{k} \sum_{j=0}^{k-1} u(\sigma^j x)$$

$$\ge \liminf_{k \to \infty} \sum_{l=0}^{\lfloor (n+M)/k \rfloor + 1} \frac{1}{n+M} \sum_{j=0}^{n-1} \left[u(\sigma^{j+l(n+M)} x) + M\bar{\lambda} \right]$$

$$\ge \lambda_u - L\varepsilon - \frac{K}{n+M} - \frac{M}{n+M} \ge \lambda_u - (L+1)\varepsilon.$$

It follows from (7.20) and these estimates that

$$d_1 + \varepsilon/4 < \dim_u S_{1,1} < d_1 + 2\varepsilon.$$

Now we construct a topological Markov chain $\sigma|S_{2,1}$ with $S_{2,1} \subset \Sigma \setminus S_{1,1}$. Since $\varepsilon < (\dim_u \Sigma - d_2)\lambda_u$, by the choice of n there exists a subset $\mathcal{C}_{2,1} \subset \mathcal{C}_n$ with

$$\operatorname{card} \mathcal{C}_{2,1} = \lfloor \exp[n(\lambda_u d_2 + d_2\varepsilon/4)] \rfloor + 1.$$

We note that there is a cylinder set $\hat{C} \subset \Sigma$ of length n with empty intersection with $S_{1,1}$. Let $m = 8n/\varepsilon$. We consider the set $S_{2,1}$ of points $x \in \Sigma$ with the property that there exists $l < m + M$ such that for each $j \in \mathbb{N}$ the cylinder set $C_{m+n}(\sigma^{l+j(m+n+M)} x)$ is of the form $C_{n,1} \cdots C_{n,m} \hat{C}$, where $C_{n,i} \in \mathcal{C}_{2,1}$ for $i = 1, \ldots, m$, and where $C_1 C_2$ is the cylinder set $C_1 \cap \sigma^{-|C_1|} C_2$. Clearly, $S_{1,1} \cap S_{2,1} = \emptyset$, and $\sigma|S_{2,1}$ is a topological Markov chain.

It follows from estimates similar to those for $S_{1,1}$ that

$$\lambda_u d_2 + 3\varepsilon/8 \le h(\sigma|S_{2,1}) \le \lambda_u d_2 + 13\varepsilon/8. \tag{7.21}$$

Furthermore, for each $x \in S_{2,1}$ we have

$$\limsup_{k \to \infty} \frac{1}{k} \sum_{j=0}^{k-1} u(\sigma^j x) \le \lambda_u + \left(L + \frac{9}{8} \right) \varepsilon$$

and

$$\liminf_{k\to\infty} \frac{1}{k} \sum_{j=0}^{k-1} u(\sigma^j x) \geq \lambda_u - \left(L + \frac{9}{8}\right)\varepsilon.$$

It follows from (7.21) and these estimates that

$$d_2 + \varepsilon/8 < \dim_u S_{2,1} < d_2 + 17\varepsilon/8.$$

Now let us assume that for a sufficiently small $\delta > 0$ and each $l \leq r$ we have constructed topological Markov chains $\sigma|S_{1,l}$ and $\sigma|S_{2,l}$ with the following properties for $i = 1, 2$:

1. $S_{i,l} \subset S_{i,l-1}$ for $l = 2, \ldots, r$;

2. for $l = 1, \ldots, r$ we have

$$\lambda_u d_i + \frac{\delta}{2^{l+1}} \leq h(\sigma|S_{i,l}) \leq \lambda_u d_i + \frac{\delta}{2^{l-2}};$$

3. for $l = 1, \ldots, r$ there exists $\lambda_{i,l} \in \mathbb{R}$ such that for every $x \in S_{i,l}$ we have

$$\limsup_{k\to\infty} \frac{1}{k} \sum_{j=0}^{k-1} u(\sigma^j x) \leq \lambda_{i,l} + \frac{\delta}{2^l}$$

and

$$\liminf_{k\to\infty} \frac{1}{k} \sum_{j=0}^{k-1} u(\sigma^j x) \geq \lambda_{i,l} - \frac{\delta}{2^l};$$

4. for $l = 1, \ldots, r$ we have

$$d_i + \frac{\delta}{2^{l+1}} < \dim_u S_{i,l} < d_i + \frac{\delta}{2^{l-2}}.$$

In particular, setting $\delta = \varepsilon$ we can assume that $\lambda_{i,l} = \lambda_u$.

Now we repeat the construction of $S_{1,1}$ in Σ replacing Σ by $S_{1,r}$ and setting $\varepsilon = \delta/2^r$. The role of the measure of maximal dimension m_u is now played by the measure of maximal dimension $\mu_{1,r}$ in $S_{1,r}$. We obtain in this manner a topological Markov chain $\sigma|S_{1,r+1}$ which satisfies properties 1–4 for $l = r+1$. Similarly, we can construct a topological Markov chain $\sigma|S_{2,r+1}$ replacing Σ by $S_{2,r}$ and proceeding as above.

We thus obtain topological Markov chains $\sigma|S_{i,r}$ for $i = 1, 2$ and $r \in \mathbb{N}$ that satisfy properties 1–4. In particular, for each i the sets $S_{i,r}$ form a nested sequence. Furthermore, for $i = 1, 2$ the sequence $\lambda_{i,r}$ converges to some number λ_i as $r \to \infty$. Set

$$S_i = \bigcap_{r \in \mathbb{N}} S_{i,r}, \quad i = 1, 2.$$

For each $x \in S_i$ we have

$$\lim_{k \to \infty} \frac{1}{k} \sum_{j=0}^{k-1} u(\sigma^j x) = \lambda_i,$$

and

$$\dim_u S_i \leq \inf_{r \in \mathbb{N}} \dim_u S_{i,r} = d_i.$$

Moreover, $S_1 \cap S_2 = \varnothing$, since $S_{1,1} \cap S_{2,1} = \varnothing$. On the other hand, any accumulation point μ_i of the sequence of measures $(\mu_{i,r})_{r \in \mathbb{N}}$ is concentrated on the set S_i. By the upper semicontinuity of the Kolmogorov–Sinai entropy we obtain

$$\dim_u S_i \geq \dim_u \mu_i \geq \liminf_{r \to \infty} \dim_u \mu_{i,r} = \liminf_{r \to \infty} \dim_u S_{i,r} = d_i.$$

This completes the proof of the lemma. $\qquad\square$

It follows from Lemma 7.5.2 that $h(\sigma|S_i) = \lambda_i d_i$ for $i = 1, 2$.

Now we construct a function $\varphi \colon \Sigma \to \mathbb{R}$ with maximum in S_1 and minimum in S_2, such that for the u-dimension spectrum of its equilibrium measure we have

$$D_u(\underline{\alpha}_u) = d_1 \quad \text{and} \quad D_u(\overline{\alpha}_u) = d_2.$$

Given real numbers $z_1 > z_2$ we consider the set $\mathcal{F}(z_1, z_2)$ of functions $\varphi \in C^\theta(\Sigma)$ such that $\varphi|S_i = z_i$ for $i = 1, 2$, and

$$\varphi|\left(\Sigma \setminus (S_1 \cup S_2)\right) \in (z_2, z_1).$$

Lemma 7.5.3. *If $d_1 = \dim_u S_1 < -z_1 < \dim_u \Sigma$ and $-z_2 > \dim_u \Sigma$, then there exists a function $\varphi_0 \in \mathcal{F}(z_1, z_2)$ such that $P(\varphi_0 u) = 0$.*

Proof of the lemma. Let $U_{1,n} \supset S_1$ and $U_{2,n} \supset S_2$ be two nested sequences of open sets converging respectively to S_1 and S_2. For each $n \in \mathbb{N}$ we consider two functions $\varphi_{1,n}$ and $\varphi_{2,n}$ in Σ such that

$$\varphi_{1,n}(x) = \begin{cases} z_1 & \text{if } x \notin U_{2,n}, \\ z_2 & \text{if } x \in S_2, \\ \in (z_2, z_1) & \text{otherwise,} \end{cases}$$

and

$$\varphi_{2,n}(x) = \begin{cases} z_1 & \text{if } x \in S_1, \\ z_2 & \text{if } x \notin U_{1,n}, \\ \in (z_2, z_1) & \text{otherwise.} \end{cases}$$

We note that these functions exist because the sets S_i are closed, and that they are in the closure of $\mathcal{F}(z_1, z_2)$. Denoting by $P_{i,n}$, $i = 1, 2$ the topological pressure of $\varphi_{i,n}u$, we obtain

$$
\begin{aligned}
P_{1,n} &= \max_{\mu \in \mathcal{M}_E} \left\{ h_\mu(\sigma) + \int_\Sigma \varphi_{1,n}u \, d\mu \right\} \\
&\geq h_u + \int_\Sigma \varphi_{1,n}u \, dm_u \\
&\geq h_u + z_2\lambda_u m_u(U_{2,n}) + z_1\lambda_u(1 - m_u(U_{2,n})).
\end{aligned}
$$

Since $\dim_u S_2 < \dim_u \Sigma$, we have $m_u(S_2) = 0$. Therefore, $m_u(U_{2,n}) \to 0$ as $n \to \infty$. Hence, by the assumptions in the lemma we obtain

$$
\liminf_{n\to\infty} P_{1,n} \geq h_u + z_1\lambda_u > 0. \tag{7.22}
$$

On the other hand,

$$
\begin{aligned}
P_{2,n} &= \max_{\mu \in \mathcal{M}_E} \left\{ h_\mu(\sigma) + \int_\Sigma \varphi_{2,n}u \, d\mu \right\} \\
&= \max \left\{ \max_{\mu \in \mathcal{M}_1} \left\{ h_\mu(\sigma) + \int_\Sigma \varphi_{2,n}u \, d\mu \right\}, \max_{\mu \in \mathcal{M}_2} \left\{ h_\mu(\sigma) + \int_\Sigma \varphi_{2,n}u \, d\mu \right\} \right\} \\
&\leq \max \left\{ h_u + \int_\Sigma \varphi_{2,n}u \, dm_u, h(\sigma|S_1) + \max_{\mu \in \mathcal{M}_2} \int_\Sigma \varphi_{2,n}u \, d\mu \right\},
\end{aligned}
$$

where

$$
\mathcal{M}_1 = \{\mu \in \mathcal{M}_E : \mu(S_1) = 0\} \quad \text{and} \quad \mathcal{M}_2 = \{\mu \in \mathcal{M}_E : \mu(S_2) = 0\}.
$$

Furthermore,

$$
h_u + \int_\Sigma \varphi_{2,n}u \, dm_u \leq h_u + z_2\lambda_u + z_1\lambda_u m_u(U_{1,n}),
$$

and

$$
h(\sigma|S_1) + \max_{\mu \in \mathcal{M}_2} \int_\Sigma \varphi_{2,n}u \, d\mu \leq d_1\lambda_1 + z_1\lambda_1 + \max_{\mu \in \mathcal{M}_2} \{z_2\mu(U_{1,n})\} \min u.
$$

Since $z_1, z_2 < 0$, this implies that

$$
\limsup_{n\to\infty} P_{2,n} \leq \max\{h_u + z_2\lambda_u, (d_1 + z_1)\lambda_1\} < 0. \tag{7.23}
$$

Finally, since the topological pressure is a continuous function in $C^\theta(\Sigma)$ and $\mathcal{F}(z_1, z_2)$ is a connected set, it follows from (7.22) and (7.23) that there exists a function $\varphi_0 \in \mathcal{F}(z_1, z_2)$ such that $P(\varphi_0 u) = 0$. □

Given a transformation $f: X \to X$, we denote by $V_f(x)$ the set of accumulation points of the sequence of measures

$$\frac{1}{n} \sum_{j=0}^{n-1} \delta_{f^j x}.$$

We need the following result of Bowen in [37].

Lemma 7.5.4. *Let* $f: X \to X$ *be a continuous transformation of a compact metric space. If*

$$Y_t = \{x \in X : h_\mu(f) \le t \text{ for some } \mu \in V_f(x)\},$$

then $h(f|Y_t) \le t$.

A simple consequence of Lemma 7.5.4 is the following.

Lemma 7.5.5. *If* $S \subset \Sigma$ *is a compact* σ-*invariant set, then the set* R *of points* $x \in \Sigma$ *such that* $V_\sigma(x)$ *contains a* σ-*invariant probability measure in* S *satisfies* $h(\sigma|R) = h(\sigma|S)$.

Proof of the lemma. Since S is compact we have $S \subset R$, and $h(\sigma|S) \le h(\sigma|R)$. On the other hand, $h_\mu(\sigma) \le h(\sigma|S)$ for every σ-invariant probability measure μ in S, and thus

$$R \subset \{x \in \Sigma : h_\mu(\sigma) \le h(\sigma|S) \text{ for some } \mu \in V_\sigma(x)\}.$$

It follows from Lemma 7.5.4 that $h(\sigma|R) \le h(\sigma|S)$. □

Let d_i, S_i, and z_i, $i = 1,2$ be as in Lemmas 7.5.2 and 7.5.3. Let also $\varphi_0 \in \mathcal{F}(z_1, z_2)$ be as in Lemma 7.5.3.

Lemma 7.5.6. *For the spectrum* $\mathcal{D}_u$ *of the equilibrium measure of* $\varphi_0 u$ *we have*

$$\underline{\alpha}_{\varphi_0 u} = -z_1, \quad \overline{\alpha}_{\varphi_0 u} = -z_2, \quad \mathcal{D}_u(\underline{\alpha}_{\varphi_0 u}) = d_1, \quad \text{and} \quad \mathcal{D}_u(\overline{\alpha}_{\varphi_0 u}) = d_2.$$

Proof of the lemma. Set $\underline{\alpha} = \underline{\alpha}_{\varphi_0 u}$ and $\overline{\alpha} = \overline{\alpha}_{\varphi_0 u}$. We show that if $x \in K_{\underline{\alpha}}$ (respectively $x \in K_{\overline{\alpha}}$), then $V_\sigma(x)$ contains a σ-invariant probability measure in S_1 (respectively S_2). Since for each $x \in S_i$ the set $V_\sigma(x)$ contains a σ-invariant probability measure in S_i, the desired statement follows from Lemma 7.5.5 because $\dim_u S_i = d_i$, and

$$d_\nu(x) = -\frac{\int_{S_i} \varphi_0 u \, d\rho}{\int_{S_i} u \, d\rho} = -z_i \tag{7.24}$$

for each $x \in S_i$, where $\rho \in V_\sigma(x)$ and ν is the equilibrium measure of the function $\varphi_0 u$. Identity (7.24) follows from Theorem 7.3.2 together with the fact that $\varphi_0|S_i = z_i$ for $i = 1,2$.

Fix $x \in \Sigma$ and assume that $V_\sigma(x)$ contains no σ-invariant probability measure in S_1. Then

$$\limsup_{n \to \infty} \frac{1}{n} \sum_{j=0}^{n-1} \varphi_0(\sigma^j x) =: w < z_1 \quad \text{and} \quad \limsup_{n \to \infty} \frac{A(n,x)}{n} < 1,$$

where $A(n,x) = \operatorname{card} G(n)$ and

$$G(n) = \{0 \le j \le n-1 : \varphi_0(\sigma^j x) \ge w\}.$$

We obtain

$$\limsup_{n \to \infty} \frac{\sum_{j=0}^{n-1} \varphi_0(\sigma^j x) u(\sigma^j x)}{\sum_{j=0}^{n-1} u(\sigma^j x)}$$

$$\le \limsup_{n \to \infty} \frac{\sum_{j \in G(n)} \varphi_0(\sigma^j x) u(\sigma^j x) + \sum_{j \notin G(n)} \varphi_0(\sigma^j x) u(\sigma^j x)}{\sum_{j=0}^{n-1} u(\sigma^j x)}$$

$$\le z_1 - (z_1 - w) \limsup_{n \to \infty} \left(1 - \frac{A(n,x)}{n}\right) \frac{\min u}{\max u} < z_1,$$

and thus $x \notin K_{\underline{\alpha}}$. The proof for $\overline{\alpha}$ is entirely analogous.   □

Now let φ_0 be as in Lemma 7.5.3 with $z_1 = -x_1$, $z_2 = -x_2$, $d_1 = y_1$, and $d_2 = y_2$. The statement in the theorem follows immediately from Lemma 7.5.6.   □

7.6   Nondegeneracy of the spectra

The following result of Schmeling in [142] shows that a typical u-dimension spectrum $\mathcal{D}_u$ is nondegenerate, that is, $\mathcal{D}_u$ is zero at the endpoints of its domain.

Theorem 7.6.1. *For each $\theta \in (0,1]$, there is a residual set $\mathcal{R} \subset C^\theta(\Sigma)$ such that $\underline{\alpha}_u < \overline{\alpha}_u$ and*

$$\mathcal{D}_u(\underline{\alpha}_u) = \mathcal{D}_u(\overline{\alpha}_u) = 0$$

for the spectrum $\mathcal{D}_u$ of the equilibrium measure of each function in $\mathcal{R}$.

Proof. We start with an auxiliary result. Set

$$\overline{M}(\varphi) = \{x \in \Sigma : \varphi(x) = \max \varphi\} \quad \text{and} \quad \underline{M}(\varphi) = \{x \in \Sigma : \varphi(x) = \min \varphi\}.$$

Lemma 7.6.2. *Let φ be a Hölder continuous function in Σ and let $\mathcal{E} = \mathcal{D}_1$ be the entropy spectrum of its equilibrium measure. If $\overline{M}(\varphi)$ and $\underline{M}(\varphi)$ are compact σ-invariant sets, then*

$$\mathcal{E}(\underline{\alpha}_1) = h(\sigma|\overline{M}(\varphi)) \quad \text{and} \quad \mathcal{E}(\overline{\alpha}_1) = h(\sigma|\underline{M}(\varphi)).$$

Proof of the lemma. Since

$$\sup_{x \in \Sigma} \limsup_{n \to \infty} \frac{1}{n} \sum_{j=0}^{n-1} \varphi(\sigma^j x) \leq \max \varphi = \frac{1}{n} \sum_{j=0}^{n-1} \varphi(\sigma^j y)$$

for every point $y \in \overline{M}(\varphi)$, we have $\overline{M}(\varphi) \subset K_{\underline{\alpha}_1}$. Similarly, $\underline{M}(\varphi) \subset K_{\overline{\alpha}_1}$. This implies that

$$\mathcal{E}(\underline{\alpha}_1) \leq h(\sigma|\overline{M}(\varphi)) \quad \text{and} \quad \mathcal{E}(\overline{\alpha}_1) \leq h(\sigma|\underline{M}(\varphi)).$$

For the reverse inequalities, we observe that

$$x \in D_{\underline{\alpha}_1}^- \quad \text{if and only if} \quad \limsup_{n \to \infty} \frac{1}{n} \sum_{j=0}^{n-1} \varphi(\sigma^j x) = \max \varphi.$$

But this is only possible if $V_\sigma(x)$, that is, the set of accumulation points of the sequence of measures

$$\frac{1}{n} \sum_{j=0}^{n-1} \delta_{\sigma^j x},$$

contains a σ-invariant probability measure in $\overline{M}(\varphi)$. It follows from Lemma 7.5.4 that $\mathcal{E}(\underline{\alpha}_1) = h(\sigma|\overline{M}(\varphi))$. The argument for $\mathcal{E}(\overline{\alpha}_1)$ is entirely analogous. □

For each $\alpha \in \mathbb{R}$ set

$$D_\alpha^- = \{x \in \Sigma : \underline{d}_\mu(x) = \alpha\} \quad \text{and} \quad D_\alpha^+ = \{x \in \Sigma : \overline{d}_\mu(x) = \alpha\}.$$

Lemma 7.6.3. *We have:*

1. $\dim_u D_{\alpha_u(q)}^- \leq \mathcal{D}_u(\alpha_u(q))$ *for each* $q \geq 0$;

2. $\dim_u D_{\alpha_u(q)}^+ \leq \mathcal{D}_u(\alpha_u(q))$ *for each* $q \leq 0$.

Proof of the lemma. Fix $q \geq 0$. Given $\varepsilon > 0$, set

$$s = \mathcal{D}_u(\alpha_u(q)) + \varepsilon = T_u(q) + q\alpha_u(q) + \varepsilon.$$

By Theorem 7.3.2, if $x \in D_{\alpha_u(q)}^-$ then

$$-\frac{\sum_{j=0}^{n_k-1} \varphi(\sigma^j x)}{\sum_{j=0}^{n_k-1} u(\sigma^j x)} \leq \alpha_u(q) + \frac{1}{k} \tag{7.25}$$

for some sequence $n_k = n_k(x) \in \mathbb{N}$ such that $n_k \nearrow \infty$ as $k \to \infty$. Therefore, proceeding as in the construction of Moran covers (see the proof of Theorem 4.1.7), we show that there exists a constant $M > 0$ such that for each $n \in \mathbb{N}$ we can find a finite set

$$\mathcal{C} = \{x_1, \ldots, x_l\} \subset D_{\alpha_u(q)}^-$$

and integers $k_1, \ldots, k_l \in \mathbb{N}$ such that:

1. $m_i = n_{k_i}(x_i) \geq n$ for $i = 1, \ldots, l$;

2. the cylinder sets $C_{m_i}(x_i)$, $i = 1, \ldots, l$ cover $D^-_{\alpha_u(q)}$;

3. each point in $D^-_{\alpha_u(q)}$ intersects at most a number M of these cylinder sets.

Since ν_q is the equilibrium measure of $-T_u(q)u + q\varphi$, it follows from (7.25) that

$$\sum_{i=1}^{l} \exp(-s(S_{m_i}u)(x_i)) = \sum_{i=1}^{l} \prod_{j=0}^{m_i-1} \exp[-(T_u(q) + q\alpha_u(q) + \varepsilon)u(\sigma^j x_i)]$$

$$\leq C_1 \sum_{i=1}^{l} \prod_{j=0}^{m_i-1} \exp[-T_u(q)u(\sigma^j x_i) + q\varphi(\sigma^j x_i)]$$

$$\leq C_2 \sum_{i=1}^{l} \nu_q(C_{m_i}(x_i)) \leq M,$$

for some constants $C_1, C_2 > 0$, provided that n is sufficiently large (so that $q/k_i < \varepsilon$ for $i = 1, \ldots, l$). This implies that

$$N(D^-_{\alpha_u(q)}, s, u, \mathcal{U}) \leq M$$

(see (7.4)), where $\mathcal{U}$ is the cover of Σ by cylinder sets of length 1. We obtain

$$\dim_u D^-_{\alpha_u(q)} \leq s = \mathcal{D}_u(\alpha_u(q)) + \varepsilon,$$

and the arbitrariness of ε yields the first statement. A similar argument establishes the second statement.   □

Given $S \subset \Sigma$ and $m \in \mathbb{N}$ we set $C_m(S) = \bigcup_{x \in S} C_m(x)$ and

$$U_m(S) = \{x \in \Sigma : \sigma^k x \in C_m(S) \text{ for every } k \in \mathbb{N}\}.$$

For each $x \in \Sigma$ we also consider the set

$$A(S, m, x) = \operatorname{card}\{0 \leq k \leq m : \sigma^k x \in S\}.$$

Lemma 7.6.4. *For any σ-invariant set $S \subset \Sigma$ and any $m \in \mathbb{N}$ we have*

$$\lim_{a \to 1} h\left(\sigma \mid \left\{x \in \Sigma : \liminf_{N \to \infty} \frac{A(C_m(S), N, x)}{N} \geq a\right\}\right) = h(\sigma|U_m(S)).$$

Proof of the lemma. Since S is σ-invariant we have $U_m(S) \neq \varnothing$. Moreover, $C_m(S)$ is a finite union of cylinder sets, and its characteristic function ψ is Hölder continuous. It is easy to verify that for the function $u = 1$ we have

$$K_\alpha = \left\{x \in \Sigma : \lim_{N \to \infty} \frac{A(U_m(S), N, x)}{N} = -\alpha + P(\psi)\right\},$$

and

$$D_{\underline{\alpha}}^- = \left\{ x \in \Sigma : \limsup_{N \to \infty} \frac{A(U_m(S), N, x)}{N} \geq -\alpha + P(\psi) \right\}.$$

We thus obtain $\underline{\alpha}_1 = P(\psi) - 1$, $\overline{\alpha}_1 = P(\psi)$, and for each $a \in (0, 1)$,

$$U_m(S) \subset K_{\underline{\alpha}_1} \subset \bigcup_{\alpha \in (\underline{\alpha}_1, \overline{\alpha}_1 + 1 - a)} D_{\underline{\alpha}}^- = D_{\overline{\alpha}_1 + 1 - a}^-$$

$$= \left\{ x \in \Sigma : \liminf_{N \to \infty} \frac{A(C_m(S), N, x)}{N} \geq a \right\}.$$

Together with Lemma 7.6.3 this yields the desired statement. $\square$

We proceed with the proof of the theorem. The following statement considers the particular case of the entropy spectrum.

Lemma 7.6.5. *For each $\theta \in (0, 1]$, there is a residual set $\Theta \subset C^\theta(\Sigma)$ such that the entropy spectrum $\mathcal{E} = \mathcal{D}_1$ of the equilibrium measure of each function in Θ is nondegenerate.*

Proof of the lemma. We find a residual set $\Theta \subset C^\theta(\Sigma)$ such that any function $\varphi \in \Theta$ satisfies

$$h(\sigma | \overline{\mathcal{M}}(\varphi)) = h(\underline{\mathcal{M}}(\varphi)) = 0. \tag{7.26}$$

The result follows then from Lemma 7.6.2. We first find an open and dense subset $\Theta_1^\varepsilon \subset C^\theta(\Sigma)$ for each $\varepsilon > 0$. Given $\varepsilon > 0$, $r \in \mathbb{N}$, and $\varphi \in C^\theta(\Sigma)$, there is a periodic point x_0 of period n_0 such that

$$M_\varphi - \frac{\varepsilon}{r} < \frac{1}{n_0} \sum_{j=0}^{n_0 - 1} \varphi(\sigma^j x_0) < M_\varphi,$$

where

$$M_\varphi := \sup_{x \in \Sigma} \limsup_{n \to \infty} \frac{1}{n} \sum_{j=0}^{n-1} \varphi(\sigma^j x).$$

We consider the set

$$C := C_{n_0}(S_{x_0}) = \bigcup_{j=0}^{n_0 - 1} C_{n_0}(\sigma^j x_0),$$

where S_{x_0} is the orbit of x_0. We observe that $U := U_{n_0}(S_{x_0}) = S_{x_0}$, and hence $h(\sigma | U) = 0$. Now we consider a function close to φ, defined by

$$\hat{\varphi}(x) = \begin{cases} \varphi(x) & \text{if } x \in C, \\ \varphi(x) - \varepsilon & \text{if } x \notin C. \end{cases}$$

Clearly, $\hat{\varphi} \in C^{\theta}(\Sigma)$ and $\|\varphi - \hat{\varphi}\|_{\theta} < \varepsilon$. We show that $h(\sigma|\overline{\mathcal{M}}(\hat{\varphi})) < \varepsilon$. For each $x \in \Sigma$ and $n \in \mathbb{N}$ we have

$$\frac{1}{n}\sum_{j=0}^{n-1} \hat{\varphi}(\sigma^j x) = \frac{1}{n}\left(\sum_{\sigma^j x \in C} \hat{\varphi}(\sigma^j x) + \sum_{\sigma^j x \notin C} \hat{\varphi}(\sigma^j x)\right)$$

$$\leq \frac{1}{n}\left(\sum_{\sigma^j x \in C} \varphi(\sigma^j x) + \sum_{\sigma^j x \notin C} \varphi(\sigma^j x) - (n - A(C,n,x))\varepsilon\right)$$

$$\leq \frac{1}{n}\sum_{j=0}^{n-1}\varphi(\sigma^j x) - \left(1 - \frac{A(C,n,x)}{n}\right)\varepsilon$$

$$\leq M_{\varphi} - \left(1 - \frac{A(C,n,x)}{n}\right)\varepsilon$$

$$\leq M_{\hat{\varphi}} + \frac{\varepsilon}{r} - \left(1 - \frac{A(C,n,x)}{n}\right)\varepsilon,$$

since

$$M_{\hat{\varphi}} \geq \frac{1}{n_0}\sum_{j=0}^{n_0-1}\varphi(\sigma^j x_0) \geq M_{\varphi} - \frac{\varepsilon}{r}.$$

Therefore,

$$\overline{\mathcal{M}}(\hat{\varphi}) \subset \left\{x \in \Sigma : \limsup_{n\to\infty} \frac{1}{n}\sum_{j=0}^{n-1}\varphi(\sigma^j x) = M_{\hat{\varphi}}\right\}$$

$$\subset \left\{x \in \Sigma : \limsup_{n\to\infty} \frac{A(C,n,x)}{n} \geq 1 - \frac{1}{r}\right\}.$$

By Lemma 7.6.4 there exists $\varepsilon > 0$ such that

$$\varepsilon > h\left(\sigma\Big|\left\{x \in \Sigma : \limsup_{n\to\infty} \frac{A(C,n,x)}{n} \geq 1 - \frac{1}{r}\right\}\right) \geq h(\sigma|\overline{\mathcal{M}}(\hat{\varphi})).$$

Now set

$$\Omega_1^{\varepsilon}(\varphi) = \{\psi \in C^{\theta}(\Sigma) : \|\hat{\varphi} - \psi\|_{\theta} < \varepsilon\}.$$

We observe that the above estimates hold with $\varepsilon' = 2\varepsilon$ for every $\psi \in \Omega_1^{\varepsilon}(\varphi)$. Therefore, for each $\varepsilon > 0$ the set

$$\Theta_1^{\varepsilon} = \bigcup_{\varphi \in C^{\theta}(\Sigma)} \Omega_1^{\varepsilon}(\varphi) \tag{7.27}$$

is open and dense in $C^{\theta}(\Sigma)$. Furthermore,

$$h(\sigma|\overline{\mathcal{M}}(\varphi)) < 2\varepsilon \quad \text{for every} \quad \varphi \in \Theta_1^{\varepsilon}.$$

Repeating the above procedure we construct a set $\Theta_2^\varepsilon \subset C^\theta(\Sigma)$ with the property that

$$h(\sigma|\underline{M}(\varphi)) < 2\varepsilon \quad \text{for every} \quad \varphi \in \Theta_2^\varepsilon.$$

Set

$$\Theta = \bigcap_{N=0}^{\infty} \bigcup_{n \geq N} \left(\Theta_1^{1/n} \cap \Theta_2^{1/n} \right).$$

Clearly, Θ is a residual subset of $C^\theta(\Sigma)$, and (7.26) holds for every $\varphi \in \Theta$. This completes the proof of the lemma. □

We have seen in the proof of Lemma 7.5.6 that the boundary values of the spectrum $\mathcal{D}_u$ of the equilibrium measure of φu coincide with those of the entropy spectrum $\mathcal{E} = \mathcal{D}_1$ of the equilibrium measure of φ. The desired statement follows from the observation that since u is positive, the transformation $\varphi \mapsto \varphi u$ is a homeomorphism in the space $C^\theta(\Sigma)$, and thus it transforms residual subsets into residual subsets. □

Chapter 8

Dimension of Irregular Sets

In Chapters 6 and 7 we described the main components of multifractal analysis for several multifractal spectra and several classes of dynamical systems. These spectra are obtained from multifractal decompositions such as the one in (7.1). In particular, we possess very detailed information from the ergodic, topological, and dimensional points of view about the level sets K_α^g in each multifractal decomposition. On the other hand, we gave no nontrivial information about the irregular set in these decompositions, that is, the set $X \setminus Y$ in (7.1). Furthermore, the irregular set is typically very small from the point of view of ergodic theory. Namely, for many "natural" multifractal decompositions it has zero measure with respect to *any* finite invariant measure. Nevertheless, it may be very large from the topological and dimensional points of view. This is the main theme of this chapter, where we also describe a general approach to the study of the u-dimension of irregular sets.

8.1 Introduction

We start by considering a model case. Birkhoff's ergodic theorem says that if $S: X \to X$ is a measurable transformation preserving a finite measure μ in X, then for each measurable function $\varphi \in L^1(X, \mu)$ the limit

$$\varphi_S(x) := \lim_{n \to \infty} \frac{1}{n} \sum_{k=0}^{n-1} \varphi(S^k x)$$

exists for μ-almost every $x \in X$. Furthermore, if μ is ergodic then

$$\varphi_S(x) = \frac{1}{\mu(X)} \int_X \varphi \, d\mu \qquad (8.1)$$

for μ-almost every $x \in X$. Of course this does not mean that identity (8.1) holds for every $x \in X$ for which $\varphi_S(x)$ is well-defined. For each $\alpha \in \mathbb{R}$ we consider the level set of Birkhoff averages

$$K_\alpha(\varphi) = \left\{ x \in X : \lim_{n \to \infty} \frac{1}{n} \sum_{k=0}^{n-1} \varphi(S^k x) = \alpha \right\}, \tag{8.2}$$

i.e., the set of points $x \in X$ such that $\varphi_S(x)$ is well-defined and is equal to α. We also consider the set

$$K(\varphi) = \left\{ x \in X : \liminf_{n \to \infty} \frac{1}{n} \sum_{k=0}^{n-1} \varphi(S^k x) < \limsup_{n \to \infty} \frac{1}{n} \sum_{k=0}^{n-1} \varphi(S^k x) \right\}. \tag{8.3}$$

We obtain the multifractal decomposition

$$X = K(\varphi) \cup \bigcup_{\alpha \in [-\infty, +\infty]} K_\alpha(\varphi).$$

By Birkhoff's ergodic theorem, the irregular set $K(\varphi)$ in (8.3) has zero measure with respect to *any* S-invariant finite measure in X. Therefore, at least from the point of view of ergodic theory, the set $K(\varphi)$ can be discarded. However, we will see that, remarkably, from the point of view of dimension theory this set is as large as the whole space. We note that if φ_1 and φ_2 are cohomologous, then $K(\varphi_1) = K(\varphi_2)$. In particular, if the function φ is cohomologous to a constant then $K(\varphi) = \varnothing$.

The following result of Barreira and Schmeling in [21] shows that if φ is not cohomologous to a constant then $K(\varphi)$ is as large as the whole space from the points of view of the topological entropy and of the Hausdorff dimension. We recall that $h(f|Z)$ denotes the topological entropy of f in the set $Z \subset X$ (see (7.3)).

Theorem 8.1.1 (Irregular sets). *Let X be a repeller of a $C^{1+\varepsilon}$ transformation f, for some $\varepsilon > 0$, such that f is conformal and topologically mixing on X, and let $\varphi \colon X \to \mathbb{R}$ be a Hölder continuous function. Then the following properties are equivalent:*

1. *$K(\varphi) \neq \varnothing$;*

2. *$K(\varphi)$ is dense in X;*

3. *φ is not cohomologous to a constant;*

4.
$$h(f|K(\varphi)) = h(f|X) \quad \text{and} \quad \dim_H K(\varphi) = \dim_H X. \tag{8.4}$$

Theorem 8.1.1 is a simple consequence of Theorem 8.5.3. Under the hypotheses of Theorem 8.1.1, we can also show (see Theorem 8.5.1) that if $\varphi_i \colon X \to \mathbb{R}$, for $i = 1, \ldots, p$, are Hölder continuous functions none of them cohomologous to a constant, then

$$h(f|K(\varphi_1) \cap \cdots \cap K(\varphi_p)) = h(f|X)$$

and

$$\dim_H(K(\varphi_1) \cap \cdots \cap K(\varphi_p)) = \dim_H X.$$

For topological Markov chains, the first identity in (8.4) was extended by Fan, Feng and Wu in [57] to arbitrary continuous functions. We note that in the case of topological Markov chains the two identities in (8.4) are in fact equivalent. This is an immediate consequence of the particular distance in (3.9). For repellers of $C^{1+\varepsilon}$ conformal transformations, the second identity in (8.4) was extended by Feng, Lau and Wu in [63] to arbitrary continuous functions. See also Theorem 8.6.1 for an appropriate version of Theorem 8.1.1 in the case of hyperbolic sets.

In order to highlight the main ideas involved in the proof of Theorem 8.1.1 we give a sketch in the particular case when $\|df\|$ is constant in X.

Sketch of the proof of Theorem 8.1.1. Let $\varphi\colon X \to \mathbb{R}$ be a Hölder continuous function which is not cohomologous to a constant. By Theorem 6.1.2, given $\delta > 0$ there exist ergodic f-invariant probability measures $\mu_1 \neq \mu_2$ in X such that

$$\int_X \varphi \, d\mu_1 \neq \int_X \varphi \, d\mu_2, \tag{8.5}$$

and

$$\min\{\dim_H \mu_1, \dim_H \mu_2\} > \dim_H X - \delta. \tag{8.6}$$

For this it is sufficient to take equilibrium measures μ_1 and μ_2 of two functions $q_1 \varphi$ and $q_2 \varphi$ with $q_1 \neq q_2$ sufficiently close to 0. Indeed, since $\|df\|$ is constant in X, it follows from (6.7) that for the dimension spectrum $\mathcal{D}$ of the equilibrium measure of φ we have

$$\alpha(q) = -T'(q) = -\frac{\int \varphi \, d\nu_q}{\log \|df\|},$$

where ν_q is the equilibrium measure of $q\varphi$. Furthermore, since φ is not cohomologous to a constant, it follows from Lemma 6.1.7 that $\alpha'(q) < 0$ for every $q \in \mathbb{R}$. Therefore, we can choose $q_1 \neq q_2$ such that the measures $\mu_1 = \nu_{q_1}$ and $\mu_2 = \nu_{q_2}$ satisfy (8.5) (we note that this is the same as $\alpha(q_1) \neq \alpha(q_2)$). Since ν_0 is the measure of maximal dimension, to obtain (8.6) we simply consider q_1 and q_2 sufficiently close to 0.

The proof of Theorem 8.1.1 starts with the juxtaposition of cylinder sets (at the level of symbolic dynamics, given by some Markov partition of X) that are alternatively typical with respect to μ_1 and μ_2. We say that a point $x \in X$ is *typical* with respect to μ_i if:

1.

$$\lim_{n\to\infty} \frac{1}{n} \sum_{k=0}^{n-1} \varphi(f^k x) = \int_X \varphi \, d\mu_i; \tag{8.7}$$

2.

$$d_{\mu_i}(x) = \dim_H \mu_i. \tag{8.8}$$

By Birkhoff's ergodic theorem and Shannon–McMillan–Breiman's theorem the identities in (8.7) and (8.8) hold for μ_i-almost every $x \in X$. We repeat the juxtaposition process indefinitely, with cylinder sets of sufficiently large (and increasing) length, centered at points whose images under the coding map are alternatively typical with respect to μ_1 and μ_2. We obtain in this manner a set I of sequences in the symbolic dynamics which is contained in $K(\varphi)$, as a consequence of (8.5).

The next step is the construction of a probability measure μ in I, essentially given by an alternate infinite product of the measures μ_1 and μ_2 (roughly speaking, we use the measure μ_i when we are on a piece of sequence obtained from a cylinder set centered at a point which is typical with respect μ_i). We note that μ is never invariant since it is concentrated on the set $K(\varphi)$. Using (8.6) we can show that

$$\liminf_{r \to 0} \frac{\log \mu(B(x,r))}{\log r} > \dim_H X - 2\delta$$

for every $x \in \chi(I)$, where χ is the coding map of the repeller (see (4.3)). Since the measure μ is concentrated on $K(\varphi)$, this implies that

$$\dim_H K(\varphi) > \dim_H X - 2\delta.$$

It follows from the arbitrariness of δ that $\dim_H K(\varphi) = \dim_H X$. $\square$

Now let $K = \bigcup_\varphi K(\varphi)$ with the union taken over all Hölder continuous functions $\varphi \colon X \to \mathbb{R}$. Under the hypotheses of Theorem 8.1.1 we have

$$h(f|K) = h(f|X) \quad \text{and} \quad \dim_H K = \dim_H X. \tag{8.9}$$

The first identity in (8.9) was established earlier by Pesin and Pitskel in [116] for the Bernoulli shift with two symbols, i.e., when $A = \left(\begin{smallmatrix} 1 & 1 \\ 1 & 1 \end{smallmatrix}\right)$ is the transition matrix. We note that their methods are different from those in [21].

8.2   Irregular sets and distinguishing measures

We now describe a general approach proposed by Barreira and Schmeling in [21] to obtain sharp estimates from below for the u-dimension of irregular sets (see Section 7.2 for the notion of u-dimension). Let $\sigma|\Sigma$ be a one-sided or two-sided topological Markov chain, where σ is the shift map in Σ. We consider sequences of functions $F_i = \{f_{i,n} \colon \Sigma \to \mathbb{R}\}_{n \in \mathbb{N}}$ for $i = 1, \ldots, m$.

Definition 8.2.1. The *irregular set* $\mathcal{F}(F_1, \ldots, F_m) \subset \Sigma$ specified by the sequences $F_1, \ldots, F_m$ is defined by

$$\left\{ x \in \Sigma \colon \liminf_{n \to \infty} f_{k,n}(x) < \limsup_{n \to \infty} f_{k,n}(x) \text{ for } k = 1, \ldots, m \right\}. \tag{8.10}$$

For example, if $m = 1$ and

$$f_{1,n} = \frac{1}{n} \sum_{k=0}^{n-1} (\varphi \circ \sigma^k)$$

for each $n \in \mathbb{N}$, then $\mathcal{F}(F_1)$ coincides with the set $K(\varphi)$ in (8.3).

We will show, under mild additional assumptions, that any irregular set carries full topological entropy and full Hausdorff dimension. Our approach is based on the notion of a distinguishing collection of measures introduced in [21].

Definition 8.2.2. A collection of measures $\mu_1, \ldots, \mu_k$ is said to be *distinguishing* for the sequences $F_1, \ldots, F_m$ if for each $i = 1, \ldots, m$ there exist distinct integers $j_1 = j_1(i)$, $j_2 = j_2(i) \in [1, k]$ and numbers $a_{i,j_1} \neq a_{i,j_2}$ such that

$$\lim_{n \to \infty} f_{i,n}(x) = a_{i,j_1} \text{ for } \mu_{j_1}\text{-almost every } x \in \Sigma,$$

and

$$\lim_{n \to \infty} f_{i,n}(x) = a_{i,j_2} \text{ for } \mu_{j_2}\text{-almost every } x \in \Sigma.$$

We note that we can always assume that $k \leq 2m$ in Definition 8.2.2. For an example of a distinguishing collection of measures, let μ_1 and μ_2 be distinct ergodic σ-invariant probability measures in Σ. Then there exists a continuous function $g \colon \Sigma \to \mathbb{R}$ such that

$$\int_\Sigma g \, d\mu_1 \neq \int_\Sigma g \, d\mu_2.$$

By Birkhoff's ergodic theorem, the measures μ_1 and μ_2 form a distinguishing collection for the sequence $\{S_n g / n\}_{n \in \mathbb{N}}$, where $S_n g$ is defined by (7.13).

Now we consider arbitrary subshifts $\sigma | \Sigma$ (which are obtained from a compact σ-invariant set Σ), and not only topological Markov chains. Let Z_Σ be the family of cylinder sets in Σ. We denote by CC' the cylinder set obtained from the juxtaposition of $C, C' \in Z_\Sigma$, in this order. That is,

$$CC' = C \cap \sigma^{-|C|} C',$$

where $|C|$ is the length of the cylinder set C. We denote by $C_n(x) \in Z_\Sigma$ the cylinder set of length n containing the point $x \in \Sigma$.

Definition 8.2.3. We say that a subshift $\sigma | \Sigma$ has the *specification property* if there exists $m \in \mathbb{N}$ such that for every $C_1, C_2 \in Z_\Sigma$ there exists a cylinder set $C \in Z_\Sigma$ of length m such that $C_1 C C_2 \in Z_\Sigma$.

With the help of distinguishing collections of measures we can obtain sharp lower bounds for the u-dimension of irregular sets. We now formulate the main result in this direction established by Barreira and Schmeling in [21].

Theorem 8.2.4. *If $\sigma|\Sigma$ is a one-sided or two-sided subshift with the specification property, $\mu_1, \ldots, \mu_k$ is a distinguishing collection of ergodic σ-invariant measures for $F_1, \ldots, F_m$, and u is a positive Hölder continuous function in Σ, then*

$$\dim_u \mathcal{F}(F_1, \ldots, F_m) \geq \min\{\dim_u \mu_1, \ldots, \dim_u \mu_k\}.$$

Proof. We first establish some auxiliary results.

Lemma 8.2.5. *If μ_1 and μ_2 are probability measures in Σ, and u is a positive Hölder continuous function in Σ, then for every $\delta > 0$ we have*

$$\mu_1(\{x \in \Sigma \colon \underline{d}_{\mu_2,u}(x) > \dim_u \mu_1 - \delta\}) > 0. \tag{8.11}$$

Proof of the lemma. If (8.11) did not hold, then the set

$$\Gamma_\delta = \{x \in \Sigma \colon \underline{d}_{\mu_2,u}(x) \leq \dim_u \mu_1 - \delta\} \tag{8.12}$$

would have full μ_1-measure. For each $x \in \Gamma_\delta$, let $\{n_k(x)\}_{k \in \mathbb{N}}$ be an increasing sequence of positive integers such that

$$-\frac{\log \mu_2(C_{n_k(x)}(x))}{S_{n_k(x)} u(x)} \leq \dim_u \mu_1 - \delta/2$$

for every $k \in \mathbb{N}$. We observe that given two cylinder sets either they are disjoint, or one of them is contained in the other. Hence, for each $L > 0$ there is a finite or countable cover $\{C_{m_i}(x_i) \colon i \in \mathbb{N}\}$ of Γ_δ formed by disjoint cylinder sets, for some points $x_i \in \Gamma_\delta$ and some integers $m_i \in \{n_k(x_i) \colon k \in \mathbb{N}\}$ such that $m_i > L$ for every $i \in \mathbb{N}$. We obtain

$$\mu_2(\Gamma_\delta) = \sum_{i=1}^{\infty} \mu_2(C_{m_i}(x_i))$$

$$\geq \sum_{i=1}^{\infty} \exp[-(\dim_u \mu_1 - \delta/2) S_{m_i} u(x_i)]$$

$$\geq c \sum_{i=1}^{\infty} \sup_{x \in C_{m_i}(x_i)} \exp[-(\dim_u \mu_1 - \delta/2) S_{m_i} u(x)],$$

where $c > 0$ is a constant depending only on the Hölder exponent of u. Since $\mu_1(\Gamma_\delta) = 1$ we obtain

$$\dim_u \mu_1 - \delta/2 \geq \dim_u \Gamma_\delta \geq \dim_u \mu_1.$$

This contradiction yields the desired result. $\square$

Lemma 8.2.6. *Let μ_1 and μ_2 be probability measures in Σ, and let u be a positive Hölder continuous function in Σ. If μ_1 is an ergodic σ-invariant measure, then*

$$\mu_1(\{x \in \Sigma \colon \underline{d}_{\mu_2,u}(x) \geq \dim_u \mu_1\}) = 1.$$

Proof of the lemma. We note that for each $\delta > 0$, the set Γ_δ in (8.12) is σ-invariant. By Lemma 8.2.5 we have $\mu_1(\Sigma \setminus \Gamma_\delta) = 1$ for every $\delta > 0$, and hence the set

$$\bigcap_{\delta > 0} (\Sigma \setminus \Gamma_\delta) = \{x \in \Sigma : \underline{d}_{\mu_2, u}(x) \geq \dim_u \mu_1\}$$

has also full μ_1-measure. □

We proceed with the proof of Theorem 8.2.4. For the sake of clarity we first present the proof when $m = 1$. The general case is discussed at the end.

Let $m = 1$. We write $f_n = f_{1,n}$ for each $n \in \mathbb{N}$, and without loss of generality we may assume that μ_1 and μ_2 form a distinguishing collection of measures for $F = \{f_n\}_{n \in \mathbb{N}}$ with $\dim_u \mu_1 \geq \dim_u \mu_2$. We also write $a_{1,j} = a_j$ for $j = 1, 2$. We may always assume that $a_j \neq 0$ for $j = 1, 2$. Otherwise we can consider the sequence of functions $F + a = \{f_n + a\}_{n \in \mathbb{N}}$, where a is a nonzero constant, since $\mathcal{F}(F + a) = \mathcal{F}(F)$.

Choose $\delta > 0$ such that

$$|a_1 - a_2| > 4\delta, \tag{8.13}$$

and for each $s \in \mathbb{N}$ set

$$p_s = \begin{cases} 1 & \text{if } s \text{ is odd,} \\ 2 & \text{if } s \text{ is even.} \end{cases}$$

For each $\ell \in \mathbb{N}$, let $\widehat{\Gamma}_1^\ell \subset \Sigma$ be the set of points $x \in \Sigma$ such that for every $n \geq \ell$ and $i = 1, 2$ we have

$$|f_n(x) - a_1| < \delta \quad \text{and} \quad -\frac{\log \mu_i(C_n(x))}{S_n u(x)} > \dim_u \mu_1 - \delta. \tag{8.14}$$

For each $\ell \in \mathbb{N}$, let also $\widehat{\Gamma}_2^\ell \subset \Sigma$ be the set of points $x \in \Sigma$ such that for every $n \geq \ell$ we have

$$|f_n(x) - a_2| < \delta \quad \text{and} \quad -\frac{\log \mu_2(C_n(x))}{S_n u(x)} > \dim_u \mu_2 - \delta. \tag{8.15}$$

Clearly, $\widehat{\Gamma}_i^{\ell+1} \supset \widehat{\Gamma}_i^\ell$ for each $\ell \in \mathbb{N}$ and $i = 1, 2$.

Now fix $\varepsilon \in (0, 1)$, and for each $s \in \mathbb{N}$ set

$$\ell_s = \min\left(\left\{\ell \in \mathbb{N} : \mu_{p_s}(\widehat{\Gamma}_{p_s}^\ell) > 1 - \varepsilon/2^{s+1}\right\} \cup \{\ell_{s-1}\}\right),$$

where $\ell_0 = \infty$. We note that $\ell_s \geq \ell_{s-1}$. It follows from Lemma 8.2.6 and Proposition 7.2.7 that $\ell_s < \infty$ for every $s \in \mathbb{N}$.

Moreover, for $j = 1, 2$, since μ_j is σ-invariant, the set of points $x \in \Sigma$ such that

$$\lim_{n \to \infty} f_n(x) = \lim_{n \to \infty} f_n(\sigma^m x) = a_j$$

for every $m \in \mathbb{N}$ has full μ_j-measure. We consider the number

$$D_{n,m}(x) = \max\left\{ |f_{n+m}(y)/f_m(x)|, |f_m(x)/f_{n+m}(z)| : y, z \in \sigma^{-n} x \right\}.$$

By Lusin's Theorem, for each $j = 1, 2$ and $\delta > 0$ there exists an integer $r_j(n, \delta) \geq n$ such that $D_{n,m}(x) < 1 + \delta$ for every $m > r_j(n, \delta)$ and every x in a set $Y_j^n(\varepsilon)$ of μ_j-measure at least $1 - \delta$.

We define inductively increasing sequences of positive integers $\{n_s\}_{s \in \mathbb{N}}$ and $\{m_s\}_{s \in \mathbb{N}}$ by $m_1 = n_1 = \ell_1$, and for each $s \geq 2$ by

$$m_s = r_{p_s}(n_{s-1} + m, \varepsilon/2^{s+1}) + \ell_{s+1}! \tag{8.16}$$

and

$$n_s = n_{s-1} + m + m_s + 1. \tag{8.17}$$

Setting

$$\Gamma_{p_s}^{\ell_s} = \widehat{\Gamma}_{p_s}^{\ell_s} \cap Y_{p_s}^{n_{s-1}}(\varepsilon/2^{s+1}), \tag{8.18}$$

we have

$$\mu_{p_s}(\Gamma_{p_s}^{\ell_s}) > 1 - \varepsilon/2^s. \tag{8.19}$$

For each $s \in \mathbb{N}$, we also consider the family of cylinder sets

$$\mathfrak{C}_s = \{ C_{m_s}(x) : x \in \Gamma_{p_s}^{\ell_s} \}. \tag{8.20}$$

Moreover, we set $\mathfrak{D}_1 = \mathfrak{C}_1$, and

$$\mathfrak{D}_s = \{ \underline{C} C \overline{C} \in Z_\Sigma : \underline{C} \in \mathfrak{D}_{s-1}, \overline{C} \in \mathfrak{C}_s \text{ and } C \in Z_\Sigma \text{ is minimal} \}, \tag{8.21}$$

where the minimality refers to the order $<$ in Z_Σ defined as follows: if the cylinder sets $C, C' \in Z_\Sigma$ are distinct, then $C < C'$ if $|C| < |C'|$, or if $|C| = |C'|$ but C is smaller than C' in the lexicographical order.

We show that for each $s \geq 2$ and $\underline{C} C \overline{C} \in \mathfrak{D}_s$ with $\underline{C} \in \mathfrak{D}_{s-1}$ and $\overline{C} \in \mathfrak{C}_s$, we have $|\underline{C}| \leq n_{s-1}$ and $|C| < m$. This is clear for $s = 2$ since $n_1 = m_1$. Using (8.16)–(8.17) and induction on $s > 2$, we obtain

$$|\underline{C} C \overline{C}| \leq n_{s-1} + m + m_s < n_s,$$

and hence, $|C'| \leq m$ for each $\underline{C'} C' \overline{C'} \in \mathfrak{D}_{s+1}$ with $\underline{C'} \in \mathfrak{D}_s$ and $\overline{C'} \in \mathfrak{C}_{s+1}$.

Now set

$$\Lambda = \bigcap_{s \geq 1} \bigcup_{C \in \mathfrak{D}_s} C. \tag{8.22}$$

We define a measure μ in Λ by $\mu(C) = \mu_1(C)$ if $C \in \mathfrak{D}_1$, by

$$\mu(\underline{C} C \overline{C}) = \mu(\underline{C}) \mu_{p_s}(\overline{C}) \tag{8.23}$$

if $\underline{C} C \overline{C} \in \mathfrak{D}_s$ for some $s > 1$, and arbitrarily for backward cylinder sets, i.e., cylinder sets with coordinates fixed in the past. We extend μ to the whole Σ by

$\mu(A) = \mu(A \cap \Lambda)$ for each measurable subset $A \subset \Sigma$. For each $s \in \mathbb{N}$ and $\underline{C} \in \mathfrak{D}_{s-1}$, it follows from (8.19) that

$$\mu \left(\bigcup_{\overline{C} \in \mathfrak{D}_s} \underline{C} \cap \overline{C} \right) \geq \mu(\underline{C}) \left(1 - \frac{\varepsilon}{2^s} \right),$$

and hence,

$$\mu(\Lambda) \geq \prod_{s=1}^{\infty} \left(1 - \frac{\varepsilon}{2^s} \right) > 0$$

for all sufficiently small ε.

Now let $x \in C \in \mathfrak{D}_s$. We have $m_s \leq |C| \leq n_s$, and

$$\sigma^{|C|-m_s} x \in \Gamma_{p_s}^{\ell_s} \quad \text{for each} \quad s \in \mathbb{N}.$$

By (8.14) and (8.15), we obtain

$$
\begin{aligned}
|f_{|C|}(x) - a_{p_s}| &\leq |f_{m_s}(\sigma^{|C|-m_s} x) - a_{p_s}| \times f_{|C|}(x)/f_{m_s}(\sigma^{|C|-m_s} x) \\
&\quad + |1 - f_{|C|}(x)/f_{m_s}(\sigma^{|C|-m_s} x)| \times |a_{p_s}| \\
&\leq D_{|C|-m_s, m_s}(x) \times |f_{m_s}(\sigma^{|C|-m_s} x) - a_{p_s}| \\
&\quad + (D_{|C|-m_s, m_s}(x) - 1) \times |a_{p_s}|.
\end{aligned}
$$

This implies that for all sufficiently large s, if $x \in C \in \mathfrak{D}_s$ then

$$|f_{|C|}(x) - a_{p_s}| < 2\delta. \tag{8.24}$$

It follows from (8.13) and (8.24) that

$$\mathcal{F}(F) \supset \Lambda. \tag{8.25}$$

Lemma 8.2.7. *If $x \in \Lambda$, then*

$$\liminf_{n \to \infty} -\frac{\log \mu(C_n(x))}{S_n u(x)} \geq \dim_u \mu_2 - 3\delta.$$

Proof of the lemma. Let $x \in \Lambda$. For each $q \in \mathbb{N}$, we choose an integer s_q such that $|C^{s_q}| \leq q < |C^{s_q+1}|$, where

$$\mathfrak{D}_{s_q+1} \ni C^{s_q+1} \subset C_q(x) \subset C^{s_q} \in \mathfrak{D}_{s_q}.$$

We first assume that

$$|C^{s_q}| \leq q \leq |C^{s_q}| + m + \ell_{s_q+1}. \tag{8.26}$$

We have $(m + \ell_{s_q+1})/|C^{s_q}| \to 0$ as $q \to \infty$, and hence

$$
\begin{aligned}
\frac{S_q u(x)}{S_{|C^{s_q}|} u(x)} &\leq \frac{S_{|C^{s_q}|+m+\ell_{s_q+1}} u(x)}{S_{|C^{s_q}|} u(x)} \\
&\leq 1 + \frac{m + \ell_{s_q+1}}{|C^{s_q}|} \times \frac{\max u}{\min u} \to 1
\end{aligned}
$$

as $q \to \infty$. Therefore, there exists $q_1 \in \mathbb{N}$ such that

$$
\begin{aligned}
-\frac{\log \mu(C_q(x))}{S_q u(x)} &\geq -\frac{\log \mu(C^{s_q})}{S_q u(x)} \\
&\geq -\frac{\log \mu(C^{s_q})}{S_{|C^{s_q}|} u(x)} \times \frac{S_{|C^{s_q}|} u(x)}{S_q u(x)} \\
&\geq \dim_u \mu_2 - 2\delta
\end{aligned}
\tag{8.27}
$$

for every $q \geq q_1$. In particular,

$$
-\frac{\log \mu(C^{s_q})}{S_{|C^{s_q}|} u(x)} \geq \dim_u \mu_2 - 2\delta
\tag{8.28}
$$

for every $q \geq q_1$. When (8.26) does not hold, we have

$$
\mu(C_q(x)) = \mu(C^{s_q}) \mu_{p_{s_q}}(\widetilde{C}) \leq \mu(C^{s_q}) \mu_{p_{s_q+1}}(\widetilde{C}),
$$

where $C_q(x) = C^{s_q} C \widetilde{C}$ and the cylinder $\widetilde{C}$ contains an element of $\mathfrak{C}_{s_q+1}$. Moreover, $|C| < m$ and $|\widetilde{C}| > \ell_{s_q+1}$. This implies that

$$
|C^{s_q}| + |\widetilde{C}| \leq q \leq |C^{s_q}| + m + |\widetilde{C}|,
$$

and

$$
\frac{S_{|C^{s_q}|} u(x) + S_{|\widetilde{C}|} u(\sigma^{q-|\widetilde{C}|} x)}{S_q u(x)} \to 1
$$

as $q \to \infty$. It follows from the definition of $\Gamma_{p_{s_q+1}}^{\ell_{s_q+1}}$ (see (8.18)) and (8.28) that there exists $q_2 \geq q_1$ such that

$$
\begin{aligned}
-\frac{\log \mu(C_q(x))}{S_q u(x)} &\geq \frac{1}{S_q u(x)} \left(-\log \mu(C^{s_q}) - \log \mu_{p_{s_q+1}}(\widetilde{C}) \right) \\
&\geq \frac{S_{|C^{s_q}|} u(x)(\dim_u \mu_2 - 2\delta) + S_{|\widetilde{C}|} u(\sigma^{q-|\widetilde{C}|} x)(\dim_u \mu_2 - \delta)}{S_q u(x)} \\
&\geq \frac{S_{|C^{s_q}|} u(x) + S_{|\widetilde{C}|} u(\sigma^{q-|\widetilde{C}|} x)}{S_q u(x)} (\dim_u \mu_2 - 2\delta) \\
&\geq \dim_u \mu_2 - 3\delta
\end{aligned}
\tag{8.29}
$$

for every $q \geq q_2$. The desired statement follows now immediately from (8.27) and (8.29). $\square$

Lemma 8.2.8. *We have*

$$
\dim_u \Lambda \geq \dim_u(\mu|\Lambda) \geq \dim_u \mu_2 - 3\delta.
\tag{8.30}
$$

Proof of the lemma. Set $\alpha = \dim_u \mu_2 - 3\delta$. We may assume that $\alpha > 0$ since otherwise there is nothing to prove. By Lemma 8.2.7, given $\varepsilon \in (0, \alpha)$, for each $x \in \Lambda$ there exists $n(x) \in \mathbb{N}$ such that if $n > n(x)$, then

$$\mu(C_n(x)) \le \exp[-(\alpha - \varepsilon)S_n u(x)]. \tag{8.31}$$

Given $m \in \mathbb{N}$, we set

$$\Lambda_m = \{x \in \Lambda : n(x) \le m\}.$$

Clearly,

$$\Lambda_{m_1} \subset \Lambda_{m_2} \quad \text{for} \quad m_1 \ge m_2, \quad \text{and} \quad \Lambda = \bigcup_{m \in \mathbb{N}} \Lambda_m.$$

Take $m \in \mathbb{N}$ with $\mu(\Lambda_m) \ge \mu(\Lambda)/2$, and let $Z \subset \Lambda$ be an arbitrary set of full $(\mu|\Lambda)$-measure. Consider any collection of cylinder sets $(C_{k_i}(x_i))_{i \in I}$ such that

$$\bigcup_{i \in I} C_{k_i}(x_i) \supset Z \cap \Lambda_m.$$

Without loss of generality we can also assume that $C_{k_i}(x_i) \cap \Lambda_m \ne \varnothing$ for every $i \in I$. Hence, there exists some point $y_i \in C_{k_i}(x_i) \cap \Lambda_m$ for each $i \in I$. Furthermore, due to the particular distances in (3.9) in the case of one-sided shifts, and in (4.24) in the case of two-sided shifts, there exists $N \in \mathbb{N}$ such that $C_{k_i}(x_i) \subset C_{k_i-N}(y_i)$ for every $i \in I$. Therefore, whenever $k_i - N > m$ for every $i \in I_i$ it follows from (8.31) that

$$\sum_{i \in I} \exp[-(\alpha - \varepsilon)S_{k_i} u(x_i)] \ge \sum_{i \in I} \exp[-(\alpha - \varepsilon)S_{k_i-N}(y_i) - (\alpha - \varepsilon)N \sup u]$$

$$\ge \exp[-(\alpha - \varepsilon)N \sup u] \sum_{i \in I} \mu(C_{k_i-N}(y_i))$$

$$\ge \exp[-(\alpha - \varepsilon)N \sup u]\mu(\Lambda_m)$$

$$\ge \exp[-(\alpha - \varepsilon)N \sup u]/2.$$

We thus obtain

$$N(Z \cap \Lambda_m, \alpha - \varepsilon, u, \mathcal{U}) \ge \exp[(\alpha - \varepsilon)N \sup u]/2$$

(see (7.4)), where $\mathcal{U}$ is the cover of Σ by cylinder sets of length 1. This implies that $\dim_u(Z \cap \Lambda_m) \ge \alpha - \varepsilon$, and it follows from the arbitrariness of ε that

$$\dim_u Z \ge \dim_u(Z \cap \Lambda_m) \ge \alpha.$$

We obtain $\dim_u(\mu|\Lambda) \ge \alpha$, and thus also (8.30). $\qquad \square$

By (8.25), it follows from Lemma 8.2.8 and the arbitrariness of δ that

$$\dim_u \mathcal{F}(F) \ge \dim_u \mu_2.$$

Since $\dim_u \mu_1 \geq \dim_u \mu_2$, this completes the proof of the theorem when $m = 1$.

Now we briefly discuss how to deal with the case when $m > 1$. For each $s \in \mathbb{N}$ set $p_s = s \pmod{k} + 1$. Without loss of generality, we may assume that

$$\dim_u \mu_{j_1(i)} \geq \dim_u \mu_{j_2(i)} \quad \text{for every} \quad 1 \leq i \leq m,$$

and that

$$\dim_u \mu_j \geq \dim_u \mu_k \quad \text{for every} \quad 1 \leq j \leq k.$$

For each $\ell \in \mathbb{N}$ and $i = 1, \ldots, m$, let $\widehat{\Gamma}^\ell_{i,j_1(i)} \subset \Sigma$ be the set of points $x \in \Sigma$ such that for every $n \geq \ell$ and $t = k, j_1(i)$ we have

$$|f_{i,n}(x) - a_{i,j_1(i)}| < \delta \quad \text{and} \quad -\frac{\log \mu_t(C_n(x))}{S_n u(x)} > \dim_u \mu_{j_1(i)} - \delta.$$

For each $\ell \in \mathbb{N}$ and $i = 1, \ldots, m$, let also $\widehat{\Gamma}^\ell_{i,j_2(i)} \subset \Sigma$ be the set of points $x \in \Sigma$ such that for every $n \geq \ell$ and $t = k, j_2(i)$ we have

$$|f_{i,n}(x) - a_{i,j_2(i)}| < \delta \quad \text{and} \quad -\frac{\log \mu_t(C_n(x))}{S_n u(x)} > \dim_u \mu_{j_2(i)} - \delta.$$

We then construct a set $\Lambda \subset \Sigma$ in a similar manner to that when $m = 1$, by juxtaposing alternatively cylinder sets whose centers are in the sets

$$\Gamma_{1,j_1(1)}, \Gamma_{1,j_2(1)}, \Gamma_{2,j_1(2)}, \Gamma_{2,j_2(2)}, \ldots, \Gamma_{m,j_1(m)}, \text{ and } \Gamma_{m,j_2(m)}$$

that are obtained as in (8.18) (although not necessarily in this order; compare with (8.20) and (8.21)). The remaining arguments are analogous. □

We remark that in the case of two-sided subshifts the cylinder sets used in the construction of the set Λ in (8.22) are forward cylinder sets, that is, they are completely determined by a finite number of symbols in the future. Moreover, for the purpose of the proof of Theorem 8.2.4, the noninvariant measure μ constructed in (8.23) can be defined arbitrarily for backward cylinder sets, essentially since we only require the statement in Lemma 8.2.7 when $n \to +\infty$. We can also consider "two-sided" irregular sets, for which there exist no limits both when $n \to +\infty$ and when $n \to -\infty$, and establish a similar statement to the one in Theorem 8.2.4 (see [21] for details).

8.3   Existence of distinguishing measures

In order to effectively use Theorem 8.2.4, we need to find distinguishing collections of measures. The following statement solves this problem for topological Markov chains. We recall that m_u and ν_q denote respectively the measure of maximal u-dimension and the full measure at $\alpha_u(q)$ (see Definition 7.3.3).

Theorem 8.3.1 (Existence of distinguishing measures [21]). *Let $\sigma|\Sigma$ be a one-sided or two-sided topologically mixing topological Markov chain, and let $\varphi_1, \ldots, \varphi_m, g, u$ be Hölder continuous functions in Σ with $g, u > 0$. If for every $i = 1, \ldots, m$ the function φ_i is not cohomologous to $\alpha_i g$, where α_i is the unique real number such that $P(\alpha_i g) = P(\varphi_i)$, then for every $\varepsilon > 0$ there exist ergodic σ-invariant measures $\mu_1, \ldots, \mu_m$ in Σ such that:*

1. *$\mu_1, \ldots, \mu_m$ are full measures for the spectrum $\mathcal{D}_u$;*

2. *$\mu_1, \ldots, \mu_m$, m_u is a distinguishing collection of measures for the sequences of functions $\{S_n\varphi_1/S_ng\}_{n\in\mathbb{N}}, \ldots, \{S_n\varphi_m/S_ng\}_{n\in\mathbb{N}}$;*

3. *$\min\{\dim_u\mu_1, \ldots, \dim_u\mu_m\} > \dim_u\Sigma - \varepsilon$.*

Proof. For each $i = 1, \ldots, m$, we have

$$\lim_{n\to\infty} -\frac{S_n\varphi_i(x)}{S_ng(x)} = -\frac{\int_\Sigma \varphi_i \, dm_u}{\int_\Sigma g \, dm_u} \quad \text{for } m_u\text{-almost every } x \in \Sigma.$$

Since φ_i is not cohomologous to $\alpha_i g$, we can show that for each $\alpha > 0$ the set of points $q \in [-\alpha, \alpha]$ such that

$$\int_\Sigma \varphi_i \, d\nu_q = \alpha_i \int_\Sigma g \, d\nu_q$$

is finite. Otherwise, by the analytic dependence of $\int_\Sigma \varphi_i \, d\nu_q$ and $\int_\Sigma g \, d\nu_q$ on q, we would have

$$\int_\Sigma \varphi_i \, d\mu = \alpha_i \int_\Sigma g \, d\mu \tag{8.32}$$

for the equilibrium measure μ of every function in a $C^\theta(\Sigma)$-open neighborhood of some φ_q (see (7.14)). Here $C^\theta(\Sigma)$ is the space of Hölder continuous functions in Σ with Hölder exponent $\theta \in (0, 1]$, equipped with the norm in (7.19). In fact this implies that (8.32) holds for every equilibrium measure μ. But this is impossible because φ_i is not cohomologous to $\alpha_i g$. Therefore, by Theorem 7.3.2, given $\varepsilon > 0$ there exists a full measure μ_i for the spectrum $\mathcal{D}_u$ such that $\dim_u\mu_i > \dim_u\Sigma - \varepsilon$, and

$$\lim_{n\to\infty} -\frac{S_n\varphi_i(x)}{S_ng(x)} \neq -\frac{\int_\Sigma \varphi_i \, dm_u}{\int_\Sigma g \, dm_u} \quad \text{for } \mu_i\text{-almost every } x \in \Sigma.$$

The collection of measures $\mu_1, \ldots, \mu_m$ has the desired properties. $\square$

8.4 Topological Markov chains

We consider in this section the particular case of topological Markov chains, and we combine the results in the former sections to show that a large class of irregular sets has full u-dimension. Given sequences of functions $F_i = \{f_{i,n} \colon \Sigma \to \mathbb{R}\}_{n\in\mathbb{N}}$ for $i = 1, \ldots, m$, we consider the irregular set in (8.10).

Theorem 8.4.1 ([21]). *Let $\sigma|\Sigma$ be a one-sided or two-sided topologically mixing topological Markov chain, and let $\varphi_1, \ldots, \varphi_m, g, u$ be Hölder continuous functions in Σ with $g, u > 0$. The following properties are equivalent:*

1. *φ_i is not cohomologous to $\alpha_i g$ for any $i = 1, \ldots, m$, where α_i is the unique real number such that $P(\alpha_i g) = P(\varphi_i)$;*

2.

$$\dim_u \mathcal{F}(\{S_n \varphi_1 / S_n g\}_{n \in \mathbb{N}}, \ldots, \{S_n \varphi_m / S_n g\}_{n \in \mathbb{N}}) = \dim_u \Sigma.$$

Proof. The statement follows readily from Theorems 8.2.4 and 8.3.1. $\square$

By using Markov partitions the statement in Theorem 8.4.1 can also be established for repellers and hyperbolic sets (see Sections 8.5 and 8.6).

We note that a priori it could happen that the cohomology assumptions in Theorem 8.4.1 were almost never satisfied, but it turns out that precisely the opposite happens. We formulate a rigorous result in the particular case when g is constant. Let $C(\Sigma)$ be the space of continuous functions in Σ, equipped with the supremum norm.

Theorem 8.4.2 ([21]). *If $\sigma|\Sigma$ is a topologically mixing topological Markov chain, then the following properties hold:*

1. *the family of Hölder continuous functions in Σ which are not cohomologous to 0 contains a dense subset of $C(\Sigma)$;*

2. *for each $\theta \in (0, 1)$, the family of functions in $C^\theta(\Sigma)$ which are not cohomologous to 0 contains an open and dense subset of $C^\theta(\Sigma)$.*

Proof. Let L be the family of nonconstant linear combinations of characteristic functions of cylinder sets (of arbitrary length). Clearly, L is a $C(\Sigma)$-dense family composed of Hölder continuous functions.

Lemma 8.4.3. *If $\sigma|\Sigma$ is a topologically mixing topological Markov chain, then the following properties hold:*

1. *the family L contains a $C(\Sigma)$-dense subset of functions which are not cohomologous to 0;*

2. *for each $\theta \in (0, 1)$, the family $L \cap C^\theta(\Sigma)$ contains a $C^\theta(\Sigma)$-dense subset of functions which are not cohomologous to 0.*

Proof of the lemma. Let $g \in L$. By Livschitz's theorem (see, for example, [84, Theorem 19.2.1]), if there exist $n \in \mathbb{N}$ and periodic points $x, y \in \Sigma$ with $\sigma^n x = x$ and $\sigma^n y = y$ such that $S_n g(x) \neq S_n g(y)$, then g is not cohomologous to 0. Given $\delta > 0$, we can find a function $h \in L$ which is δ-close to g (with respect to the supremum norm) simply by changing slightly the value of g in a sufficiently small cylinder set containing the orbit of only one of the points x and y, such that $S_n h(x) \neq S_n h(y)$. Again by Livschitz's theorem, h is not cohomologous to 0.

We conclude that in any $C(\Sigma)$-open neighborhood of a function in L there exist functions in L which are not cohomologous to 0. Furthermore, if $g \in C^{\theta}(\Sigma)$ is not cohomologous to 0, then any sufficiently small $C^{\theta}(\Sigma)$-open neighborhood of g contains only functions which are not cohomologous to 0. This completes the proof of the lemma. $\qquad\square$

The statement in the theorem is an immediate consequence of the lemma. $\quad\square$

8.5 Repellers

We show in this section how to use Markov partitions and Theorem 8.4.1 to obtain corresponding results in the case of repellers. See Section 8.6 for the case of hyperbolic sets.

Theorem 8.5.1 ([21]). *Let J be a repeller of a $C^{1+\varepsilon}$ transformation f, for some $\varepsilon > 0$, such that f is conformal and topologically mixing on J. If $\varphi_1, \ldots, \varphi_m, g$ are Hölder continuous functions in J with $g > 0$, then the following properties are equivalent:*

1. *φ_i is not cohomologous to $\alpha_i g$ for every $i = 1, \ldots, m$, where α_i is the unique real number such that $P(\alpha_i g) = P(\varphi_i)$;*

2.

$$h(f|\mathcal{F}(\{S_n\varphi_1/S_ng\}_{n\in\mathbb{N}}, \ldots, \{S_n\varphi_m/S_ng\}_{n\in\mathbb{N}})) = h(f|J), \qquad (8.33)$$

and

$$\dim_H \mathcal{F}(\{S_n\varphi_1/S_ng\}_{n\in\mathbb{N}}, \ldots, \{S_n\varphi_m/S_ng\}_{n\in\mathbb{N}}) = \dim_H J.$$

Proof. Set

$$\mathcal{F} := \mathcal{F}(\{S_n\varphi_1/S_ng\}_{n\in\mathbb{N}}, \ldots, \{S_n\varphi_m/S_ng\}_{n\in\mathbb{N}}).$$

Repeating arguments in the proof of Theorem 6.1.2 (see Lemma 6.1.6) we find that $\mathcal{F}$ coincides with the image under the coding map χ (see (4.3)) of the corresponding irregular set $\mathcal{F}' \subset \Sigma$, for the symbolic dynamics $\sigma|\Sigma$ associated to some Markov partition $\mathcal{R}$.

In the case of the topological entropy, identity (8.33) can be obtained as follows. We start with an auxiliary result.

Lemma 8.5.2. *If $\mathcal{R}$ is a Markov partition of J, then $h(f|\partial\mathcal{R}) < h(f|J)$.*

Proof of the lemma. The partition $\mathcal{R}$ is a generating partition, and hence the diameter of the image under χ of cylinder sets tends (uniformly) to zero as the length of the cylinder sets tends to infinity. Therefore, there exist $n \in \mathbb{N}$ and $C \in \bigvee_{k=0}^{n} f^{-k}\mathcal{R}$ such that $C \cap \partial\mathcal{R} = \varnothing$. Since $f(\partial\mathcal{R}) \subset \partial\mathcal{R}$, the coding of the boundary in the symbolic dynamics does not contain at least the cylinder set corresponding to C. This implies that $h(f|\partial\mathcal{R}) < h(f|J)$. $\qquad\square$

Since the topological entropy of a set coincides with the topological entropy of its invariant hull, it follows from Lemma 8.5.2 that

$$h(f| \bigcup_{n=1}^{\infty} f^{-n}\partial\mathcal{R}) < h(f|J).$$

Since the coding map χ is a homeomorphism on the set $J \setminus \bigcup_{n=1}^{\infty} f^{-n}\partial\mathcal{R}$, if $A \subset \Sigma$ is such that $h(\sigma|A) = h(\sigma|\Sigma)$, then

$$h(f|\chi(A)) = h(f|J) = h(\sigma|\Sigma).$$

Identity (8.33) is now an immediate consequence of Theorem 8.4.1.

Now we consider the Hausdorff dimension. For each sufficiently small $r > 0$, let $\mathcal{U}_r$ be the associated Moran cover of J. We recall that the cover is composed of images under χ of cylinder sets, such that given $x \in X$ and a sufficiently small $r > 0$ the number of elements in $\mathcal{U}_r$ that intersect the ball $B(x, r)$ is bounded from above by a constant $\kappa > 0$ independent of x and r.

We equip Σ with the unique distance such that each cylinder set C of length n has diameter $\sup_{x \in C}(\|d_{\chi(x)}f^n\|^{-1})$. Setting $u = \log \|df\| \circ \chi$, the Hausdorff dimension associated to this distance coincides with the u-dimension in Σ. Now let $\mathcal{U}$ be a cover of $\mathcal{F}$ by open balls. For each $B \in \mathcal{U}$ of radius r there are at most κ cylinder sets (not necessarily all with the same length) such that their images under χ are the elements of the Moran cover $\mathcal{U}_r$ that intersect B. Collecting the cylinder sets for every $B \in \mathcal{U}$, we obtain a family $\mathcal{V}$ of cylinder sets in Σ which form a cover of $\mathcal{F}'$. Furthermore,

$$\sum_{U \in \mathcal{U}}(\operatorname{diam} U)^s \leq \sum_{C \in \mathcal{V}}(\operatorname{diam} C)^s \leq \kappa \sup \|df\|^s \sum_{U \in \mathcal{U}}(\operatorname{diam} U)^s.$$

This implies that $\dim_u \mathcal{F}' = \dim_H \mathcal{F}$. Furthermore, repeating this argument with $\mathcal{F}$ replaced by J we conclude that $\dim_u \Sigma = \dim_H J$. By Theorem 8.4.1, we have $\dim_u \mathcal{F}' = \dim_u \Sigma$, and hence $\dim_H \mathcal{F} = \dim_H J$. □

In particular, Theorem 8.5.1 indicates that the boundaries of Markov partitions have no influence in the study of the topological entropy and the Hausdorff dimension of irregular sets of repellers.

Now we describe several irregular sets which carry full topological entropy and full Hausdorff dimension. Let

$$\mathcal{B} = \left\{ x \in \Sigma : \liminf_{n \to \infty} \frac{1}{n} S_n g(x) < \limsup_{n \to \infty} \frac{1}{n} S_n g(x) \text{ for some } g \in C(J) \right\},$$

where $C(J)$ is the space of continuous functions in J. We also set $\mathcal{B}_f = \chi(\mathcal{B})$. Note that

$$\mathcal{B}_f \supset \left\{ x \in J : \liminf_{n \to \infty} \frac{1}{n} S_n g(x) < \limsup_{n \to \infty} \frac{1}{n} S_n g(x) \text{ for some } g \in C(J) \right\}.$$

We define the *irregular set for the Lyapunov exponents of f* by

$$\mathcal{L}_f = \left\{ x \in J : \liminf_{n \to \infty} \frac{1}{n} \log \|d_x f^n\| < \limsup_{n \to \infty} \frac{1}{n} \log \|d_x f^n\| \right\}.$$

Given a probability measure in Σ we set

$$\mathcal{H}(\mu) = \left\{ x \in \Sigma : \liminf_{n \to \infty} -\frac{\log \mu(C_n(x))}{n} < \limsup_{n \to \infty} -\frac{\log \mu(C_n(x))}{n} \right\},$$

and given a probability measure μ in J, we define the *irregular set for the local entropies of μ* by

$$\mathcal{H}_f(\mu) = \chi(\mathcal{H}(\mu \circ \chi)),$$

and the *irregular set for the pointwise dimensions of μ* by

$$\mathcal{D}(\mu) = \{ x \in J : \underline{d}_\mu(x) < \overline{d}_\mu(x) \}.$$

Finally, we denote by μ_D and m_E respectively the measures of maximal dimension and maximal entropy.

Theorem 8.5.3 ([21]). *If J is a repeller of a $C^{1+\varepsilon}$ transformation f, for some $\varepsilon > 0$, such that f is conformal and topologically mixing on J, then the following properties hold:*

1. $h(f|\mathcal{B}_f) = h(f|J)$ *and* $\dim_H \mathcal{B}_f = \dim_H J$;

2. $m_D \neq m_E$ *if and only if* $h(f|\mathcal{L}_f) = h(f|J)$ *and* $\dim_H \mathcal{L}_f = \dim_H J$.

If, in addition, μ is the equilibrium measure of a Hölder continuous function in J, then:

3. $\mu \neq m_D$ *if and only if* $h(f|\mathcal{D}(\mu)) = h(f|J)$ *and* $\dim_H \mathcal{D}(\mu) = \dim_H J$;

4. $\mu \neq m_E$ *if and only if* $h(f|\mathcal{H}_f(\mu)) = h(f|J)$ *and* $\dim_H \mathcal{H}_f(\mu) = \dim_H J$;

5. *the three measures μ, m_D, and m_E are distinct if and only if*

$$h(f|\mathcal{D}(\mu) \cap \mathcal{H}_f(\mu) \cap \mathcal{L}_f) = h(f|J)$$

and

$$\dim_H(\mathcal{D}(\mu) \cap \mathcal{H}_f(\mu) \cap \mathcal{L}_f) = \dim_H J.$$

Proof. Let μ be the equilibrium measure of a Hölder continuous function φ in J with $P(\varphi) = 0$. All the identities follow from Theorem 8.5.1 taking respectively for each statement the sequences of functions:

1. $\{S_n g/n\}_{n \in \mathbb{N}}$, where g is any Hölder continuous function which is not cohomologous to 0, since

$$\mathcal{B}_f \supset \chi(\mathcal{B}(g)) = \chi(\mathcal{F}(\{S_n g/n\}_{n \in \mathbb{N}}));$$

2. $\{S_n \log \|df\|/n\}_{n \in \mathbb{N}}$;

3. $\{S_n \varphi / S_n \log \|df\|\}_{n \in \mathbb{N}}$;

4. $\{S_n \varphi / n\}_{n \in \mathbb{N}}$;

5. $\{S_n \varphi / S_n \log \|df\|\}_{n \in \mathbb{N}}$, $\{S_n \varphi / n\}_{n \in \mathbb{N}}$, and $\{S_n \log a / n\}_{n \in \mathbb{N}}$.

This completes the proof of the theorem.   □

8.6   Hyperbolic sets

We describe in this section several irregular subsets of hyperbolic sets which carry full topological entropy and full Hausdorff dimension. Given a hyperbolic set Λ of a diffeomorphism f, we denote by $\mathcal{M}_D$ the set of f-invariant probability Borel measures in Λ of full dimension (see Definition 5.1.3). We note that $\mathcal{M}_D$ may be empty (see the discussion after Definition 5.1.3). We continue to denote by m_E the measure of maximal entropy.

Theorem 8.6.1 ([21]). *If Λ is a locally maximal hyperbolic set of a $C^{1+\varepsilon}$ diffeomorphism f, for some $\varepsilon > 0$, such that f is conformal and topologically mixing on Λ, then the following properties hold:*

1. $h(f|\mathcal{B}_f) = h(f|\Lambda)$ *and* $\dim_H \mathcal{B}_f = \dim_H \Lambda$;

2. $\log \|df|E^u\|$ *is not cohomologous to 0 if and only if*

$$h(f|\mathcal{L}_f) = h(f|\Lambda) \quad and \quad \dim_H \mathcal{L}_f = \dim_H \Lambda;$$

3. $\log \|df|E^s\|$ *is not cohomologous to 0 if and only if*

$$h(f|\mathcal{L}_{f^{-1}}) = h(f|\Lambda) \quad and \quad \dim_H \mathcal{L}_{f^{-1}} = \dim_H \Lambda.$$

If, in addition, μ is the equilibrium measure of a Hölder continuous function in Λ, then:

4. $\mu \notin \mathcal{M}_D$ *if and only if*

$$h(f|\mathcal{D}(\mu)) = h(f|\Lambda) \quad for \quad \dim_H \mathcal{D}(\mu) = \dim_H \Lambda;$$

5. $\mu \neq m_E$ *if and only if*

$$h(f|\mathcal{H}_f(\mu)) = h(f|\Lambda) \quad for \quad \dim_H \mathcal{H}_f(\mu) = \dim_H \Lambda;$$

6. $\mu \neq m_E$ *and* $\mu \notin \mathcal{M}_D$ *if and only if*

$$h(f|\mathcal{D}(\mu) \cap \mathcal{H}_f(\mu)) = h(f|\Lambda) \quad and \quad \dim_H(\mathcal{D}(\mu) \cap \mathcal{H}_f(\mu)) = \dim_H \Lambda.$$

Proof. The case of hyperbolic sets can be reduced to the case of repellers in the following manner. By Proposition 4.2.11, if g is a continuous function in a two-sided topological Markov chain Σ_A, then there exists a cohomologous function g_+ such that

$$g_+(\cdots i_{-1} i_0 i_1 \cdots) = g_+(\cdots i'_{-1} i'_0 i'_1 \cdots)$$

whenever $i_k = i'_k$ for every $k \geq 0$. Let Σ_A^+ be the one-sided topological Markov chain with the same transition matrix A. Then the irregular set $\mathcal{B}(g_+)$ with respect to Σ_A^+ coincides with $\mathcal{B}(g)$. This establishes the first three statements.

For statement 4 we decompose Λ into local stable and unstable manifolds. For μ-almost every $x \in \Lambda$ we can define conditional measures ν_x^s and ν_x^u in the local stable and unstable manifolds at x. These coincide μ-almost everywhere with the measures μ_x^s and μ_x^u constructed in Section 4.2.3 (see (4.38)). By Proposition 4.2.13, there exists a constant $\kappa > 0$ such that

$$\kappa^{-1}\mu(A) < (\mu_x^s \times \mu_x^u)(A) < \kappa\mu(A)$$

for every measurable set A in a small rectangle. Furthermore, the measures μ_x^s and μ_x^u are equilibrium measures of some Hölder continuous functions φ_x^s and φ_x^u (see Section 4.2.3). Since $\mu \notin \mathcal{M}_D$, the measures μ_x^s and μ_x^u cannot be both equivalent respectively to the measures of maximal dimension in the local stable and unstable manifolds at x. Without loss of generality we assume that μ_x^u is not equivalent to the measure of maximal dimension in the local unstable manifold.

Let $\mathcal{D}_x^u(\mu)$ be the set of points in $V^u(x)$ for which the pointwise dimension of μ_x^u does not exist, and set $g = \log \|df|E^u\|$. We note that $\mathcal{D}_x^u(\mu)$ is the image under the coding map χ (see (4.25)) of the set of points $y \in \Sigma_A$ such that $(S_n\varphi_x^u/S_ng)(y)$ does not converge. Proceeding as in the proof of Theorem 8.5.3, it follows from Theorem 8.5.1 that

$$h(f|\mathcal{D}_x^u(\mu)) = h(f|\mathcal{F}(\{S_n\varphi_x^u/S_ng\}_{n\in\mathbb{N}})) = h(f|V^u(x) \cap \Lambda),$$

and

$$\dim_H \mathcal{D}_x^u(\mu) = \dim_g \mathcal{F}(\{S_n\varphi_x^u/S_ng\}_{n\in\mathbb{N}}) = \dim_H(V^u(x) \cap \Lambda).$$

Now let $m_{D,x}^s$ be the measure of maximal dimension in $V^s(x)$. This is the equilibrium measure of $t^s \log \|df|E^s\|$ (see (4.42)). One can easily verify that $\mathcal{D}_y^u(\mu) \subset \mathcal{D}(\mu)$ for every $y \in V^s(x)$ in a set $G_{D,x}$ of full $m_{D,x}^s$-measure. Therefore, the set $\bigcup_{y\in G_{D,x}} \mathcal{D}_y^u(\mu)$ is contained in $\mathcal{D}(\mu)$, and has full stable and unstable dimensions. We thus obtain the second identity in statement 4.

For the first identity in statement 4, let $m_{E,x}^s$ be the measure of maximal entropy in $V^s(x)$, and let $G_{E,x}$ be a set of full $m_{E,x}^s$-measure such that $\mathcal{D}_y^u(\mu) \subset \mathcal{D}(\mu)$ for every $y \in G_{E,x}$. The set

$$B = \bigcup_{y\in G_{E,x}} \mathcal{D}_y^u(\mu)$$

has full topological entropy with respect to f, and hence,

$$h(f|\mathcal{D}(\mu)) \geq h(f|B) = h(f|\Lambda).$$

This completes the proof of statement 4.

The proofs of the remaining statements are analogous.   □

Shereshevsky proved earlier in [146] that for a generic C^2 surface diffeomorphism with a locally maximal hyperbolic set Λ, and an equilibrium measure μ of a Hölder continuous C^0-generic function,

$$\dim_H Y = \left\{ x \in \Lambda : \liminf_{r \to 0} \frac{\log \mu(B(x,r))}{\log r} < \limsup_{r \to 0} \frac{\log \mu(B(x,r))}{\log r} \right\} > 0.$$

This statement can easily be recovered from Theorem 8.6.1 (see [21] for details).

Chapter 9

Variational Principles in Multifractal Analysis

Following the general concept of multifractal analysis introduced in Section 7.1, one can consider several multifractal spectra. In particular, we showed in Chapters 6 and 7 that the spectra $\mathcal{D}_D$ and $\mathcal{E}_E$ are analytic in several contexts. Furthermore, they coincide with the Legendre transform of certain functions defined in terms of the topological pressure, and this allows us to show that they are always convex (in fact, in a certain sense, they are "generically" strictly convex; see Theorem 8.4.2). A priori it is unclear whether it is possible to effect a similar analysis in the case of the mixed multifractal spectra $\mathcal{D}_E$ and $\mathcal{E}_D$ which combine local and global characteristics of distinct nature. This is precisely the main theme of this chapter. In particular, we show that the mixed spectra are analytic in several contexts. The analyticity follows from a conditional variational principle for the u-dimension which is also established in this chapter, and which is important in its own right. On the other hand, we show that there are many nonconvex mixed spectra.

9.1 Conditional variational principle

Let f be a continuous transformation in the compact metric space X. We denote by $C(X)$ the space of continuous functions $\varphi\colon X \to \mathbb{R}$. Given φ, $\psi \in C(X)$ with $\psi > 0$, we set

$$K_\alpha = K_\alpha(\varphi, \psi) = \left\{ x \in X : \lim_{n \to \infty} \frac{\varphi_n(x)}{\psi_n(x)} = \alpha \right\}, \tag{9.1}$$

where

$$\varphi_n(x) = \sum_{k=0}^{n-1} \varphi(f^k x) \quad \text{and} \quad \psi_n(x) = \sum_{k=0}^{n-1} \psi(f^k x).$$

Moreover, given a positive function $u \in C(X)$ we continue to denote by $\dim_u Z$ the u-dimension of the set $Z \subset X$ (see Definition 7.2.2). For example, if $u = 1$, then $\dim_u Z$ coincides with the topological entropy $h(f|Z)$.

Definition 9.1.1. The function $\mathcal{F}_u = \mathcal{F}_{u,(\varphi,\psi)}$ defined by

$$\mathcal{F}_u(\alpha) = \dim_u K_\alpha(\varphi, \psi)$$

is called the *u-dimension spectrum of the pair* (φ, ψ).

We denote by $D(X) \subset C(X)$ the family of continuous functions with a unique equilibrium measure, and we denote by P the topological pressure in X. We collect in the following statement a few properties of the pressure.

Theorem 9.1.2. *If the Kolmogorov–Sinai entropy is upper semicontinuous, that is, if the map $\mu \mapsto h_\mu(f)$ is upper semicontinuous, then the following properties hold:*

1. *each function $\varphi \in C(X)$ has an equilibrium measure;*

2. *given $\varphi \in C(X)$, the function $q \mapsto P(\varphi + q\psi)$ is differentiable at $q = 0$ for every $\psi \in C(X)$ if and only if $\varphi \in D(X)$; furthermore, the unique equilibrium measure μ_φ is ergodic, and given $\psi \in C(X)$ we have*

$$\frac{d}{dq}P(\varphi + q\psi)\Big|_{q=0} = \int_X \psi \, d\mu_\varphi; \qquad (9.2)$$

3. *if φ, $\psi \in C(X)$ are such that $\mathrm{span}\{\varphi, \psi\} \subset D(X)$, then the function $q \mapsto P(\varphi + q\psi)$ is of class C^1 in $\mathbb{R}$;*

4. *the family $D(X)$ is dense in $C(X)$.*

We refer to [132, 86] for details (in particular, see [86, Theorem 4.2.11] for statement 3 and [132, Theorem 6.14] for statement 4).

For example, when f is a one-sided or two-sided topological Markov chain, or an expansive homeomorphism, then the entropy is upper semicontinuous (see, for example, [86, Theorem 4.5.6]). Furthermore, if f is a one-sided or two-sided topologically mixing topological Markov chain, or an expansive homeomorphism with the specification property, and $\varphi \in C_f(X)$ (see Definition 9.1.3), then φ has a unique equilibrium measure, that is, $C_f(X) \subset D(X)$ (see [84] for details).

Definition 9.1.3. We denote by $C_f(X) \subset C(X)$ the family of continuous functions $\varphi \colon X \to \mathbb{R}$ for which there exist constants $\varepsilon > 0$ and $\kappa > 0$ such that

$$\left| \sum_{k=0}^{n-1} \varphi(f^k x) - \sum_{k=0}^{n-1} \varphi(f^k y) \right| < \kappa$$

whenever $d(f^k x, f^k y) < \varepsilon$ for every $k = 0, \ldots, n-1$.

On the other hand, there exist many transformations without the specification property for which the entropy is upper semicontinuous. For example, all β-shifts are expansive, and thus their entropy map is upper semicontinuous (see [86] for details), but it was shown by Schmeling in [141] that for β in a residual set of full Lebesgue measure (although the complement has full Hausdorff dimension) the corresponding β-shift does not have the specification property. On the other hand, it follows from work of Walters in [162] that for every β-shift the family of Lipschitz functions is contained in $D(X)$.

Now we present a conditional variational principle for the multifractal spectrum $\mathcal{F}_u$, obtained by Barreira and Saussol in [16]. We denote by $\mathcal{M}$ the family of f-invariant probability Borel measures in the compact metric space X, and by $\mathcal{M}_E \subset \mathcal{M}$ the subset of all ergodic measures. Let also

$$\underline{\alpha} = \underline{\alpha}(\varphi, \psi) = \inf \left\{ \frac{\int_X \varphi \, d\mu}{\int_X \psi \, d\mu} : \mu \in \mathcal{M} \right\}, \tag{9.3}$$

and

$$\overline{\alpha} = \overline{\alpha}(\varphi, \psi) = \sup \left\{ \frac{\int_X \varphi \, d\mu}{\int_X \psi \, d\mu} : \mu \in \mathcal{M} \right\}. \tag{9.4}$$

Theorem 9.1.4. *Assume that the Kolmogorov–Sinai entropy is upper semicontinuous. If φ, ψ, $u \in C(X)$ with $\psi, u > 0$ are such that $\mathrm{span}\{\varphi, \psi, u\} \subset D(X)$, then the following properties hold:*

1. *if $\alpha \notin [\underline{\alpha}, \overline{\alpha}]$, then $K_\alpha = \varnothing$;*

2. *if $\alpha \in (\underline{\alpha}, \overline{\alpha})$, then $K_\alpha \neq \varnothing$ and*

$$\mathcal{F}_u(\alpha) = \max \left\{ \frac{h_\mu(f)}{\int_X u \, d\mu} : \mu \in \mathcal{M} \text{ and } \frac{\int_X \varphi \, d\mu}{\int_X \psi \, d\mu} = \alpha \right\}; \tag{9.5}$$

3. *the function $\mathcal{F}_u$ is continuous in $(\underline{\alpha}, \overline{\alpha})$;*

4.
$$\inf_{q \in \mathbb{R}} P(q\varphi - q\alpha\psi - \mathcal{F}_u(\alpha)u) = 0;$$

5. *if $\Delta(p, q)$ is the unique real number such that*

$$P(q\varphi - p\psi - \Delta(p, q)u) = 0,$$

then

$$\mathcal{F}_u(\alpha) = \inf_{q \in \mathbb{R}} \Delta(q\alpha, q). \tag{9.6}$$

Proof. We start with some preparatory lemmas.

Lemma 9.1.5. *If $\alpha \in \mathbb{R}$, then*

$$\inf_{q \in \mathbb{R}} P(q\varphi - q\alpha\psi - \mathcal{F}_u(\alpha)u) \geq 0.$$

Proof of the lemma. By Proposition 7.2.3, the number $\mathcal{F}_u(\alpha)$ coincides with the unique root δ of the equation $P_{K_\alpha}(-\delta u) = 0$, where P_{K_α} is the topological pressure in the set K_α (see Definition 7.2.1). Given $\delta > 0$ and $\tau \in \mathbb{N}$, we consider the set

$$L_{\delta,\tau} = \{x \in X : |\varphi_n(x) - \alpha\psi_n(x)| < \delta n \text{ for every } n \geq \tau\}.$$

Since $\psi > 0$ we can easily show that

$$K_\alpha \subset \bigcap_{\delta > 0} \bigcup_{\tau \in \mathbb{N}} L_{\delta,\tau}.$$

Now let $\mathcal{U}$ be an open cover of X with sufficiently small diameter so that if $n \in \mathbb{N}$ is sufficiently large, $\mathbf{U} \in \bigcup_{k \geq n} W_k(\mathcal{U})$, and $x \in X(\mathbf{U})$, then

$$|\varphi(\mathbf{U}) - \varphi_{m(\mathbf{U})}(x)| \leq \delta m(\mathbf{U}) \quad \text{and} \quad |\psi(\mathbf{U}) - \psi_{m(\mathbf{U})}(x)| \leq \delta m(\mathbf{U}).$$

This implies that if $\mathbf{U} \in \bigcup_{k \geq n} W_k(\mathcal{U})$ and $X(\mathbf{U}) \cap L_{\delta,\tau} \neq \varnothing$, then

$$|\varphi(\mathbf{U}) - \alpha\psi(\mathbf{U})| < (2 + |\alpha|)\delta m(\mathbf{U}),$$

and we obtain

$$P_{L_{\delta,\tau}}(-\mathcal{F}_u(\alpha)u, \mathcal{U}) \leq P_{L_{\delta,\tau}}(q\varphi - q\alpha\psi - \mathcal{F}_u(\alpha)u, \mathcal{U}) + (2 + |\alpha|)\delta|q|.$$

Letting $\operatorname{diam} \mathcal{U} \to 0$ yields

$$P_{L_{\delta,\tau}}(-\mathcal{F}_u(\alpha)u) \leq P(q\varphi - q\alpha\psi - \mathcal{F}_u(\alpha)u) + (2 + |\alpha|)\delta|q|,$$

and hence,

$$0 \leq P_{\bigcup_{\tau \in \mathbb{N}} L_{\delta,\tau}}(-\mathcal{F}_u(\alpha)u) = \sup_{\tau \in \mathbb{N}} P_{L_{\delta,\tau}}(-\mathcal{F}_u(\alpha)u)$$

$$\leq P(q\varphi - q\alpha\psi - \mathcal{F}_u(\alpha)u) + (2 + |\alpha|)\delta|q|.$$

Since δ is arbitrary, we obtain

$$\inf_{q \in \mathbb{R}} P(q\varphi - q\alpha\psi - \mathcal{F}_u(\alpha)u) \geq 0.$$

This completes the proof of the lemma.   □

We denote by $\zeta = \zeta_{q,\alpha,\delta}$ the unique equilibrium measure of $q\varphi - q\alpha\psi - \delta u$, which is well-defined in our setting (see Theorem 9.1.2).

Lemma 9.1.6. *For each $\delta \in \mathbb{R}$ and $\alpha \in (\underline{\alpha}, \overline{\alpha})$ there exists $q = q(\delta, \alpha)$ such that $\int_X \varphi \, d\zeta / \int_X \psi \, d\zeta = \alpha$.*

Proof of the lemma. Given $\delta \in \mathbb{R}$ and $\alpha \in (\underline{\alpha}, \overline{\alpha})$, we define a function $S \colon \mathbb{R} \to \mathbb{R}$ by

$$S(q) := \int_X \varphi \, d\zeta - \alpha \int_X \psi \, d\zeta = \frac{d}{dq} P(q\varphi - q\alpha\psi - \delta u). \tag{9.7}$$

By the upper semicontinuity of the entropy and Theorem 9.1.2, the function

$$q \mapsto P(q\varphi - q\alpha\psi - \delta u)$$

is of class C^1, and thus S is continuous.

Now we prove that $S(q) > 0$ for all sufficiently large $q > 0$. Since ζ is an equilibrium measure, if $q > 0$ then

$$
\begin{aligned}
S(q) &= \frac{1}{q}\left[P(q\varphi - q\alpha\psi - \delta u) + \delta \int_X u \, d\zeta - h_\zeta(f) \right] \\
&= \sup_{\mu \in \mathcal{M}} \left[\int_X \varphi \, d\mu - \alpha \int_X \psi \, d\mu + \frac{\delta(\int_X u \, d\zeta - \int_X u \, d\mu) + h_\mu(f) - h_\zeta(f)}{q} \right] \\
&\geq \sup_{\mu \in \mathcal{M}} \left[\int_X \varphi \, d\mu - \overline{\alpha} \int_X \psi \, d\mu + \frac{\delta(\int_X u \, d\zeta - \int_X u \, d\mu) + h_\mu(f) - h_\zeta(f)}{q} \right] \\
&\quad + (\overline{\alpha} - \alpha) \inf \psi.
\end{aligned}
$$

Since the functions ψ, u and the entropies are bounded, and $(\overline{\alpha} - \alpha) \inf \psi > 0$, we conclude that $S(q) > 0$ for all sufficiently large $q > 0$. A similar argument shows that $S(q) < 0$ for all sufficiently small $q < 0$. The desired result follows from the continuity of S. □

Lemma 9.1.7. *Given $\alpha \in (\underline{\alpha}, \overline{\alpha})$, if $\delta \in \mathbb{R}$ is such that*

$$P(q\varphi - q\alpha\psi - \delta u) \geq 0 \quad \text{for every} \quad q \in \mathbb{R}, \tag{9.8}$$

then there exists $\mu \in \mathcal{M}_E$ such that

$$\int_X \varphi \, d\mu / \int_X \psi \, d\mu = \alpha \quad \text{and} \quad \dim_u \mu \geq \delta.$$

Proof of the lemma. Let $q = q(\delta, \alpha)$ be as in Lemma 9.1.6. We continue to write $\zeta = \zeta_{q,\alpha,\delta}$. By (9.8) we have

$$h_\zeta(f) - \delta \int_X u \, d\zeta = h_\zeta(f) + q \int_X \varphi \, d\zeta - q\alpha \int_X \psi \, d\zeta - \delta \int_X u \, d\zeta \geq 0.$$

It follows from Proposition 7.2.7 that $\delta \leq h_\zeta(f) / \int_X u \, d\zeta = \dim_u \zeta$. □

Lemma 9.1.8. *If $\alpha \in (\underline{\alpha}, \overline{\alpha})$, then $K_\alpha \neq \varnothing$ and*

$$\mathcal{F}_u(\alpha) = \sup \left\{ \dim_u \mu : \mu \in \mathcal{M}_E \text{ and } \frac{\int_X \varphi \, d\mu}{\int_X \psi \, d\mu} = \alpha \right\}. \tag{9.9}$$

Proof of the lemma. By Lemma 9.1.5 we can apply Lemma 9.1.7 with $\delta = \mathcal{F}_u(\alpha)$ to obtain

$$\mathcal{F}_u(\alpha) \le \sup\left\{\dim_u\mu : \mu \in \mathcal{M}_E \text{ and } \frac{\int_X \varphi\, d\mu}{\int_X \psi\, d\mu} = \alpha\right\}.$$

Now let $\mu \in \mathcal{M}_E$ be a measure such that

$$\int_X \varphi\, d\mu \Big/ \int_X \psi\, d\mu = \alpha.$$

Birkhoff's ergodic theorem implies that $\mu(K_\alpha) = 1$, and it follows from Proposition 7.2.7 that

$$\mathcal{F}_u(\alpha) \ge \dim_u\mu = \frac{h_\mu(f)}{\int_X u\, d\mu}.$$

This completes the proof of the lemma.   □

Lemma 9.1.9. *If $\alpha \in (\underline{\alpha}, \overline{\alpha})$, then*

$$\inf_{q\in\mathbb{R}} P(q\varphi - q\alpha\psi - \mathcal{F}_u(\alpha)u) = 0.$$

Proof of the lemma. Since u is positive, it follows from Lemma 9.1.5 and the continuity of the topological pressure (in the supremum norm) that there exists $\delta^* \ge \mathcal{F}_u(\alpha)$ such that

$$\inf_{q\in\mathbb{R}} P(q\varphi - q\alpha\psi - \delta^*u) = 0.$$

On the other hand, by Lemma 9.1.7 there exists $\mu \in \mathcal{M}_E$ such that

$$\int_X \varphi\, d\mu \Big/ \int_X \psi\, d\mu = \alpha \quad\text{and}\quad \dim_u\mu \ge \delta^*.$$

It follows from Lemma 9.1.8 that $\mathcal{F}_u(\alpha) \ge \delta^*$.   □

We now establish the continuity of the spectrum.

Lemma 9.1.10. *The function $\mathcal{F}_u$ is continuous in $(\underline{\alpha}, \overline{\alpha})$.*

Proof of the lemma. We first show that $\mathcal{F}_u$ is upper semicontinuous. Given $\alpha \in (\underline{\alpha}, \overline{\alpha})$, let $\alpha_n \in (\underline{\alpha}, \overline{\alpha})$ be any sequence converging to α. By Lemma 9.1.6, for each $n \in \mathbb{N}$ there exists $q_n \in \mathbb{R}$ such that

$$\int_X \varphi\, d\mu_n = \alpha_n \int_X \psi\, d\mu_n,$$

where $\mu_n = \zeta_{q_n,\alpha_n,\mathcal{F}_u(\alpha_n)}$. By Lemma 9.1.9, the function

$$q \mapsto P(q\varphi - q\alpha_n\psi - \mathcal{F}_u(\alpha_n)u)$$

attains its infimum at $q = q_n$ (and the infimum is equal to zero). Since μ_n is an equilibrium measure, we have

$$h_{\mu_n}(f) = \mathcal{F}_u(\alpha_n) \int_X u \, d\mu_n.$$

Now let $\beta = \limsup_{n \to \infty} \mathcal{F}_u(\alpha_n)$. Taking a subsequence, if necessary, we may assume that $\mathcal{F}_u(\alpha_n)$ converges to β, and that the sequence of measures μ_n converges weakly to some measure μ. Since the entropy is upper semicontinuous, we obtain

$$h_{\mu}(f) \geq \limsup_{n \to \infty} h_{\mu_n}(f) = \limsup_{n \to \infty} \mathcal{F}_u(\alpha_n) \int_X u \, d\mu. \qquad (9.10)$$

Since $\mu_n \rightharpoonup \mu$ and $\alpha_n \to \alpha$, we have

$$\int_X \varphi \, d\mu - \alpha \int_X \psi \, d\mu = 0,$$

and hence,

$$P(q\varphi - q\alpha\psi - \mathcal{F}_u(\alpha)u) \geq h_{\mu}(f) - \mathcal{F}_u(\alpha) \int_X u \, d\mu$$

for every $q \in \mathbb{R}$. Taking the infimum over q, it follows from Lemma 9.1.9 that

$$\mathcal{F}_u(\alpha) \int_X u \, d\mu \geq h_{\mu}(f).$$

Since $u > 0$, we conclude from (9.10) that $\mathcal{F}_u$ is upper semicontinuous.

Now we show that $\mathcal{F}_u$ is lower semicontinuous. Let $\alpha_* \in (\underline{\alpha}, \overline{\alpha})$. We consider the functions

$$\chi_q = q\varphi - q\alpha_*\psi - \mathcal{F}_u(\alpha_*)u \quad \text{and} \quad F(q) = P_X(\chi_q).$$

By Lemma 9.1.5 we have $F(q) \geq 0$ for every $q \in \mathbb{R}$. On the other hand, by Theorem 9.1.2 the function F is of class C^1 and

$$S(q) := F'(q) = \int_X \varphi \, d\mu_q - \alpha_* \int_X \psi \, d\mu_q, \qquad (9.11)$$

where μ_q is the equilibrium measure of χ_q. We note that S is increasing, by the convexity of the topological pressure. By Lemma 9.1.6, there exists $q_* = q_*(\alpha_*) \in \mathbb{R}$ such that $S(q_*) = 0$. Furthermore, it is shown in the proof of Lemma 9.1.6 that $S(q) > 0$ for all sufficiently large $q > 0$, and that $S(q) < 0$ for all sufficiently small $q < 0$. Therefore, we can always choose q_* so that $S(q) > 0$ for every $q > q_*$.

Let $\varepsilon > 0$. Since S is continuous, there exists $\delta \in (0, \varepsilon)$ such that

$$\sup\{|S(q)| : q \in (q_* - \delta, q_* + \delta)\} \leq \varepsilon \inf u.$$

Since μ_q is the equilibrium measure of χ_q and $F(q) \geq 0$, it follows from Proposition 7.2.7 that

$$\dim_u \mu_q = \frac{h_{\mu_q}(f)}{\int_X u \, d\mu_q} = \mathcal{F}_u(\alpha_*) + \frac{F(q) - qS(q)}{\int_X u \, d\mu_q}$$

$$\geq \mathcal{F}_u(\alpha_*) - \frac{qS(q)}{\int_X u \, d\mu_q} \qquad (9.12)$$

$$\geq \mathcal{F}_u(\alpha_*) - (|q_*| + \varepsilon)\varepsilon.$$

For every sufficiently small $\alpha > \alpha_*$ it follows from the continuity of S and the choice of q_* that there exists $q = q(\alpha) \in (q_*, q_* + \delta)$ such that

$$\alpha = \alpha_* + S(q) / \int_X \psi \, d\mu_q, \qquad (9.13)$$

and hence, by (9.11),

$$\int_X \varphi \, d\mu_q = \alpha \int_X \psi \, d\mu_q. \qquad (9.14)$$

By Lemma 9.1.8 and (9.12) we conclude that

$$\mathcal{F}_u(\alpha) \geq \dim_u \mu_q \geq \mathcal{F}_u(\alpha_*) - (|q_*| + \varepsilon)\varepsilon. \qquad (9.15)$$

Since ε is arbitrary, this establishes the right lower semicontinuity of $\mathcal{F}_u$.

The left lower semicontinuity of $\mathcal{F}_u$ can be established in a similar manner. If necessary, we first rechoose q_* so that $S(q) < 0$ for every $q < q_*$. For every sufficiently small $\alpha < \alpha_*$ there exists $q = q(\alpha) \in (q_* - \delta, q_*)$ with α as in (9.13), and hence (9.14) holds. By Lemma 9.1.8 and (9.12), we conclude that (9.15) holds, and the arbitrariness of ε implies that $\mathcal{F}_u$ is left lower semicontinuous. This completes the proof of the lemma. $\qquad \square$

Now we proceed with the proof of Theorem 9.1.4. Setting $\delta = \mathcal{F}_u(\alpha)$ in Lemma 9.1.6 we conclude that for $q = q(\delta, \alpha)$ the equilibrium measure $\zeta = \zeta_{q,\alpha,\delta}$ satisfies

$$P(q\varphi - q\alpha\psi - \mathcal{F}_u(\alpha)u) = h_\zeta(f) - \mathcal{F}_u(\alpha) \int_X u \, d\zeta,$$

and hence,

$$\mathcal{F}_u(\alpha) = \frac{h_\zeta(f)}{\int_X u \, d\zeta} = \dim_u \zeta.$$

This shows that in (9.9) we can replace the supremum by the maximum.

Now let $\mu \in \mathcal{M}$ be a measure such that $\int_X \varphi \, d\mu / \int_X \psi \, d\mu = \alpha$ (we note that μ is not necessarily ergodic). To complete the proof of statement 2 it is sufficient to show that

$$\mathcal{F}_u(\alpha) \geq \frac{h_\mu(f)}{\int_X u \, d\mu}. \qquad (9.16)$$

We first observe that

$$P(q\varphi - q\alpha\psi - \mathcal{F}_u(\alpha)u)$$
$$\geq h_\mu(f) + q \int_X \varphi \, d\mu - q\alpha \int_X \psi \, d\mu - \mathcal{F}_u(\alpha) \int_X u \, d\mu$$
$$= h_\mu(f) - \mathcal{F}_u(\alpha) \int_X u \, d\mu.$$

Taking the infimum over $q \in \mathbb{R}$, it follows from Lemma 9.1.9 that

$$0 \geq h_\mu(f) - \mathcal{F}_u(\alpha) \int_X u \, d\mu,$$

and hence (9.16) holds. This establishes statement 2. Statement 1 is an immediate consequence of statement 2.

The continuity of $\mathcal{F}_u$ is established in Lemma 9.1.10. Statement 4 is the content of Lemma 9.1.9. Statement 5 is an immediate consequence of statement 4. This completes the proof of the theorem. □

For a repeller of a $C^{1+\varepsilon}$ conformal expanding map f, when $u = \log \|df\|$ (that is, in the case of the Hausdorff dimension) the identity in (9.5) was established by Feng, Lau and Wu in [63] for arbitrary continuous functions.

Definition 9.1.11. Identity (9.5) is called a *conditional variational principle* for the u-dimension spectrum $\mathcal{F}_u$.

Identity (9.6) shows that the spectrum $\mathcal{F}_u$ is given by a formula that may remind a Legendre transform of the topological pressure. However, in general the spectrum is not convex (see Proposition 9.2.2).

Taking $u = 1$ in Theorem 9.1.4 we obtain a conditional variational principle for the entropy spectrum.

Theorem 9.1.12 ([16]). *Assume that the Kolmogorov–Sinai entropy is upper semicontinuous. If φ, $\psi \in C(X)$ with $\psi > 0$ are such that $\mathrm{span}\{\varphi, \psi\} \subset D(X)$, and $\alpha \in (\underline{\alpha}, \overline{\alpha})$, then*

$$h(f|K_\alpha) = \max \left\{ h_\mu(f) : \mu \in \mathcal{M} \text{ and } \frac{\int_X \varphi \, d\mu}{\int_X \psi \, d\mu} = \alpha \right\}. \tag{9.17}$$

When $\psi = 1$, identity (9.17) (with the maximum replaced by the supremum) was established by Takens and Verbitski in [154], under the assumptions that f is a continuous transformation with the specification property, and φ is an arbitrary continuous function (which thus may have more than one equilibrium measure). We note that they use a different approach. In [121] the authors discuss a problem with the proof in [154] but they also give the appropriate correction.

In [18], Barreira and Saussol obtained conditional variational principles for hyperbolic flows.

9.2 Topological Markov chains

Now we consider the particular case of one-sided and two-sided topological Markov chains (see Definitions 4.1.3 and 4.2.7).

Theorem 9.2.1 ([16]). *If $\sigma|\Sigma$ is a topologically mixing topological Markov chain, and φ, ψ, $u \in C_\sigma(\Sigma)$ with $\psi, u > 0$, then the following properties hold:*

1. *if $\alpha \notin [\underline{\alpha}, \overline{\alpha}]$, then $K_\alpha = \varnothing$;*

2. *if $\alpha \in (\underline{\alpha}, \overline{\alpha})$, then $K_\alpha \neq \varnothing$ and*

$$\mathcal{F}_u(\alpha) = \max \left\{ \frac{h_\mu(\sigma)}{\int_\Sigma u \, d\mu} : \mu \in \mathcal{M} \text{ and } \frac{\int_\Sigma \varphi \, d\mu}{\int_\Sigma \psi \, d\mu} = \alpha \right\}. \tag{9.18}$$

Proof. It is well known that for topologically mixing topological Markov chains and functions in $C_\sigma(\Sigma)$ the assumptions in Theorem 9.1.4 are satisfied. The desired statement is thus an immediate consequence of that theorem. $\qquad\square$

In the case of the full shift and for $\psi = u = 1$, the identity in (9.18) was first established by Olivier [105, 106], for the more general class of g-measures. This class, introduced by Keane in [85], is composed of equilibrium measures of a class of continuous functions that need not be Hölder continuous. It is known that any Gibbs measure is a g-measure (see [106] for details).

Now we provide an explicit example for which the spectrum $\mathcal{F}_u$ is not convex. This strongly contrasts with what happens with the multifractal spectra studied in Chapters 6 and 7, which are always Legendre transforms of the topological pressure, and which thus are always convex.

Proposition 9.2.2 ([16]). *Let $\sigma|\Sigma_2^+$ be the full shift on two symbols. There exist Hölder continuous functions φ, ψ, and u in Σ_2^+ such that the spectrum $\mathcal{F}_u$ is not convex.*

Proof. We have $\Sigma_2^+ = \{1, 2\}^{\mathbb{N}}$. Set $\varphi(x_1 x_2 \ldots) = \varphi_{x_1}$, with $\varphi_1 = 0$ and $\varphi_2 = 1$, $\psi = 1$, and $u(x_1 x_2 \ldots) = u_{x_1}$, for some positive numbers u_1 and u_2. Writing

$$\beta_\mu = \mu(\{(x_1 x_2 \ldots) \in \Sigma_2^+ : x_1 = 1\}),$$

it follows from Theorem 9.2.1 that

$$\mathcal{F}_u(\alpha) = \max_{\mu \in \mathcal{M}} \left\{ \frac{h_\mu(\sigma)}{v_1(1 - \beta_\mu) + v_2 \beta_\mu} : \beta_\mu = \alpha \right\}$$

$$= \frac{1}{-\alpha(v_1 - v_2) + v_1} \max_{\mu \in \mathcal{M}} \{h_\mu(\sigma) : \beta_\mu = \alpha\}.$$

Now let ξ be the partition of Σ_2^+ into cylinder sets of length 1. For each $\mu \in \mathcal{M}$ with $\beta_\mu = \alpha$ we have

$$h_\mu(\sigma) = \inf_{n \in \mathbb{N}} \frac{1}{n} H_\mu \left(\bigvee_{k=0}^{n-1} \sigma^{-k} \xi \right) \le H_\mu(\xi)$$

$$= -\alpha \log \alpha - (1 - \alpha) \log(1 - \alpha),$$

with equality if μ is the Bernoulli measure in Σ_2^+ with $\beta_\mu = \alpha$. Therefore,

$$\mathcal{F}_u(\alpha) = \frac{\alpha \log \alpha + (1 - \alpha) \log(1 - \alpha)}{\alpha(u_1 - u_2) - u_1}.$$

For example, taking $u_1 = 1$ and $u_2 = 100$ the function $\mathcal{F}_u$ is not convex. $\square$

See Figure 9.1 for the graph of a nonconvex spectrum $\mathcal{F}_u$, as constructed in the proof of Proposition 9.2.2.

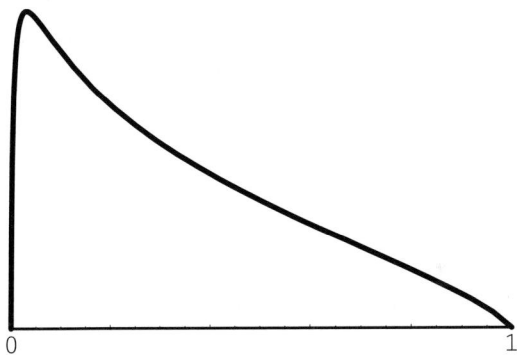

Figure 9.1: A nonconvex u-dimension spectrum

We also show that for topological Markov chains the multifractal spectrum $\mathcal{F}_u$ is analytic, thus substantially improving statement 3 in Theorem 9.1.4.

Theorem 9.2.3 ([16]). *If $\sigma|\Sigma$ is a topologically mixing topological Markov chain, and φ, ψ, and u are Hölder continuous functions in X with $\psi, u > 0$, then the following properties hold:*

1. *if φ is cohomologous to some multiple of ψ, then $\underline{\alpha} = \overline{\alpha}$;*

2. *if φ is not cohomologous to any multiple of ψ, then the function $\mathcal{F}_u$ is analytic in the nonempty interval $(\underline{\alpha}, \overline{\alpha})$.*

Proof. We assume first that there exist $\beta \in \mathbb{R}$ and a continuous function $\chi \colon \Sigma \to \mathbb{R}$ such that $\varphi - \beta\psi = \chi \circ \sigma - \chi$. Then

$$\frac{\varphi_n}{\psi_n} - \beta = \frac{\chi \circ \sigma^n - \chi}{\psi_n} \to 0 \quad \text{as} \quad n \to \infty,$$

where

$$\varphi_n = \sum_{k=0}^{n-1} \varphi \circ \sigma^k \quad \text{and} \quad \psi_n = \sum_{k=0}^{n-1} \psi \circ \sigma^k.$$

By Birkhoff's ergodic theorem, if $\mu \in \mathcal{M}$ is ergodic, then

$$\frac{\int_\Sigma \varphi \, d\mu}{\int_\Sigma \psi \, d\mu} = \beta,$$

and hence, $\underline{\alpha} = \overline{\alpha} = \beta$. This establishes the first property.

We continue with an auxiliary lemma.

Lemma 9.2.4. *If $\varphi - \beta\psi$ is not cohomologous to a constant for every $\beta \in \mathbb{R}$, then the spectrum $\mathcal{F}_u$ is analytic in the nonempty interval $(\underline{\alpha}, \overline{\alpha})$.*

Proof of the lemma. For $\alpha \in (\underline{\alpha}, \overline{\alpha})$ we set

$$F(q, \delta, \alpha) = P(q\varphi - q\alpha\psi - \delta u).$$

By Theorem 9.1.4, the number $\mathcal{F}_u(\alpha)$ coincides with the unique $\delta \in \mathbb{R}$ such that

$$\inf_{q \in \mathbb{R}} F(q, \delta, \alpha) = 0.$$

It is well known that in the present context the topological pressure is analytic in the space of Hölder continuous functions (with a given Hölder exponent), and thus F is analytic in all variables. In addition, due to the cohomology assumption the function $q \mapsto F(q, \delta, \alpha)$ is strictly convex. Hence, if there exists $q \in \mathbb{R}$ such that the derivative $\partial F/\partial q$ vanishes, then the minimum is attained at q. But

$$\frac{\partial F}{\partial q}(q, \delta, \alpha) = S(q),$$

with $S(q)$ as in (9.7). By Lemma 9.1.6, there exists $q = q(\delta, \alpha) \in \mathbb{R}$ such that $S(q) = 0$ (we note that such a q is necessarily unique, by the strict convexity of $q \mapsto F(q, \delta, \alpha)$). The u-dimension $\mathcal{F}_u(\alpha)$ is then given by the unique root $\delta = \delta(\alpha)$ of the system of equations

$$F(q(\delta, \alpha), \delta, \alpha) = \frac{\partial F}{\partial q}(q(\delta, \alpha), \delta, \alpha) = 0.$$

We want to apply the implicit function theorem in order to obtain the regularity of the functions $q(\alpha) = q(\delta(\alpha), \alpha)$ and $\delta(\alpha)$. Writing

$$G(q, \delta, \alpha) = \begin{pmatrix} F(q, \delta, \alpha) \\ \partial F/\partial q(q, \delta, \alpha) \end{pmatrix},$$

it is sufficient to show that

$$\det \left[\left(\frac{\partial G}{\partial q}, \frac{\partial G}{\partial \delta} \right) \right] = \frac{\partial F}{\partial q} \cdot \frac{\partial^2 F}{\partial \delta \partial q} - \frac{\partial^2 F}{\partial q^2} \cdot \frac{\partial F}{\partial \delta}$$

does not vanish when $\delta = \delta(\alpha)$ and $q = q(\delta(\alpha), \alpha)$. Since $\partial F/\partial q = 0$ at $q(\delta, \alpha)$, it is sufficient to show that $\partial^2 F/\partial q^2$ and $\partial F/\partial \delta$ are nonzero. We observe that

$$\int_\Sigma (\varphi - \alpha\psi) \, d\zeta_{q,\alpha,\delta} = 0$$

when $q = q(\delta(\alpha), \alpha)$ and $\delta = \delta(\alpha)$. Since $\varphi - \alpha\psi$ is not cohomologous to a constant, $\partial F/\partial q^2$ does not vanish (see [132]). Finally,

$$\frac{\partial F}{\partial \delta} = -\int_\Sigma u \, d\zeta_{q,\alpha,\delta} \le -\inf u < 0.$$

This shows that $\delta(\alpha)$ and $q(\alpha)$ are analytic. $\qquad\square$

By Lemma 9.2.4, to prove the second statement in the theorem it remains to consider the case when there exist β, $c \in \mathbb{R}$ with $c \ne 0$, and a continuous function $\chi \colon \Sigma \to \mathbb{R}$ such that

$$\varphi - \beta\psi = c + \chi \circ \sigma - \chi, \qquad (9.19)$$

but φ is not cohomologous to $\beta'\psi$ for every $\beta' \in \mathbb{R}$. We can easily verify that $x \in K_\alpha(\varphi, \psi)$ if and only if $x \in K_{c/(\alpha-\beta)}(\psi, 1)$. Furthermore, it follows from (9.19) that

$$\frac{\varphi_n}{\psi_n} - \beta = \frac{cn}{\psi_n} + \frac{\chi \circ \sigma^n - \chi}{\psi_n}.$$

Since $\psi > 0$ and $c \ne 0$, we conclude that $\beta \ne \alpha$ for every $\alpha \in \mathbb{R}$ such that $K_\alpha(\varphi, \psi) \ne \varnothing$. This shows that the function $\alpha \mapsto c/(\alpha - \beta)$ is real analytic in $(\underline{\alpha}, \overline{\alpha})$.

Furthermore, we observe that ψ cannot be cohomologous to a constant, say $\gamma \in \mathbb{R}$. Otherwise the function φ would be cohomologous to $\beta\psi + c = (\beta + c/\gamma)\psi$ (since $\psi > 0$, the constant γ would be positive), which contradicts the hypothesis that φ is not cohomologous to $\beta'\psi$ for every $\beta' \in \mathbb{R}$. Therefore, we can apply Lemma 9.2.4 to the pair of functions $(\psi, 1)$ to conclude that the spectrum $\mathcal{F}_{u,(\psi,1)}$ is analytic in the nonempty interval $(\underline{\beta}, \overline{\beta})$, where

$$\underline{\beta} = \inf\left\{ \int_\Sigma \psi \, d\mu : \mu \in \mathcal{M} \right\} \quad \text{and} \quad \overline{\beta} = \sup\left\{ \int_\Sigma \psi \, d\mu : \mu \in \mathcal{M} \right\}.$$

Since $\psi > 0$, we have $\underline{\beta} > 0$.

Since $\mathcal{F}_{u,(\varphi,\psi)}$ is the composition of the analytic functions $\mathcal{F}_{u,(\psi,1)}$ and $\alpha \mapsto c/(\alpha - \beta)$, we conclude that it is also analytic. Furthermore,

$$(\underline{\alpha}, \overline{\alpha}) = \begin{cases} (\beta + c/\overline{\beta}, \beta + c/\underline{\beta}) & \text{when } c > 0, \\ (\beta + c/\underline{\beta}, \beta + c/\overline{\beta}) & \text{when } c < 0. \end{cases}$$

This completes the proof of the theorem. $\qquad\square$

9.3   Dimension of irregular sets

Let $\sigma|\Sigma$ be a subshift, obtained from a compact σ-invariant set Σ. Given functions $\varphi, \psi \in C(\Sigma)$ with $\psi > 0$, we obtain the *multifractal decomposition*

$$\Sigma = \bigcup_{\alpha \in [\underline{\alpha}, \overline{\alpha}]} K_\alpha(\varphi, \psi) \cup I(\varphi, \psi),$$

where $K_\alpha(\varphi, \psi)$, $\underline{\alpha}$, and $\overline{\alpha}$ are defined by (9.1), (9.3), and (9.4) (with $X = \Sigma$), and where

$$I(\varphi, \psi) = \left\{ x \in \Sigma : \liminf_{n \to \infty} \frac{\varphi_n(x)}{\psi_n(x)} < \limsup_{n \to \infty} \frac{\varphi_n(x)}{\psi_n(x)} \right\}.$$

Definition 9.3.1. The set $I(\varphi, \psi)$ is called the *irregular set* for the pair (φ, ψ).

By Birkhoff's ergodic theorem, $I(\varphi, \psi)$ has zero measure with respect to any finite invariant measure.

The following result of Barreira and Saussol [16] shows that from the point of view of dimension theory the irregular set $I(\varphi, \psi)$ is as large as the whole space. The proof uses the techniques developed in [21] (see Chapter 8).

Theorem 9.3.2. *Let $\sigma|\Sigma$ be a subshift with the specification property. If $\varphi, \psi, u \in C(\Sigma)$ with $\psi, u > 0$ are such that $\operatorname{span}\{\varphi, \psi, u\} \subset D(\Sigma)$, and $\underline{\alpha} < \overline{\alpha}$, then*

$$\dim_u I(\varphi, \psi) = \dim_u \Sigma.$$

Proof. By Theorem 9.1.4, the function $\mathcal{F}_u$ is continuous in $(\underline{\alpha}, \overline{\alpha})$. Now we show that $\mathcal{F}_u$ is continuous where it attains its maximum even if this occurs at $\alpha \in \{\underline{\alpha}, \overline{\alpha}\}$. Let m_u be the measure of maximal dimension, that is, the equilibrium measure of $-\dim_u \Sigma \cdot u$. Clearly, $\mathcal{F}_u(\alpha) \leq \dim_u m_u$ for every α. Furthermore, we can easily verify that if

$$\alpha = \int_\Sigma \varphi \, dm_u / \int_\Sigma \psi \, dm_u,$$

then $\mathcal{F}_u(\alpha) = \dim_u m_u$. Therefore, when $\alpha \in \{\underline{\alpha}, \overline{\alpha}\}$ we can use a similar argument to that in the proof of Lemma 9.1.10 to establish the continuity at α.

We also need the following statement, which is an immediate consequence of Theorem 8.2.4.

Lemma 9.3.3. *If $\sigma|\Sigma$ is a subshift with the specification property, and $\mu_1, \mu_2 \in \mathcal{M}_E$ are such that*

$$\frac{\int_\Sigma \varphi \, d\mu_1}{\int_\Sigma \psi \, d\mu_1} \neq \frac{\int_\Sigma \varphi \, d\mu_2}{\int_\Sigma \psi \, d\mu_2},$$

then

$$\dim_u I(\varphi, \psi) \geq \min\{\dim_u \mu_1, \dim_u \mu_2\}.$$

The continuity of $\mathcal{F}_u$ at the point $\alpha = \int_\Sigma \varphi \, dm_u / \int_\Sigma \psi \, dm_u$ guarantees that for each $\varepsilon > 0$ there exists an equilibrium measure $\mu \in \mathcal{M}_E$ such that

$$\frac{\int_\Sigma \varphi \, d\mu}{\int_\Sigma \psi \, d\mu} \neq \alpha \quad \text{and} \quad \dim_u \mu > \dim_u \Sigma - \varepsilon.$$

By Lemma 9.3.3, we conclude that

$$\dim_u I(\varphi, \psi) \geq \min\{\dim_u \mu_u, \dim_u \mu\} > \dim_u \Sigma - \varepsilon.$$

The arbitrariness of ε implies the desired result. □

Using the techniques described in Chapter 8, when $\sigma|\Sigma$ is a subshift with the specification property, we can show that if the functions φ_i, ψ_i, $u \in C(\Sigma)$ with $\psi_i, u > 0$ are such that $\text{span}\{\varphi_i, \psi_i, u\} \subset D(\Sigma)$ and $\underline{\alpha}(\varphi_i, \psi_i) < \overline{\alpha}(\varphi_i, \psi_i)$ for $i = 1$, ..., k, then

$$\dim_u \bigcap_{i=1}^{k} I(\varphi_i, \psi_i) = \dim_u \Sigma.$$

In the particular case of topologically mixing topological Markov chains and Hölder continuous functions, the statement in Theorem 9.3.2 was obtained by Barreira and Schmeling in [21].

9.4 Repellers and mixed spectra

Now we consider the case of repellers. Using Markov partitions we establish a conditional variational principle for the u-dimension, as well as the analyticity of the spectrum $\mathcal{F}_u$. We also obtain versions of the remaining results in the former sections. We continue to denote by $\mathcal{M}$ the family of f-invariant probability Borel measures, and by $\mathcal{M}_E \subset \mathcal{M}$ the subset of all ergodic measures.

Theorem 9.4.1 ([16]). *Let J be a repeller of a $C^{1+\varepsilon}$ transformation f, for some $\varepsilon > 0$, such that f is conformal and topologically mixing on J, and let φ, ψ, and u be Hölder continuous functions in J with $\psi, u > 0$. If φ is not cohomologous to any multiple of ψ, then:*

1. *the function $\mathcal{F}_u$ is analytic in $(\underline{\alpha}, \overline{\alpha})$;*

2. *if $\alpha \in (\underline{\alpha}, \overline{\alpha})$, then*

$$\mathcal{F}_u(\alpha) = \max \left\{ \frac{h_\mu(f)}{\int_J u \, d\mu} : \mu \in \mathcal{M} \text{ and } \frac{\int_J \varphi \, d\mu}{\int_J \psi \, d\mu} = \alpha \right\} = \inf_{q \in \mathbb{R}} \Delta(q\alpha, q),$$

where $\Delta(p, q)$ is the unique real number such that

$$P(q\varphi - p\psi - \Delta(p, q)u) = 0;$$

3. *for every $\alpha \in (\underline{\alpha}, \overline{\alpha})$ we have*

$$\inf_{q \in \mathbb{R}} P(q\varphi - q\alpha\psi - \mathcal{F}_u(\alpha)u) = 0;$$

4. $\dim_u I(\varphi, \psi) = \dim_u J$.

Proof. Let $R_1, \ldots, R_p$ be the elements of a Markov partition of J, and let $\chi \colon \Sigma \to J$ be the associated coding map. The following result is due to Schmeling [143].

Lemma 9.4.2. *We have $\dim_{u \circ \chi} B = \dim_u \chi(B)$ for every set $B \subset \Sigma$.*

Proof of the lemma. Clearly,

$$\dim_u \chi(B) \le \dim_{u \circ \chi} B \quad \text{for every set} \quad B \subset \Sigma.$$

For the reverse inequality, let $\mathcal{U}$ be a finite open cover of J, and consider a collection $\Gamma \subset \bigcup_{n \in \mathbb{N}} W_n(\mathcal{U})$ covering $\chi(B)$. For each $\mathbf{U} \in \Gamma$, let $C_i(\mathbf{U}) \subset \Sigma$, $i = 1, \ldots, N(\mathbf{U})$ be the cylinder sets of length $m(\mathbf{U})$ such that $\chi(C_i(\mathbf{U}))$ intersects $\chi(\mathbf{U})$. Since f is conformal on J, we have $M := \sup_{\mathbf{U} \in \Gamma} N(\mathbf{U}) < \infty$. The set

$$\mathcal{V} = \left\{ C_i(\mathbf{U}) : \mathbf{U} \in \Gamma \text{ and } i = 1, \ldots, N(\mathbf{U}) \right\}$$

is a cover of B, and

$$\sum_{C \in \mathcal{V}} \exp\left[-\alpha \inf_{x \in C} S_{|C|} u(\chi(x)) \right] \le M \sum_{\mathbf{U} \in \Gamma} \exp[-\alpha u(\mathbf{U})]. \tag{9.20}$$

Note that due to the uniform continuity of u in J (and thus of $u \circ \chi$ in Σ), we can replace the supremum in (7.2) by the infimum (see (3.40)). Therefore, it follows from (9.20) that $\dim_{u \circ \chi} B \le \dim_u \chi(B)$. $\quad\square$

The statement in the theorem is now a simple consequence of Theorems 9.1.4, 9.2.1, 9.2.3, and 9.3.2, together with Lemma 9.4.2 (in a similar manner to that in the proof of Theorem 6.1.2; see Lemma 6.1.6). $\quad\square$

Let J be a repeller of f. Given a Hölder continuous function $\varphi \colon J \to \mathbb{R}$, we consider the sets

$$\mathfrak{D} = \left\{ -\frac{\int_J \varphi \, d\mu}{\int_J \log \|df\| \, d\mu} : \mu \in \mathcal{M} \right\},$$

$$\mathfrak{E} = \left\{ -\int_J \varphi \, d\mu : \mu \in \mathcal{M} \right\}, \quad \mathfrak{L} = \left\{ \int_J \log \|df\| \, d\mu : \mu \in \mathcal{M} \right\}.$$

Using Theorem 9.4.1 we can obtain conditional variational principles for each of the multifractal spectra introduced in Section 7.1.

Theorem 9.4.3 ([16]). *Let J be a repeller of a $C^{1+\varepsilon}$ transformation f, for some $\varepsilon > 0$, such that f is conformal and topologically mixing on J. If ν is the equilibrium measure of a Hölder continuous function φ in J with $P(\varphi) = 0$, then:*

1. *the set $\mathfrak{D}$ is an interval or a point, it coincides with the domains of the functions $\mathcal{D}_{D,\nu}$ and $\mathcal{D}_{E,\nu}$, and if $\alpha \in \text{int } \mathfrak{D}$ then*

$$\mathcal{D}_{D,\nu}(\alpha) = \max \left\{ \dim_H \mu : \mu \in \mathcal{M}_E \text{ and } -\frac{\int_J \varphi \, d\mu}{\int_J \log \|df\| \, d\mu} = \alpha \right\},$$

$$\mathcal{D}_{E,\nu}(\alpha) = \max \left\{ h_\mu(f) : \mu \in \mathcal{M}_E \text{ and } -\frac{\int_J \varphi \, d\mu}{\int_J \log \|df\| \, d\mu} = \alpha \right\};$$

2. *the set $\mathfrak{E}$ is an interval or a point, it coincides with the domains of the functions $\mathcal{E}_{D,\nu}$ and $\mathcal{E}_{E,\nu}$, and if $\alpha \in \text{int } \mathfrak{E}$ then*

$$\mathcal{E}_{D,\nu}(\alpha) = \max \left\{ \dim_H \mu : \mu \in \mathcal{M}_E \text{ and } -\int_J \varphi \, d\mu = \alpha \right\},$$

$$\mathcal{E}_{E,\nu}(\alpha) = \max \left\{ h_\mu(f) : \mu \in \mathcal{M}_E \text{ and } -\int_J \varphi \, d\mu = \alpha \right\};$$

3. *the set $\mathfrak{L}$ is an interval or a point, it coincides with the domains of the functions $\mathcal{L}_D$ and $\mathcal{L}_E$, and if $\alpha \in \text{int } \mathfrak{L}$ then*

$$\mathcal{L}_D(\alpha) = \max \left\{ \dim_H \mu : \mu \in \mathcal{M}_E \text{ and } \int_J \log \|df\| \, d\mu = \alpha \right\},$$

$$\mathcal{L}_E(\alpha) = \max \left\{ h_\mu(f) : \mu \in \mathcal{M}_E \text{ and } \int_J \log \|df\| \, d\mu = \alpha \right\}.$$

Proof. Set $\psi = \log \|df\|$. We have

$$d_\nu(x) = \lim_{n \to \infty} -\frac{\varphi_n(x)}{\psi_n(x)}, \quad h_\nu(x) = \lim_{n \to \infty} -\frac{\varphi_n(x)}{n},$$

$$\lambda(x) = \lim_{n \to \infty} \frac{\psi_n(x)}{n},$$

whenever the corresponding limits exist. Therefore, by Theorem 9.4.1:

1. setting $u = \log \|df\|$ (see Example 7.2.5) we obtain each of the first identities in statements 1, 2, and 3 in the theorem, considering respectively the pairs of functions $(\varphi, -\psi)$, $(\varphi, -1)$, and $(\psi, 1)$;

2. setting $u = 1$ (see Example 7.2.4) we obtain each of the second identities in statements 1, 2, and 3 in the theorem, considering respectively the pairs of functions $(\varphi, -\psi)$, $(\varphi, -1)$, and $(\psi, 1)$.

This completes the proof of the theorem. $\square$

We remark that the conformality of f on J is essential to all but the spectrum $\mathcal{E}_{E,\nu}$. Indeed, we can formulate the following stronger statement for this spectrum.

Theorem 9.4.4 ([16]). *Let J be a repeller of a $C^{1+\varepsilon}$ transformation f, for some $\varepsilon > 0$, such that f is topologically mixing on J. If ν is the equilibrium measure of a Hölder continuous function φ in J with $P(\varphi) = 0$, then*

$$\mathcal{E}_{E,\nu}(\alpha) = \max\left\{ h_\mu(f) : \mu \in \mathcal{M}_E \text{ and } -\int_J \varphi\, d\mu = \alpha \right\}.$$

Proof. Considering the functions $u = 1$ and $\psi = -1$, the statement is an immediate consequence of Theorem 9.2.1 and Lemma 9.4.2. $\square$

For a repeller of a conformal map, let m_D and m_E be respectively the measures of maximal dimension and maximal entropy. The following is an immediate consequence of statement 2 in Theorem 9.2.3.

Theorem 9.4.5 ([16]). *If J is a repeller of a $C^{1+\varepsilon}$ transformation f, for some $\varepsilon > 0$, such that f is conformal and topologically mixing on J, and ν is the equilibrium measure of a Hölder continuous function in J, then:*

1. *if $\nu \neq m_D$, then the functions $\mathcal{D}_{D,\nu}$ and $\mathcal{D}_{E,\nu}$ are analytic;*

2. *if $\nu \neq m_E$, then the functions $\mathcal{E}_{D,\nu}$ and $\mathcal{E}_{E,\nu}$ are analytic;*

3. *if $m_D \neq m_E$, then the functions $\mathcal{L}_D$ and $\mathcal{L}_E$ are analytic.*

Part III

Multifractal Analysis: Further Developments

Chapter 10

Multidimensional Spectra and Number Theory

In the theory of dynamical systems we are sometimes interested in more than one local quantity at the same time. Examples include Lyapunov exponents, local entropy, and pointwise dimension. However, the theory of multifractal analysis described in the former chapters only considers separately each of these local quantities. This led Barreira, Saussol and Schmeling to develop in [20] a multi-dimensional version of the theory of multifractal analysis. For example, we can consider intersections of level sets of Birkhoff averages of different functions, and describe their multifractal properties, including their "size" in terms of topological entropy and of Hausdorff dimension. It turns out that the corresponding multi-dimensional multifractal spectra exhibit several nontrivial phenomena that are absent in the one-dimensional case. A unifying element continues to be the use of the thermodynamic formalism.

10.1 Conditional variational principle

As an illustration we first formulate a rigorous statement in the case of topological Markov chains.

Let $\mathcal{M}$ be the family of σ-invariant probability Borel measures in a topological Markov chain Σ. Given continuous functions $\varphi, \psi \colon \Sigma \to \mathbb{R}$, we consider the intersections of the level sets of Birkhoff averages

$$K_{\alpha,\beta} = K_\alpha(\varphi) \cap K_\beta(\psi), \tag{10.1}$$

where

$$K_\alpha(\varphi) = \left\{ x \in \Sigma : \lim_{n \to \infty} \frac{1}{n} \sum_{k=0}^{n-1} \varphi(\sigma^k x) = \alpha \right\},$$

and

$$K_\beta(\psi) = \left\{ x \in \Sigma : \lim_{n\to\infty} \frac{1}{n} \sum_{k=0}^{n-1} \psi(\sigma^k x) = \beta \right\}.$$

We also consider the set

$$\mathfrak{D} = \left\{ \left(\int_\Sigma \varphi\,d\mu, \int_\Sigma \psi\,d\mu \right) \in \mathbb{R}^2 : \mu \in \mathfrak{M} \right\}.$$

The following result is a conditional variational principle for the sets $K_{\alpha,\beta}$ in (10.1). It was established by Barreira, Saussol and Schmeling in [20].

Theorem 10.1.1. *Let $\sigma|\Sigma$ be a topologically mixing topological Markov chain, and let φ and ψ be Hölder continuous functions in Σ. Then for each $(\alpha, \beta) \in \operatorname{int}\mathfrak{D}$ we have $K_{\alpha,\beta} \neq \varnothing$, and*

$$h(\sigma|K_{\alpha,\beta}) = \sup\left\{ h_\mu(\sigma) : \mu \in \mathfrak{M} \text{ and } \left(\int_\Sigma \varphi\,d\mu, \int_\Sigma \psi\,d\mu \right) = (\alpha, \beta) \right\}$$

$$= \inf\left\{ P(p(\varphi - \alpha) + q(\psi - \beta)) : (p,q) \in \mathbb{R}^2 \right\}. \qquad (10.2)$$

Theorem 10.1.1 is a particular case of Theorem 10.1.4 below. The first identity in (10.2) was established independently by Fan, Feng and Wu [57], also for arbitrary continuous functions φ and ψ. We will show that the second identity in (10.2) can be applied with success to several problems in number theory (see Section 10.7).

The following result was also established in [20].

Theorem 10.1.2. *If $\sigma|\Sigma$ is a topologically mixing topological Markov chain, and φ and ψ are Hölder continuous functions in Σ, then the following properties hold:*

1. *if $(\alpha,\beta) \notin \overline{\mathfrak{D}}$, then $K_{\alpha,\beta} = \varnothing$;*

2. *if for every $(p,q) \in \mathbb{R}^2$ the function $p\varphi + q\psi$ is not cohomologous to a constant, then $\mathfrak{D} = \operatorname{int}\overline{\mathfrak{D}}$;*

3. *the function $(\alpha,\beta) \mapsto h(\sigma|K_{\alpha,\beta})$ is analytic in $\operatorname{int}\mathfrak{D}$;*

4. *there is an ergodic equilibrium measure $\mu_{\alpha,\beta} \in \mathfrak{M}$ with*

$$\int_\Sigma \varphi\,d\mu = \alpha \quad \text{and} \quad \int_X \psi\,d\mu = \beta,$$

   such that

$$\mu_{\alpha,\beta}(K_{\alpha,\beta}) = 1 \quad \text{and} \quad h_{\mu_{\alpha,\beta}}(\sigma) = h(\sigma|K_{\alpha,\beta}).$$

Theorem 10.1.2 is also a consequence of Theorem 10.1.4 below. In particular, statement 2 gives a condition which guarantees that the identities in Theorem 10.1.1 are valid for a dense set of pairs $(\alpha,\beta) \in \mathfrak{D}$. We note that this is automatic in the case of one-dimensional spectra, since then $\mathfrak{D}$ is an interval (see Chapters 6, 7, and 9).

Now let f be a continuous transformation in the compact metric space X. We continue to denote by $C(X)$ the space of continuous functions $\varphi \colon X \to \mathbb{R}$. Given $d \in \mathbb{N}$, consider a pair of vectors $(\Phi, \Psi) \in C(X)^d \times C(X)^d$, and write

$$\Phi = (\varphi_1, \ldots, \varphi_d) \quad \text{and} \quad \Psi = (\psi_1, \ldots, \psi_d).$$

We always assume in this chapter that $\psi_i > 0$ for $i = 1, \ldots, d$. Given $\alpha = (\alpha_1, \ldots, \alpha_d) \in \mathbb{R}^d$ we set

$$K_\alpha = K_\alpha(\Phi, \Psi) = \bigcap_{i=1}^{d} \left\{ x \in X : \lim_{n \to \infty} \frac{\varphi_{i,n}(x)}{\psi_{i,n}(x)} = \alpha_i \right\}, \tag{10.3}$$

where for each i,

$$\varphi_{i,n}(x) = \sum_{k=0}^{n-1} \varphi_i(f^k x) \quad \text{and} \quad \psi_{i,n}(x) = \sum_{k=0}^{n-1} \psi_i(f^k x).$$

We continue to denote by $\mathfrak{M}$ the family of f-invariant probability Borel measures in X, and we define a function $\mathcal{P} = \mathcal{P}_{(\Phi, \Psi)} \colon \mathfrak{M} \to \mathbb{R}^d$ by

$$\mathcal{P}(\mu) = \left(\frac{\int_X \varphi_1 \, d\mu}{\int_X \psi_1 \, d\mu}, \ldots, \frac{\int_X \varphi_d \, d\mu}{\int_X \psi_d \, d\mu} \right). \tag{10.4}$$

Since $\mathfrak{M}$ is compact and connected, and $\mathcal{P}$ is continuous, the image $\mathcal{P}(\mathfrak{M})$ is also compact and connected.

Let $u \in C(X)$ be a positive function.

Definition 10.1.3. The function $\mathcal{F}_u = \mathcal{F}_{u,(\Phi,\Psi)}$ defined by

$$\mathcal{F}_u(\alpha) = \dim_u K_\alpha(\Phi, \Psi) \tag{10.5}$$

is called the *u-dimension spectrum* of the pair (Φ, Ψ).

Now we formulate a conditional variational principle for the spectrum $\mathcal{F}_u$, that was established by Barreira, Saussol and Schmeling in [20]. Given a vector $\alpha = (\alpha_1, \ldots, \alpha_d) \in \mathbb{R}^d$ we write

$$\alpha * \Phi = (\alpha_1 \varphi_1, \ldots, \alpha_d \varphi_d) \in C(X)^d \quad \text{and} \quad \langle \alpha, \Phi \rangle = \sum_{i=1}^{d} \alpha_i \varphi_i \in C(X).$$

Theorem 10.1.4 (Multidimensional conditional variational principle). *Assume that the Kolmogorov–Sinai entropy of f is upper semicontinuous, and that*

$$\operatorname{span}\{\varphi_1, \psi_1, \ldots, \varphi_d, \psi_d, u\} \subset D(X).$$

If $\alpha \notin \mathcal{P}(\mathfrak{M})$, then $K_\alpha = \varnothing$. Furthermore, if $\alpha \in \operatorname{int} \mathcal{P}(\mathfrak{M})$, then $K_\alpha \neq \varnothing$ and the following properties hold:

1.

$$\mathcal{F}_u(\alpha) = \max\left\{\frac{h_\mu(f)}{\int_X u\,d\mu} : \mu \in \mathcal{M} \text{ and } \mathcal{P}(\mu) = \alpha\right\}; \qquad (10.6)$$

2. *we have*

$$\mathcal{F}_u(\alpha) = \inf\{T_u(q) : q \in \mathbb{R}^d\},$$

where $T_u(q)$ is the unique real number such that

$$P(\langle q, \Phi - \alpha * \Psi\rangle - T_u(q)u) = 0;$$

3. *there exists an ergodic equilibrium measure $\mu_\alpha \in \mathcal{M}$ with $\mathcal{P}(\mu_\alpha) = \alpha$ and $\mu_\alpha(K_\alpha) = 1$ such that*

$$\dim_u \mu_\alpha = \frac{h_{\mu_\alpha}(f)}{\int_X u\,d\mu_\alpha} = \mathcal{F}_u(\alpha). \qquad (10.7)$$

Proof. We begin with some preparatory lemmas. Let $\|q\| = |p| + \cdots + |q_d|$ be the norm of a vector $q \in \mathbb{R}^d$.

Lemma 10.1.5. *If $\alpha \in \mathcal{P}(\mathcal{M})$, then*

$$\inf_{q\in\mathbb{R}^d} P(\langle q, \Phi - \alpha * \Psi\rangle - \mathcal{F}_u(\alpha)u) \geq 0.$$

Proof of the lemma. We first assume that $\mathcal{F}_u(\alpha) = 0$. By the definition of $\mathcal{P}$, there exists $\mu \in \mathcal{M}$ such that

$$\int_X \Phi\,d\mu = \int_X \alpha * \Psi\,d\mu.$$

We obtain

$$P(\langle q, \Phi - \alpha * \Psi\rangle) \geq h_\mu(f) + \left\langle q, \int_X (\Phi - \alpha * \Psi)\,d\mu\right\rangle = h_\mu(f) \geq 0.$$

Now we imitate the argument in the proof of Lemma 9.1.5, using the notion of topological pressure in Section 7.2. Assume that $\mathcal{F}_u(\alpha) > 0$. By Proposition 7.2.3, the number $\mathcal{F}_u(\alpha)$ is equal to the unique root δ of the equation $P_{K_\alpha}(-\delta u) = 0$. Given $\delta > 0$ and $\tau \in \mathbb{N}$, we consider the sets

$$L_{\delta,\tau} = \{x \in X : \|\Phi_n(x) - \alpha\Psi_n(x)\| < \delta n \text{ for every } n \geq \tau\},$$

where

$$\Phi_n = \sum_{k=0}^{n-1} \Phi \circ f^k \quad \text{and} \quad \Psi_n = \sum_{k=0}^{n-1} \Psi \circ f^k.$$

Since all components of Ψ are positive, we can easily show that

$$K_\alpha \subset \bigcap_{\delta>0}\bigcup_{\tau\in\mathbb{N}} L_{\delta,\tau}.$$

Now let $\mathfrak{U}$ be a finite open cover of X with sufficiently small diameter such that if $n \in \mathbb{N}$ is sufficiently large, $\mathbf{U} \in \bigcup_{k \geq n} W_k(\mathfrak{U})$, and $x \in X(\mathbf{U})$, then

$$\|\Phi(\mathbf{U}) - \Phi_{m(\mathbf{U})}(x)\| \leq \delta m(\mathbf{U}) \quad \text{and} \quad \|\Psi(\mathbf{U}) - \Psi_{m(\mathbf{U})}(x)\| \leq \delta m(\mathbf{U}),$$

where

$$\Phi(\mathbf{U}) = (\varphi_1(\mathbf{U}), \dots, \varphi_d(\mathbf{U})) \quad \text{and} \quad \Psi(\mathbf{U}) = (\psi_1(\mathbf{U}), \dots, \psi_d(\mathbf{U})).$$

This implies that if $\mathbf{U} \in \bigcup_{k \geq n} W_k(\mathfrak{U})$ and $X(\mathbf{U}) \cap L_{\delta,\tau} \neq \varnothing$, then

$$\|\Phi(\mathbf{U}) - \alpha \Psi(\mathbf{U})\| < (2 + |\alpha|)\delta m(\mathbf{U}).$$

Therefore,

$$P_{L_{\delta,\tau}}(-\mathcal{F}_u(\alpha)u, \mathfrak{U}) \leq P_{L_{\delta,\tau}}(\langle q, \Phi - \alpha * \Psi \rangle - \mathcal{F}_u(\alpha)u, \mathfrak{U}) + (2 + |\alpha|)\delta|q|.$$

Letting $\operatorname{diam} \mathfrak{U} \to 0$ yields

$$P_{L_{\delta,\tau}}(-\mathcal{F}_u(\alpha)u) \leq P(\langle q, \Phi - \alpha * \Psi \rangle - \mathcal{F}_u(\alpha)u) + (2 + |\alpha|)\delta|q|,$$

and hence,

$$0 \leq P_{\bigcup_{\tau \in \mathbb{N}} L_{\delta,\tau}}(-\mathcal{F}_u(\alpha)u) = \sup_{\tau \in \mathbb{N}} P_{L_{\delta,\tau}}(-\mathcal{F}_u(\alpha)u)$$
$$\leq P(\langle q, \Phi - \alpha * \Psi \rangle - \mathcal{F}_u(\alpha)u) + (2 + |\alpha|)\delta|q|.$$

The arbitrariness of δ implies the desired result. $\qquad \square$

Lemma 10.1.6. *If $\alpha \in \operatorname{int} \mathcal{P}(\mathcal{M})$, then*

$$\inf_{q \in \mathbb{R}^d} P(\langle q, \Phi - \alpha * \Psi \rangle - \mathcal{F}_u(\alpha)u) = 0,$$

and there exists an ergodic equilibrium measure $\mu_\alpha \in \mathcal{M}$ with $\mathcal{P}(\mu_\alpha) = \alpha$ and $\mu_\alpha(K_\alpha) = 1$ such that $\dim_u \mu_\alpha = \mathcal{F}_u(\alpha)$.

Proof of the lemma. Let

$$r = \inf\{\|\alpha - \beta\| : \beta \in \mathbb{R}^d \setminus \mathcal{P}(\mathcal{M})\} > 0.$$

We claim that the infimum over $q \in \mathbb{R}^d$ of the function

$$F(q) = P(\langle q, \Phi - \alpha * \Psi \rangle - \mathcal{F}_u(\alpha)u)$$

is attained inside the ball of radius

$$R = \frac{\dim_u X \cdot \sup u + F(0)}{r \min_i \inf \psi_i}$$

centered at zero. Take $q \in \mathbb{R}^d$ such that $\|q\| \geq R$. We show that $F(q) \geq F(0)$. Let $a \in (0,1)$ and $\beta \in \mathbb{R}^d$ such that $\beta_i = \alpha_i + ar \operatorname{sgn} q_i$. Clearly, $\beta \in \mathcal{P}(\mathcal{M})$, and hence there exists $\mu \in \mathcal{M}$ such that

$$\int_X \Phi \, d\mu = \int_X \beta * \Psi \, d\mu.$$

We obtain

$$F(q) \geq h_\mu(f) + \left\langle q, \int_X (\Phi - \alpha * \Psi) \, d\mu \right\rangle - \mathcal{F}_u(\alpha) \int_X u \, d\mu$$
$$\geq \left\langle q, \int_X (\beta - \alpha) * \Psi \, d\mu \right\rangle - \dim_u X \cdot \sup u$$
$$\geq \|q\| ar \min_{i=1,\dots,d} \inf \psi_i - \dim_u X \cdot \sup u$$
$$> a \dim_u X \cdot \sup u + F(0) - \dim_u X \cdot \sup u.$$

We obtain the claim letting $a \to 1$.

Since F is of class C^1 its minimum is attained at a point $q = q(\alpha)$ with $\|q(\alpha)\| < R$ such that $\nabla F(q(\alpha)) = 0$. Let μ_α be the equilibrium measure of the function

$$\langle q(\alpha), \Phi - \alpha * \Psi \rangle - \mathcal{F}_u(\alpha)u.$$

Then

$$\int_X (\Phi - \alpha * \Psi) \, d\mu_\alpha = \nabla F(q(\alpha)) = 0,$$

and hence $\mathcal{P}(\mu_\alpha) = \alpha$. Furthermore,

$$F(q(\alpha)) = h_{\mu_\alpha}(f) - \mathcal{F}_u(\alpha) \int_X u \, d\mu_\alpha.$$

By Lemma 10.1.5, we have $F(q(\alpha)) \geq 0$, and thus,

$$\dim_u \mu_\alpha = \frac{h_{\mu_\alpha}(f)}{\int_X u \, d\mu_\alpha} \geq \mathcal{F}_u(\alpha).$$

On the other hand, since μ_α is ergodic and

$$\int_X \Phi \, d\mu_\alpha = \int_X \alpha * \Psi \, d\mu_\alpha,$$

it follows from Birkhoff's ergodic theorem that $\mu_\alpha(K_\alpha) = 1$. Therefore,

$$\mathcal{F}_u(\alpha) \geq \dim_u \mu_\alpha,$$

and hence $\dim_u \mu_\alpha = \mathcal{F}_u(\alpha)$. This completes the proof of the lemma. $\square$

We proceed with the proof of the theorem. Let $\alpha \in \mathbb{R}^d$ with $K_\alpha \neq \varnothing$, and take $x \in K_\alpha$. The sequence of measures

$$\mu_n = \frac{1}{n} \sum_{k=0}^{n-1} \delta_{f^k x}$$

has an accumulation point, say μ, which is invariant. Moreover, for $i = 1, \ldots, d$ we have $\int_X \varphi_i \, d\mu_n / \int_X \psi_i \, d\mu_n \to \alpha_i$ when $n \to \infty$. This implies that

$$\int_X \varphi_i \, d\mu / \int_X \psi_i \, d\mu = \alpha_i$$

for $i = 1, \ldots, d$, and hence $\alpha \in \mathcal{P}(\mathcal{M})$. This proves the first statement.

Now let $\alpha \in \operatorname{int} \mathcal{P}(\mathcal{M})$. For each $\mu \in \mathcal{M}$ with $\mathcal{P}(\mu) = \alpha$, it follows from Lemma 10.1.6 that

$$0 = \inf_{q \in \mathbb{R}^d} P(\langle q, \Phi - \alpha * \Psi \rangle) - \mathcal{F}_u(\alpha)u) \geq h_\mu(f) - \mathcal{F}_u(\alpha) \int_X u \, d\mu.$$

Therefore,

$$h_\mu(f) / \int_X u \, d\mu \leq \mathcal{F}_u(\alpha).$$

On the other hand, again by Lemma 10.1.6 there exists an ergodic measure μ_α such that $\mu_\alpha(K_\alpha) = 1$, $\mathcal{P}(\mu_\alpha) = \alpha$, and

$$\mathcal{F}_u(\alpha) = \dim_u \mu_\alpha = \frac{h_{\mu_\alpha}(f)}{\int_X u \, d\mu_\alpha}$$

(using ergodicity and Proposition 7.2.7). This establishes the identities in (10.6) and (10.7). Statement 2 is an immediate consequence of Lemma 10.1.6. This completes the proof of the theorem. $\square$

When $\psi_1 = \cdots = \psi_d = 1$ and $u = 1$ the identity in (10.6) (with the maximum replaced by the supremum) was established by Takens and Verbitski in [154] under the assumptions that f is a continuous transformation with the specification property, and $\varphi_1, \ldots, \varphi_d$ are arbitrary continuous functions (see also [121]). See [81] for related results.

We refer to [20] for the study of irregular sets of multidimensional spectra.

10.2 Geometry of the domains

Theorem 10.1.4 gives very detailed information about the multifractal spectrum inside $\operatorname{int} \mathcal{P}(\mathcal{M})$. Therefore, it is important to discuss the properties of this set, and in particular to give conditions under which it is nonempty.

For each q, $\alpha \in \mathbb{R}^d$, we consider the function $S_{q\alpha} : \mathbb{R} \to \mathbb{R}$ defined by

$$S_{q\alpha}(t) = P_X(t\langle q, \Phi - \alpha * \Psi \rangle).$$

For example, when $f|X$ is a topological Markov chain, and the components of $\Phi - \alpha * \Psi$ are Hölder continuous, the function $S_{q\alpha}$ has the following interpretation. Let

$$\mathcal{E}_{q\alpha}(\beta) = h\left(f\Big| \left\{ x \in X : \lim_{n\to\infty} \frac{1}{n} \sum_{k=0}^{n-1} \langle q, \Phi - \alpha * \Psi \rangle (f^k x) = \beta \right\}\right).$$

Using (9.2) we obtain

$$\beta_{q\alpha}(t) := -S'_{q\alpha}(t) = -\int_X \langle q, \Phi - \alpha * \Psi \rangle \, d\mu_{t\langle q, \Phi - \alpha * \Psi \rangle},$$

and by Theorem 7.3.2,

$$\mathcal{E}_{q\alpha}(\beta_{q\alpha}(t)) = S_{q\alpha}(t) + t\beta_{q\alpha}(t)$$

for every $t \in \mathbb{R}$. In other words, $S_{q\alpha}$ is the Legendre transform of $\mathcal{E}_{q\alpha}$.

The following statement gives a characterization of the points in the set $\operatorname{int} \mathcal{P}(\mathcal{M})$ which is optimal in a certain sense.

Theorem 10.2.1 ([20]). *Assume that $f|X$ has finite topological entropy, and that $\Phi, \Psi \in C(X)^d$. For each $\alpha \in \mathcal{P}(\mathcal{M})$ we have:*

1. *if $S_{q\alpha}$ is constant for no $q \in \mathbb{R}^d$, then $\alpha \in \overline{\operatorname{int} \mathcal{P}(\mathcal{M})}$;*

2. *if $S_{q\alpha}$ is constant for some $q \in \mathbb{R}^d$, then $\alpha \notin \operatorname{int} \mathcal{P}(\mathcal{M})$.*

Proof. Replacing if necessary Φ by $\Phi - \alpha * \Psi$, we may assume without loss of generality that $\alpha = 0$. Note that this corresponds to a translation of the set $\mathcal{P}(\mathcal{M})$ by the vector $-\alpha$.

Now we establish the first statement. Since $\alpha = 0 \in \mathcal{P}(\mathcal{M})$, there exists a measure $m_0 \in \mathcal{M}$ with $\int_X \Phi \, dm_0 = 0$. Moreover, the map $m \mapsto \int_X \Phi \, dm$ is affine in the convex set $\mathcal{M}$, and hence

$$\mathcal{M}(\Phi) := \left\{ \int_X \Phi \, dm : m \in \mathcal{M} \right\}$$

is also convex. We show that it has nonempty interior. If $\operatorname{int} \mathcal{M}(\Phi) = \varnothing$, then $\mathcal{M}(\Phi)$ is contained in some hyperplane, and hence there exists $q \in \mathbb{R}^d$ such that $\langle q, \int_X \Phi \, dm \rangle = 0$ for every $m \in \mathcal{M}$. This implies that for any real number t we have

$$P(t\langle q, \Phi \rangle) = \sup_{m \in \mathcal{M}} \left(h_m(f) + t \int_X \langle q, \Phi \rangle \, dm \right)$$
$$= \sup_{m \in \mathcal{M}} h_m(f) = P(0),$$

which contradicts the hypotheses in the theorem. Therefore, $\operatorname{int} \mathcal{M}(\Phi) \neq \varnothing$, and there exist measures $m_1, \ldots, m_d \in \mathcal{M}$ such that the vectors

$$\int_X \Phi \, dm_1, \int_X \Phi \, dm_2, \ldots, \int_X \Phi \, dm_d \tag{10.8}$$

form a basis of $\mathbb{R}^d$. Consider the set

$$\Delta = \{ p \in \mathbb{R}^d : p_i \geq 0 \text{ for } i = 1, \ldots, d, \text{ and } p_1 + \cdots + p_d \leq 1 \}.$$

For each $p \in \Delta$ let

$$\mu_p = p_1 m_1 + \cdots + p_d m_d + \left(1 - \sum_{i=1}^{d} p_i \right) m_0 \in \mathcal{M}.$$

We define a map $\beta \colon \Delta \to \mathbb{R}^d$ by

$$\beta(p) = \left(\frac{\int_X \varphi_1 \, d\mu_p}{\int_X \psi_1 \, d\mu_p}, \ldots, \frac{\int_X \varphi_d \, d\mu_p}{\int_X \psi_d \, d\mu_p} \right).$$

Since $\int_X \Phi \, dm_0 = 0$ we have

$$\begin{aligned}
\frac{\partial}{\partial p_j} \left(\frac{\int_X \varphi_i \, d\mu_p}{\int_X \psi_i \, d\mu_p} \right) \bigg|_{p=0} &= \left. \frac{\int_X \varphi_i \, dm_j - \int_X \varphi_i \, dm_0}{\int_X \psi_i \, d\mu_p} \right|_{p=0} \\
&\quad - \left. \frac{\left(\int_X \psi_i \, dm_j - \int_X \psi_i \, dm_0 \right) \int_X \varphi_i \, d\mu_p}{\left(\int_X \psi_i \, d\mu_p \right)^2} \right|_{p=0} \\
&= \frac{\int_X \varphi_i \, dm_j}{\int_X \psi_i \, dm_0}.
\end{aligned}$$

Therefore, the map β is of class C^1, and its derivative at $p = 0$ is given by

$$d_0 \beta = \begin{bmatrix} \frac{\int_X \varphi_1 \, dm_1}{\int_X \psi_1 \, dm_0} & \cdots & \frac{\int_X \varphi_1 \, dm_d}{\int_X \psi_1 \, dm_0} \\ \vdots & \ddots & \vdots \\ \frac{\int_X \varphi_d \, dm_1}{\int_X \psi_d \, dm_0} & \cdots & \frac{\int_X \varphi_d \, dm_d}{\int_X \psi_d \, dm_0} \end{bmatrix}.$$

Denoting by $M = (M_{ij})$ the $d \times d$ matrix with entries $M_{ij} = \int_X \varphi_j \, dm_i$, we obtain

$$\det d_0 \beta = \left(\prod_{j=1}^{d} \int_X \psi_j \, dm_0 \right)^{-1} \det M.$$

Since the vectors in (10.8) are linearly independent, the matrix M is invertible, and thus β is a local diffeomorphism at 0. Hence, there exist open sets $U \subset \Delta$

and $D = \beta(U)$ such that $0 \in \overline{U}$, and β is a diffeomorphism from U onto D. In particular, we have $0 \in \overline{D}$, and

$$\alpha = 0 \in \overline{\text{int } \beta(\Delta)} \subset \overline{\text{int } \mathcal{P}(\mathcal{M})}.$$

Now we prove the second statement. We continue to assume that $\alpha = 0$. There exists $q \in \mathbb{R}^d$ such that $P(t\langle q, \Phi\rangle) = P(0)$ for every $t \in \mathbb{R}$. We show that

$$\{sq : s \in \mathbb{R}\} \cap \mathcal{P}(\mathcal{M}) = \{0\},$$

which immediately implies the statement in the theorem. Let $s \neq 0$. If $sq \in \mathcal{P}(\mathcal{M})$, then there exists $\mu \in \mathcal{M}$ such that

$$\int_X \Phi \, d\mu = sq * \int_X \Psi \, d\mu.$$

For each $t > 0$ we obtain

$$P(0) = P(t\langle sq, \Phi\rangle) \geq h_\mu(f) + t\left\langle sq, \int_X \Phi \, d\mu \right\rangle \geq t|sq|^2 \inf_{i=1,\dots,d} \inf \psi_i.$$

But this is impossible when t is sufficiently large. Therefore, $\alpha = 0 \notin \text{int } \mathcal{P}(\mathcal{M})$. This completes the proof of the theorem.   □

A noteworthy consequence of Theorem 10.2.1 is that if the topological pressure is strictly convex, that is, if for any $q \in \mathbb{R}^d$ and $\alpha \in \mathcal{P}(\mathcal{M})$ the function $S_{q\alpha}$ is strictly convex, then $\mathcal{P}(\mathcal{M}) = \overline{\text{int } \mathcal{P}(\mathcal{M})}$.

We note that if $f|X$ is a subshift with the specification property, and φ_i, $\psi_i \in C_f(X)$ for $i = 1, \dots, d$ (see Definition 9.1.3), then for each $\alpha \in \mathcal{P}(\mathcal{M})$ the following properties are equivalent:

1. the function $q \mapsto P(\langle q, \Phi - \alpha * \Psi\rangle)$ is strictly convex;

2. the function $S_{q\alpha}$ is constant for no $q \in \mathbb{R}^d$;

3. the functions $\varphi_i - \alpha_i\psi_i$, $i = 1, \dots, d$ are linearly independent as cohomology classes.

For each $q \in S^{2d-1} := \{x \in \mathbb{R}^{2d} : \|x\| = 1\}$ we set

$$\Gamma(q) = \partial\left\{\mathcal{P}(\mu_{t\langle q, (\Phi, \Psi)\rangle}) : t \in \mathbb{R}\right\},$$

where ∂A denotes the boundary of the set A.

Theorem 10.2.2 ([20]). *If $f|X$ is a subshift with the specification property, and φ_i, $\psi_i \in C_f(X)$ for $i = 1, \dots, d$, then*

$$\partial\mathcal{P}(\mathcal{M}) \subset \bigcup_{q \in S^{2d-1}} \Gamma(q).$$

If, in addition, the functions 1, φ_1, $\dots$, φ_d, ψ_1, $\dots$, ψ_d are linearly independent as cohomology classes, then $\mathcal{P}(\mathcal{M}) = \overline{\text{int } \mathcal{P}(\mathcal{M})}$.

Proof. The second statement follows immediately from Theorem 10.2.1. Now let

$$\mathcal{E}_{2d}(\mu) = -\left(\int_X \Phi \, d\mu, \int_X \Psi \, d\mu\right),$$

and

$$\widehat{\mathcal{P}} = \left\{\left(\frac{\alpha_1}{\beta_1}, \ldots, \frac{\alpha_d}{\beta_d}\right) : (\alpha, \beta) \in \partial\mathcal{E}_{2d}(\mathcal{M})\right\}.$$

We show that $\mathcal{E}_{2d}(\mathcal{M}) = \overline{\text{int } \mathcal{E}_{2d}(\mathcal{M})}$. The proof consists of three claims.

Claim. We have $\partial\mathcal{P}(\mathcal{M}) \subset \widehat{\mathcal{P}}$.

Let $(\alpha, \beta) \in \text{int } \mathcal{E}_{2d}(\mathcal{M})$. This means that $(\alpha, \beta) + \varepsilon \in \mathcal{E}_{2d}(\mathcal{M})$ for all sufficiently small $\varepsilon \in \mathbb{R}^{2d}$, and thus, by Theorem 10.1.4, there exists an ergodic measure $\mu_\varepsilon \in \mathcal{M}$ with $\mathcal{P}(\mu_\varepsilon) = (\alpha, \beta) + \varepsilon$. Hence, for all sufficiently small $\delta \in \mathbb{R}^d$ the δ-neighborhood of $(\alpha_1/\beta_1, \ldots, \alpha_d/\beta_d)$ is contained in $\mathcal{P}(\mathcal{M})$. This establishes the claim.

Claim. The set $\mathcal{E}_{2d}(\mathcal{M})$ is convex.

Since the function $\mathcal{E}_{2d}$ is linear, the claim follows immediately from the convexity of $\mathcal{M}$.

Claim. For each $(\alpha, \beta) \in \partial\mathcal{E}_{2d}(\mathcal{M})$ there exists a vector $q \in S^{2d-1}$ such that $\langle(\alpha, \beta), q\rangle \in \partial W_q$, where

$$W_q = -\left\{\int_X \langle q, (\Phi, \Psi)\rangle \, d\mu : \mu \in \mathcal{M}\right\}.$$

Since $\mathcal{E}_{2d}(\mathcal{M})$ is a convex set each of its boundary points has a supporting plane. Let $(\alpha, \beta) \in \partial\mathcal{E}_{2d}(\mathcal{M})$, and denote by Q the orthogonal projection of $\mathcal{E}_{2d}(\mathcal{M})$ onto the normal to the supporting plane at (α, β). The point (α, β) is mapped by Q to a boundary point of the interval $Q(\mathcal{E}_{2d}(\mathcal{M}))$. The orthogonal projection of a point (α, β) onto a line in the direction of a normal vector $q \in S^{2d-1}$ is given by $\langle(\alpha, \beta), q\rangle$. This establishes the claim since W_q is the image of $\mathcal{E}_{2d}(\mathcal{M})$ under this projection.

Now we are ready to prove the theorem. Let $(\alpha, \beta) \in \mathbb{R}^{2d}$ be such that

$$(\alpha_1/\beta_1, \ldots, \alpha_d/\beta_d) \in \widehat{\mathcal{P}}.$$

This means that $(\alpha, \beta) \in \partial\mathcal{E}_{2d}(\mathcal{M})$. Therefore, there exists $q \in S^{2d-1}$ such that $\langle(\alpha, \beta), q\rangle \in \partial W_q$. This completes the proof. $\square$

10.3 Regularity of the multifractal spectra

Now we give conditions to obtain the analyticity of the spectrum, as an application of Theorems 10.1.4 and 10.2.1.

Theorem 10.3.1 ([20]). *Assume that:*

1. *the Kolmogorov–Sinai entropy of f is upper semicontinuous;*

2. *the topological pressure of f is analytic.*

If $\alpha \in \operatorname{int} \mathcal{P}(\mathcal{M})$ is such that the second derivative of $q \mapsto P(\langle q, \Phi - \alpha * \Psi \rangle)$ is a positive definite bilinear form for each $q \in \mathbb{R}^d$, then $\mathcal{F}_u$ is analytic in some open neighborhood of α.

Proof. Let $\alpha \in \operatorname{int} \mathcal{P}(\mathcal{M})$ and consider the function

$$Q(\delta, q, \alpha) = P(\langle q, \Phi - \alpha * \Psi \rangle - \delta u).$$

Proceeding as in the proof of Lemma 10.1.6, we show that there exist $q(\alpha) \in \mathbb{R}^d$ and an ergodic equilibrium measure μ_α such that $q \mapsto Q(\mathcal{F}_u(\alpha), q, \alpha)$ attains a minimum at $q = q(\alpha)$, and thus

$$\frac{\partial Q}{\partial q}(\mathcal{F}_u(\alpha), q(\alpha), \alpha) = \int_X (\Phi - \alpha * \Psi)\, d\mu_\alpha = 0.$$

By Lemma 10.1.6 we have

$$Q(\mathcal{F}_u(\alpha), q(\alpha), \alpha) = 0.$$

Consider the system of equations

$$Q(\delta, q, \alpha) = 0 \quad \text{and} \quad \frac{\partial Q}{\partial q}(\delta, q, \alpha) = 0.$$

We want to apply the implicit function theorem to establish the uniqueness of the solution $(\delta, q) = (\mathcal{F}_u(\alpha), q(\alpha))$ of this system, and to obtain its regularity in α. In particular, this will establish the regularity of the spectrum. Let

$$G(q, \delta, \alpha) = \left(Q(\delta, q, \alpha), \frac{\partial Q}{\partial p}(\delta, q, \alpha), \dots, \frac{\partial Q}{\partial q_d}(\delta, q, \alpha) \right). \tag{10.9}$$

It is sufficient to show that the matrix

$$(\partial/\partial\delta, \partial/\partial p, \dots, \partial/\partial q_d)^t G = \begin{bmatrix} \frac{\partial Q}{\partial \delta} & \frac{\partial^2 Q}{\partial \delta \partial p} & \cdots & \frac{\partial^2 Q}{\partial \delta \partial q_d} \\ \frac{\partial Q}{\partial p} & \frac{\partial^2 Q}{\partial p \partial p} & \cdots & \frac{\partial^2 Q}{\partial p \partial q_d} \\ \vdots & \vdots & \ddots & \vdots \\ \frac{\partial Q}{\partial q_d} & \frac{\partial^2 Q}{\partial q_d \partial p} & \cdots & \frac{\partial^2 Q}{\partial q_d \partial q_d} \end{bmatrix} \tag{10.10}$$

is invertible at $(q(\alpha), \mathcal{F}_u(\alpha), \alpha)$. For each $i = 1, \dots, d$ we have

$$\frac{\partial Q}{\partial q_i}(\mathcal{F}_u(\alpha), q(\alpha), \alpha) = 0,$$

and $\delta = \mathcal{F}_u(\alpha)$. This shows that the first column of the matrix in (10.10) is zero at $(q(\alpha), \mathcal{F}_u(\alpha), \alpha)$, with the exception of the first term which is

$$\frac{\partial Q}{\partial \delta}(\mathcal{F}_u(\alpha), q(\alpha), \alpha) = -\int_X u\, d\mu_{q,\delta,\alpha} < 0,$$

where $\mu_{q,\delta,\alpha}$ is the unique equilibrium measure of the function $\langle q, \Phi - \alpha * \Psi \rangle - \delta u$. Therefore, it is sufficient to show that the right lower $d \times d$ matrix, say H, is invertible. The second derivative of the topological pressure at $q(\alpha)$ is a bilinear symmetric form $B\colon \mathbb{R}^d \times \mathbb{R}^d \to \mathbb{R}$, and we have

$$\frac{\partial^2 Q}{\partial q_i \partial q_j}(\mathcal{F}_u(\alpha), q(\alpha), \alpha) = B(e_i, e_j)$$

where $e_1, \ldots, e_d$ is the canonical basis of $\mathbb{R}^d$. By hypothesis B is positive definite. If H was not invertible, then some nontrivial linear combination of its columns would be equal to zero, and thus there would exist $\lambda \in \mathbb{R}^d \setminus \{0\}$ such that

$$\sum_{j=1}^d \lambda_j B(e_i, e_j) = 0$$

for $i = 1, \ldots, d$. Setting $g = \sum_{i=1}^d \lambda_i e_i$ we obtain

$$B(g, g) = \sum_{i=1}^d \lambda_i \sum_{j=1}^d \lambda_j B(e_i, e_j) = 0.$$

Since $g \neq 0$ this contradicts the positive definiteness of B. Thus H is invertible. By the implicit function theorem, the functions $\delta(\alpha)$ and $q(\alpha)$ are as regular as the analytic function G in (10.9). This completes the proof of the theorem. □

10.4 New phenomena in multidimensional spectra

When $d = 1$, the connectedness of $\mathcal{P}(\mathcal{M})$ implies that only one of the following two exclusive alternatives can occur:

1. The spectrum is *degenerate*: in this case $\mathcal{P}(\mathcal{M}) = \{a\}$ for some $a \in \mathbb{R}$. Furthermore, $K_a = X$, and $K_\alpha = \varnothing$ for every $\alpha \neq a$.

2. The spectrum is *nondegenerate*: in this case $\mathcal{P}(\mathcal{M}) = [\underline{a}, \overline{a}]$ for some real numbers $\underline{a} < \overline{a}$. In particular, $\mathcal{P}(\mathcal{M})$ has nonempty interior, and $\mathcal{P}(\mathcal{M}) = \operatorname{int} \mathcal{P}(\mathcal{M})$.

When $d > 1$, that is, for multidimensional multifractal spectra, several new phenomena can occur. Namely:

1. $\mathcal{P}(\mathcal{M})$ may not be convex;

2. int $\mathcal{P}(\mathcal{M})$ may have more than one connected component;

3. $\mathcal{P}(\mathcal{M})$ may have empty interior, but still be uncountable.

See Examples 10.4.1 and 10.4.2 for explicit constructions.

When $d = 1$ the existence of a cohomology relation between the functions φ and ψ immediately implies that the domain of the spectrum $\mathcal{F}_u$ consists of a single point. This is not the case for multidimensional spectra, as the following example illustrates.

Example 10.4.1. Consider a topological Markov chain $f|X$ and set $d = 2$. Let $\varphi, \psi \in C_f(X)$ be such that φ, ψ, and 1 are linearly independent as cohomology classes. We assume that $\int_X \varphi \, d\mu = 0$ for some measure $\mu \in \mathcal{M}$, and that $\psi > 0$.

Setting

$$\varphi_1 = \varphi, \quad \varphi_2 = \varphi, \quad \psi_1 = 1, \quad \text{and} \quad \psi_2 = \psi$$

we obtain $0 \in \mathcal{P}(\mathcal{M})$ (since $\int_X \varphi \, d\mu = 0$), and

$$(\varphi_1 - 0 \cdot \psi_1) - (\varphi_2 - 0 \cdot \psi_2) = 0.$$

On the other hand, it is easy to verify that $\varphi_1 - \alpha_1 \psi_1$ and $\psi_2 - \alpha_2 \psi_2$ are linearly independent as cohomology classes whenever $\alpha = (\alpha_1, \alpha_2) \neq 0$, and hence $\mathcal{P}^* := \mathcal{P}(\mathcal{M}) \setminus \{0\}$ is nonempty, by Theorem 10.2.1. Moreover, it follows from Theorem 10.2.1 that $\mathcal{P}^* \subset \text{int}\, \mathcal{P}(\mathcal{M})$. Since $\mathcal{P}(\mathcal{M})$ is closed, we conclude that $\mathcal{P}(\mathcal{M}) = \text{int}\, \mathcal{P}(\mathcal{M})$. This shows that even though there exists a cohomology relation, the set $\mathcal{P}(\mathcal{M})$ is uncountable. Furthermore, it has nonempty interior.

The first drawing in Figure 10.1 is an explicit example of a set $\mathcal{P}(\mathcal{M})$ when f is the Bernoulli shift on three symbols. Namely, we took the functions

$$\varphi = \chi_1 - \chi_2 \quad \text{and} \quad \psi = \chi_1 + \chi_2 + 2\chi_3,$$

where χ_i is the characteristic function of the cylinder set C_i of length 1. Observe that in this particular case int $\mathcal{P}(\mathcal{M})$ has two connected components. Furthermore, the set $\mathcal{P}(\mathcal{M})$ is not convex, but each connected component of int $\mathcal{P}(\mathcal{M})$ is still convex.

The second drawing in Figure 10.1 is also obtained from the Bernoulli shift on three symbols, now with the functions

$$\varphi_1 = -4\chi_1 + 4\chi_2 + 8\chi_3 \quad \text{and} \quad \varphi_2 = -6\chi_1 - 3\chi_2 + 5\chi_3,$$

$$\psi_1 = 2\chi_1 + 9\chi_2 + 2\chi_3 \quad \text{and} \quad \psi_2 = 6\chi_1 + \chi_2 + 2\chi_3.$$

Again the set int $\mathcal{P}(\mathcal{M})$ has two connected components, but one of them is not convex. □

Now we illustrate that $\mathcal{P}(\mathcal{M})$ may have empty interior, but still be uncountable.

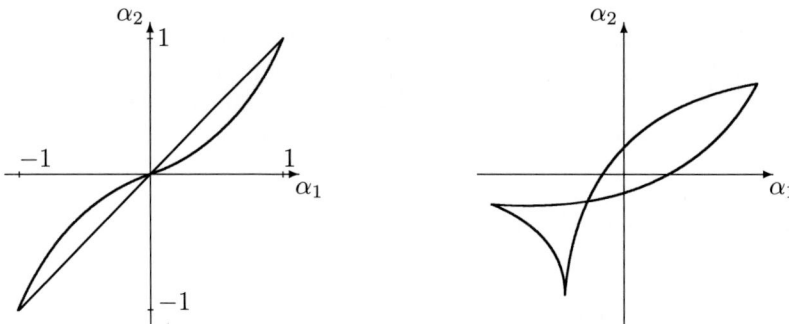

Figure 10.1: Two examples of the set $\mathcal{P}(\mathcal{M})$ for which the interior has two connected components, due to the presence of a cohomology relation. The curves represent the boundary of $\mathcal{P}(\mathcal{M})$.

Example 10.4.2. Consider the Bernoulli shift on two symbols and set $d = 2$. In a similar manner to that in Example 10.4.1 we define functions

$$\varphi_1 = a_1\chi_1 + b_1\chi_2, \quad \varphi_2 = a_2\chi_1 + b_2\chi_2, \quad \text{and} \quad \psi_1 = \psi_2 = u = 1.$$

We assume that $a_1b_2 - b_1a_2 = 1$. The case when $a_1b_2 - b_1a_2 \neq 1$ can be treated in a similar manner. Observe that

$$b_2\varphi_1 - b_1\varphi_2 = \chi_1 \quad \text{and} \quad a_1\varphi_2 - a_2\varphi_1 = \chi_2.$$

Since $\chi_1 + \chi_2 = 1$ we obtain (see (8.2))

$$K_{(\alpha_1,\alpha_2)} = K_{b_2\alpha_1 - b_1\alpha_2}(\chi_1) = K_{a_1\alpha_2 - a_2\alpha_1}(\chi_2),$$

where we must have
$$b_2\alpha_1 - b_1\alpha_2 + a_1\alpha_2 - a_2\alpha_1 = 1. \tag{10.11}$$

It follows from Theorem 10.1.4 and (10.11) that

$$
\begin{aligned}
h(f|K_{(\alpha_1,\alpha_2)}) &= \sup\{h_\mu(f) : \mu(C_1) = b_2\alpha_1 - b_1\alpha_2\} \\
&= -(b_2\alpha_1 - b_1\alpha_2)\log(b_2\alpha_1 - b_1\alpha_2) \\
&\quad - (a_1\alpha_2 - a_2\alpha_1)\log(a_1\alpha_2 - a_2\alpha_1).
\end{aligned}
$$

Furthermore, the domain of the spectrum $(\alpha_1, \alpha_2) \mapsto h(f|K_{(\alpha_1,\alpha_2)})$ is a segment contained in the straight line defined by (10.11). □

We remark that in some sense the situation described in Example 10.4.2 should be considered degenerate. Indeed, Theorem 10.2.1 implies that the degeneracy in Example 10.4.2, with the domain of the spectrum contained in a straight line, is due to the presence of cohomology relations. When this happens we can replace the $2d$ components of the vectors Φ and Ψ by a maximal set of functions that are independent as cohomology classes, without changing the level sets (up to a change of variables), and in such a way that the domain of the spectrum with respect to the new functions has nonempty interior.

10.5   Topological Markov chains

We show in this section that for topological Markov chains the spectrum $\mathcal{F}_u$ is often analytic and nondegenerate, that is, $\mathcal{F}_u(\alpha) = 0$ for every $\alpha \in \partial \mathcal{P}(\mathcal{M})$.

Theorem 10.5.1 ([20]). *Let $f|X$ be a topologically mixing topological Markov chain. If the functions (Φ, Ψ) and u are Hölder continuous, then the spectrum $\mathcal{F}_u$ is analytic in* int $\mathcal{P}(\mathcal{M})$.

Proof. We consider the functions

$$G = \Phi - \alpha * \Psi \quad \text{and} \quad F(q) = P(\langle q, G \rangle).$$

Ruelle's formula for the second derivative of the topological pressure (see (6.24) and (6.25)) shows that for every $p \in \mathbb{R}^d$ we have

$$d_q^2 F(p, p) = \int_X \langle p, G \rangle^2 \, d\mu_q + 2 \sum_{n=1}^{\infty} \int_X \langle p, G \rangle \cdot \langle p, G \circ f^n \rangle \, d\mu_q \geq 0,$$

where μ_q is the unique equilibrium measure of the function $\langle q, G \rangle$.
     We prove that $d_q^2 F(p, p) > 0$ whenever $p \neq 0$. Assume on the contrary that $d_q^2 F(p, p)$ is zero. Then the function $\langle p, G \rangle$ must be cohomologous to a constant, say c. Since $\alpha \in \mathcal{P}(\mathcal{M})$ there exists a measure $\mu \in \mathcal{M}$ such that $\int_X G \, d\mu = 0$. This implies that $c = 0$. Since $\langle p, G \rangle$ is cohomologous to zero we conclude that $t \mapsto P(t\langle p, G \rangle)$ is a constant function. By Theorem 10.2.1 this never happens when $\alpha \in$ int $\mathcal{P}(\mathcal{M})$, and thus $d_q^2 F(p, p) > 0$. The analyticity of the spectrum follows now immediately from Theorem 10.3.1. $\square$

The following result shows that typically the spectrum $\mathcal{F}_u$ is nondegenerate. We continue to denote by $C^\theta(X)$ the space of Hölder continuous functions $\varphi \colon X \to \mathbb{R}$ with Hölder exponent θ, equipped with the norm in (7.19).

Theorem 10.5.2 ([20]). *Let $f|X$ be a topologically mixing topological Markov chain. There exists a residual subset $\Theta \subset C^\theta(X)^d \times C^\theta(X)^d$ such that if $(\Phi, \Psi) \in \Theta$ and $u \in C^\theta(X)$ with $u > 0$, then:*

1. $\mathcal{P}(\mathcal{M}) = \overline{\text{int } \mathcal{P}(\mathcal{M})}$;

2. $\mathcal{F}_u(\alpha) = 0$ *for every $\alpha \in \partial \mathcal{P}(\mathcal{M})$.*

Proof. We need an auxiliary statement.

Lemma 10.5.3. *The set of vectors $(\varphi_1, \ldots, \varphi_d) \in C^\theta(X)^d$ such that $\varphi_1, \ldots, \varphi_d$ are linearly independent as cohomology classes is residual in $C^\theta(X)^d$.*

Proof of the lemma. Consider d distinct periodic points $x_1, \ldots, x_d$ respectively with periods $n_1, \ldots, n_d$, and set

$$S_{ij} = \frac{1}{n_i} \sum_{k=0}^{n_i-1} \varphi_j(f^k x_i).$$

By Livschitz's theorem (see, for example, [84, Theorem 19.2.1]), if the $d \times d$ matrix S with entries S_{ij} has rank d, then the functions $\varphi_1, \ldots, \varphi_d$ are linearly independent as cohomology classes. The desired statement follows now from the fact that the condition rank $S = d$ is generic. □

Set
$$\underline{u} = \min\{u(x) : x \in X\} \quad \text{and} \quad \overline{u} = \max\{u(x) : x \in X\}.$$

It follows from the definition of $N(Z, \alpha, u, \mathcal{U})$ in (7.5) that

$$N(Z, \alpha\overline{u}, 1, \mathcal{U}) \leq N(Z, \alpha, u, \mathcal{U}) \leq N(Z, \alpha\underline{u}, 1, \mathcal{U}).$$

Therefore,

$$\frac{\dim_{1,\mathfrak{u}} Z}{\overline{u}} \leq \dim_{u,\mathfrak{u}} Z \leq \frac{\dim_{1,\mathfrak{u}} Z}{\underline{u}},$$

and hence

$$h(f|Z)/\overline{u} \leq \dim_u Z \leq h(f|Z)/\underline{u}. \tag{10.12}$$

This shows that $\mathcal{F}_1(\alpha) = 0$ if and only if $\mathcal{F}_u(\alpha) = 0$. Hence, it is sufficient to prove that the topological entropy vanishes at the boundary of $\mathcal{P}(\mathcal{M})$.

We first reduce the problem to a one-dimensional problem. Set $H_\theta = C^\theta(X)^d$. For each $q, \alpha \in \mathbb{R}^d$ and $(\Phi, \Psi) \in H_\theta \times H_\theta$ we consider the function

$$\chi_{\Phi\Psi} = \langle q, \Phi - \alpha * \Psi \rangle.$$

It is shown in the proof of Lemma 7.6.5 (see (7.27)) that there exists an open and dense subset $\Theta^\varepsilon \subset C^\theta(X)$ such that if $\chi \in \Theta^\varepsilon$, then

$$h\left(f|\left\{x \in X : \lim_{n\to\infty} \frac{1}{n} \sum_{k=0}^{n-1} \chi(f^n x) \in \{\underline{\beta}, \overline{\beta}\}\right\}\right) < \varepsilon, \tag{10.13}$$

where

$$\underline{\beta} = \inf\left\{\int_X \chi \, d\mu : \mu \in \mathcal{M}\right\} \quad \text{and} \quad \overline{\beta} = \sup\left\{\int_X \chi \, d\mu : \mu \in \mathcal{M}\right\}.$$

Therefore, for each fixed $q, \alpha \in \mathbb{R}^d$ the set

$$\Theta^\varepsilon_{q\alpha} = \{(\Phi, \Psi) \in H_\theta \times H_\theta : \chi_{\Phi\Psi} \in \Theta^\varepsilon\}$$

is open and dense in $H_\theta \times H_\theta$.

It follows from Lemma 10.5.3 that by changing $\Theta^\varepsilon_{q\alpha}$ slightly while still leaving it open and dense we may assume that there is no cohomology relation between the functions $\varphi_i - \alpha_i \psi_i$, $i = 1, \ldots, d$. This implies that the vector (q, α) has an open neighborhood $U(q, \alpha) \subset \mathbb{R}^d \times \mathbb{R}^d$ with the property that there is an open and dense subset $\Theta^\varepsilon_{q\alpha} \subset H_\theta \times H_\theta$ without cohomology relations such that $(\Phi, \Psi) \in \Theta^\varepsilon_{q'\alpha'}$ for every $(q', \alpha') \in U(q, \alpha)$. Now we choose a sequence (q_n, α_n) such that $\bigcup_{n=1}^\infty U(q_n, \alpha_n) = \mathbb{R}^d \times \mathbb{R}^d$, and we set

$$\Theta = \bigcap_{m=1}^\infty \bigcup_{n=1}^\infty \Theta^{1/m}_{q_n \alpha_n}. \tag{10.14}$$

By construction the set Θ is residual, and for each $(\Phi, \Psi) \in \Theta$ there are no cohomology relations (see Theorem 10.2.1 and the discussion after this theorem). This establishes the first statement in the theorem.

By Theorem 10.2.2, for each $(\Phi, \Psi) \in \Theta$ the boundary points of $\mathcal{P}(\mathcal{M})$ are contained in $\bigcup_{q \in S^{2d-1}} \Gamma(q)$. By construction of the set Θ (see (10.13) and (10.14)) the spectrum $\mathcal{F}_1$ vanishes at these points (and by (10.12) the same happens for $\mathcal{F}_u$). This completes the proof of the theorem. $\qquad\square$

When $d = 1$ the statement in Theorem 10.5.2 was established by Schmeling in [142] (see Theorem 7.6.1). We note that when $d > 1$ the set $\partial\mathcal{P}(\mathcal{M})$ may be uncountable (see Examples 10.4.1 and 10.4.2), and by property 1 in Theorem 10.5.2 this is the generic situation.

10.6   Finer structure of the spectrum

Now we have a closer look at the finer structure of the level sets K_α in (10.3). In particular, we show that the u-dimension of K_α is entirely carried by a certain level set of another set of functions, which is strictly inside K_α.

Let $f|X$ be a topologically mixing topological Markov chain. Given functions $(\Phi, \Psi) \in C^\theta(X)^d \times C^\theta(X)^d$, we define $\mathcal{P}(\mu)$ and $\mathcal{F}_u(\alpha)$ as in (10.4) and (10.5). For each $(p, q) \in \mathbb{R}^d \times \mathbb{R}^d$, we consider the unique real number $T(p, q)$ such that

$$P(\langle p, \Phi \rangle + \langle q, \Psi \rangle - T(p, q)u) = 0,$$

and we denote by $\mu_{p,q}$ the equilibrium measure of the function

$$\langle p, \Phi \rangle + \langle q, \Psi \rangle - T(p, q)u.$$

Set

$$\beta(p, q) := \nabla_p T(p, q) \quad \text{and} \quad \gamma(p, q) := \nabla_q T(p, q).$$

For each $(\beta, \gamma) \in \mathbb{R}^d \times \mathbb{R}^d$, we consider the set $K_{\beta,\gamma}$ of points $x \in X$ such that

$$\lim_{n \to \infty} \frac{\sum_{k=0}^n \varphi_i(f^k x)}{\sum_{k=0}^n u(f^k x)} = \beta_i \quad \text{and} \quad \lim_{n \to \infty} \frac{\sum_{k=0}^n \psi_i(f^k x)}{\sum_{k=0}^n u(f^k x)} = \gamma_i.$$

for every $i = 1, \ldots, d$, where $\beta = (\beta_1, \ldots, \beta_d)$ and $\gamma = (\gamma_1, \ldots, \gamma_d)$. Now we establish the relation between the d-dimensional spectrum $\mathcal{F}_u$ and the $2d$-dimensional spectrum

$$\mathcal{H}_u(\beta, \gamma) = \dim_u K_{\beta, \gamma}.$$

Theorem 10.6.1 ([20]). *Let $f|X$ be a topologically mixing topological Markov chain. If the functions (Φ, Ψ) and u are Hölder continuous, then the following properties hold:*

1. $\mu_{p,q}(K_{\beta(p,q),\gamma(p,q)}) = 1$ *and*

$$\mathcal{H}_u(\beta(p,q), \gamma(p,q)) = \dim_u \mu_{p,q}$$
$$= T(p,q) - \langle p, \beta(p,q) \rangle - \langle q, \gamma(p,q) \rangle;$$

2. *if $\alpha \in \operatorname{int} \mathcal{P}(\mathcal{M})$, then there exists $\gamma \in \mathbb{R}^d$ such that*

$$\mathcal{F}_u(\alpha) = \mathcal{H}_u(\alpha * \gamma, \gamma).$$

Proof. Applying Theorem 10.1.4 with $\Psi = (1, \ldots, 1)$, we obtain

$$\mu_{p,q}(K_{\beta(p,q),\gamma(p,q)}) = 1 \quad \text{and} \quad \mathcal{H}_u(\beta(p,q), \gamma(p,q)) = \dim_u \mu_{p,q}.$$

Furthermore,

$$\beta(p,q) = \frac{\int_X \Phi \, d\mu_{p,q}}{\int_X u \, d\mu_{p,q}} \quad \text{and} \quad \gamma(p,q) = \frac{\int_X \Psi \, d\mu_{p,q}}{\int_X u \, d\mu_{p,q}}.$$

We obtain

$$\dim_u \mu_{p,q} = \frac{h_{\mu_{p,q}}(f)}{\int_X u \, d\mu_{p,q}}$$
$$= \frac{-\int_X (\langle p, \Phi \rangle + \langle q, \Psi \rangle - T(p,q)u) \, d\mu_{p,q}}{\int_X u \, d\mu_{p,q}}$$
$$= T(p,q) - \langle p, \beta(p,q) \rangle - \langle q, \gamma(p,q) \rangle.$$

This completes the proof of the first statement in the theorem.

Now we establish the second statement. For each $\alpha \in \operatorname{int} \mathcal{P}(\mathcal{M})$, let μ_α be the measure in Theorem 10.1.4. For μ_α-almost every $x \in X$ there exist the limits

$$\lim_{n \to \infty} \frac{\sum_{k=0}^{n} \Phi(f^k x)}{\sum_{k=0}^{n} u(f^k x)} = \beta(\alpha) = \frac{\int_X \Phi \, d\mu_\alpha}{\int_X u \, d\mu_\alpha}$$

and

$$\lim_{n \to \infty} \frac{\sum_{k=0}^{n} \Psi(f^k x)}{\sum_{k=0}^{n} u(f^k x)} = \gamma(\alpha) = \frac{\int_X \Psi \, d\mu_\alpha}{\int_X u \, d\mu_\alpha},$$

and $\beta(\alpha) = \alpha * \gamma(\alpha)$. Therefore,

$$\mu_\alpha(K_{\beta(\alpha),\gamma(\alpha)}) = \mu_\alpha(K_{\alpha*\gamma(\alpha),\gamma(\alpha)}) = 1.$$

This implies that

$$\dim_u K_{\alpha*\gamma(\alpha),\gamma(\alpha)} \geq \dim_u \mu_\alpha = \dim_u K_\alpha.$$

On the other hand,

$$K_{\alpha*\gamma,\gamma} \subset K_\alpha \quad \text{for every} \quad \gamma \in \mathbb{R}^d, \tag{10.15}$$

and thus

$$\dim_u K_{\alpha*\gamma,\gamma} \leq \dim_u K_\alpha.$$

We conclude that

$$\dim_u K_\alpha = \sup\{\dim_u K_{\alpha*\gamma,\gamma} : \gamma \in \mathbb{R}^d\} = \dim_u K_{\alpha*\gamma(\alpha),\gamma(\alpha)}.$$

This completes the proof of the theorem. □

We observe that $\mathcal{H}_u$ is the Legendre transform of the function T.

By (10.15), the second statement in Theorem 10.6.1 says that the u-dimension of the set K_α is entirely carried by some subset $K_{\alpha*\gamma,\gamma}$ of K_α. In particular, the spectrum $\mathcal{F}_u$ can be obtained from the spectrum $\mathcal{H}_u$ by

$$\mathcal{F}_u(\alpha) = \max\{\mathcal{H}_u(\alpha * \gamma, \gamma) : \gamma \in \mathbb{R}^d\}$$

for each $\alpha \in \operatorname{int} \mathcal{P}(\mathcal{M})$. This conclusion is particularly remarkable since the inclusion $\bigcup_\gamma K_{\alpha*\gamma,\gamma} \subset K_\alpha$ may be proper, and since the u-dimension of an uncountable union $\bigcup_\gamma I_\gamma$ may be strictly larger than $\sup_\gamma \dim_u I_\gamma$.

10.7   Applications to number theory

The multifractal analysis of multidimensional spectra has several applications to number theory. These were described by Barreira, Saussol and Schmeling in [19]. Instead of formulating general statements we describe explicit examples that illustrate well the nature of the results.

Given $m \in \mathbb{N}$, for each $x \in [0,1]$ we denote by $x = 0.x_1x_2\cdots$ the base-m representation of x. Note that the representation is unique except for countably many points. Since countable sets have zero Hausdorff dimension, the nonuniqueness of the representation does not affect the study of the dimensional properties.

For each $k \in \{0, \ldots, m-1\}$, $x \in [0,1]$, and $n \in \mathbb{N}$ we define

$$\tau_k(x,n) = \operatorname{card}\{i \in \{1,\ldots,n\} : x_i = k\}.$$

When the limit

$$\tau_k(x) = \lim_{n \to \infty} \frac{\tau_k(x, n)}{n}$$

exists we call it the *frequency* of the number k in the base-m representation of x. A classical result of Borel in [33] says that for Lebesgue-almost every $x \in [0, 1]$ we have

$$\tau_k(x) = 1/m \quad \text{for} \quad k = 0, \ldots, m - 1. \tag{10.16}$$

Furthermore, for $m = 2$, Hardy and Littlewood showed in [70] that for Lebesgue-almost every $x \in [0, 1]$, $k = 0, 1$, and all sufficiently large $n \in \mathbb{N}$ we have

$$\left| \frac{\tau_k(x, n)}{n} - \frac{1}{2} \right| < \sqrt{\frac{\log n}{n}}.$$

Of course, these remarkable results do not imply that the set of numbers in $[0, 1]$ for which (10.16) fails to hold is empty. In particular, consider the set

$$F_m(\alpha_0, \ldots, \alpha_{m-1}) = \{x \in [0, 1] : \tau_k(x) = \alpha_k \text{ for } k = 0, \ldots, m - 1\},$$

with $\alpha_i \in [0, 1]$ for each i, and $\alpha_0 + \cdots + \alpha_{m-1} = 1$. It was shown by Eggleston in [50] that

$$\dim_H F_m(\alpha_0, \ldots, \alpha_{m-1}) = -\sum_{k=0}^{m-1} \alpha_k \log_m \alpha_k. \tag{10.17}$$

This statement can also be obtained as a application of Theorem 10.6.1 (see Proposition 10.7.3 below). A former result concerning the dimension of these sets is due to Besicovitch [28]. For $m = 2$ he showed that if $\alpha \in (0, 1/2)$, then

$$\dim_H \left\{ x \in [0, 1] : \limsup_{n \to \infty} \frac{\tau_1(x, n)}{n} \leq \alpha \right\} = -\frac{\alpha \log \alpha + (1 - \alpha) \log(1 - \alpha)}{\log 2}.$$

Identity (10.17) follows from Theorem 6.1.2 when $m = 2$, from Theorem 10.1.1 when $m = 3$, and from Theorem 10.1.4 when $m \geq 4$ (see Theorem 10.7.2 and Proposition 10.7.3). This provides a new proof of Eggleston's result

Now we consider sets of more complicated nature. Let $m = 3$ and define

$$F = \{x \in [0, 1] : \tau_1(x) = 5\tau_0(x)\}. \tag{10.18}$$

We emphasize that the frequency of the number 2 is arbitrary. The following result is a consequence of work of Barreira, Saussol and Schmeling in [19].

Proposition 10.7.1. *If F is given by (10.18), then*

$$\dim_H F = \frac{\log(1 + 6/5^{5/6})}{\log 3}. \tag{10.19}$$

In order to explain how this result is obtained, and its relation to the multi-fractal analysis of multidimensional spectra, we first observe that

$$F = \bigcup_{\alpha \in [0,1/6]} F_3(\alpha, 5\alpha, 1 - 6\alpha). \tag{10.20}$$

We show that the constant in (10.19) is a lower bound for $\dim_H F$. It follows from (10.17) and (10.20) that, since $F \supset F_3(\alpha, 5\alpha, 1 - 6\alpha)$ for every $\alpha \in [0, 1/6]$,

$$\dim_H F \geq \max_{\alpha \in [0,1/6]} -\frac{\alpha \log \alpha + 5\alpha \log(5\alpha) + (1 - 6\alpha) \log(1 - 6\alpha)}{\log 3}. \tag{10.21}$$

The maximum in (10.21) is attained at $\alpha = 1/(5^{5/6} + 6)$, and it is easy to verify that it is equal to the constant in (10.19). This establishes a lower bound for the Hausdorff dimension.

The corresponding upper bound is more delicate, essentially because the union in (10.20) is composed of an *uncountable* number of pairwise disjoint sets. Consider the map $g_m \colon [0,1] \to [0,1]$ defined by $g_m x = mx \pmod 1$. We observe that if $0.x_1 x_2 \cdots$ is a base-m representation of $x \in [0,1]$, then $g_m x = 0.x_2 x_3 \cdots$. Now consider a function $\varphi \colon [0,1] \to \mathbb{R}$ such that

$$\varphi(0.x_1 x_2 \cdots) = a_{x_1 \cdots x_\kappa}$$

for some constants $a_{i_1 \cdots i_\kappa} \in \mathbb{R}$ for $i_1, \ldots, i_\kappa \in \{0, \ldots, m-1\}$, and some fixed $\kappa \in \mathbb{N}$. The function φ is called a *κ-locally constant function*. It follows easily from (3.12) that if φ is a 1-locally constant function, then

$$P(\varphi) = \lim_{n \to \infty} \frac{1}{n} \log \sum_{i_1 \cdots i_n} \prod_{j=1}^{n} \exp a_{i_j} = \log \sum_{k=0}^{m-1} \exp a_k. \tag{10.22}$$

Taking $\psi_i = 1$ for $i = 1, \ldots, d$ in Theorem 10.1.4, for each $\alpha = (\alpha_1, \ldots, \alpha_d) \in \mathbb{R}^d$ the set K_α in (10.3) is given by

$$K_\alpha = \left\{ x \in [0,1] : \lim_{n \to \infty} \sum_{j=0}^{n} \varphi_i(g_m^j x) = \alpha_i \text{ for } i = 1, \ldots, d \right\}.$$

Let also $\mathcal{M}$ be the family of g_m-invariant probability Borel measures in $[0,1]$. For each $\mu \in \mathcal{M}$ set

$$\mathcal{Q}(\mu) = \left(\int_0^1 \varphi_1 \, d\mu, \ldots, \int_0^1 \varphi_d \, d\mu \right).$$

Theorem 10.7.2 ([19]). *Let $\varphi_i \colon [0,1] \to \mathbb{R}$ be 1-locally constant functions for $i = 1, \ldots, d$. If $\alpha = (\alpha_1, \ldots, \alpha_d) \in \operatorname{int} \mathcal{Q}(\mathcal{M})$, then*

$$\dim_H K_\alpha = \inf \left\{ \log_m \sum_{k=0}^{m-1} \exp \sum_{i=1}^{d} q_i(\varphi_{ik} - \alpha_i) : (q_1, \ldots, q_d) \in \mathbb{R}^d \right\},$$

where $\varphi_{ik} = \varphi_i([k/m, (k+1)/m))$.

Proof. By (10.22) we have

$$P\left(\sum_{k=1}^{d} q_k(\varphi_k - \alpha_k \psi_k)\right) = \log \sum_{k=0}^{m-1} \exp \sum_{i=1}^{d} q_i(\varphi_{ik} - \alpha_i).$$

Applying Theorem 10.1.4 we obtain the desired statement.   □

A first consequence of Theorem 10.7.2 is the result of Eggleston in [50].

Proposition 10.7.3. *For each* $(\alpha_0, \ldots, \alpha_{m-1}) \in (0,1)^m$ *with* $\sum_{k=0}^{m-1} \alpha_i = 1$ *the identity in* (10.17) *holds.*

Proof. Set $d = m$ and $\varphi_i = \chi_{[i/m,(i+1)/m]}$ for $i = 1, \ldots, d$. For each $\alpha_0, \ldots, \alpha_{m-1}$ as in the proposition, we have $K_\alpha = F_m(\alpha_0, \ldots, \alpha_{m-1})$, and by Theorem 10.7.2,

$$\dim_H F_m(\alpha_0, \ldots, \alpha_{m-1}) = \inf_{q \in \mathbb{R}^d} \log_m \sum_{k=0}^{m-1} e^{q_{k+1} - \sum_{i=1}^{d} q_i \alpha_{i-1}}.$$

Computing the gradient of the logarithm we obtain

$$\frac{1}{\log m} \nabla \left(\log \sum_{k=0}^{m-1} e^{q_{k+1}} - \sum_{i=1}^{d} q_i \alpha_{i-1} \right)$$

$$= \frac{1}{\log m} \left(\frac{e^{q_1}}{\sum_{k=0}^{m-1} e^{q_{k+1}}} - \alpha_0, \ldots, \frac{e^{q_d}}{\sum_{k=0}^{m-1} e^{q_{k+1}}} - \alpha_{d-1} \right).$$

This is the zero vector provided that

$$\frac{e^{q_1}}{\alpha_0} = \cdots = \frac{e^{q_d}}{\alpha_{d-1}} = \sum_{k=0}^{m-1} e^{q_{k+1}},$$

which implies that there exists $c > 0$ such that $e^{q_k} = \alpha_{k-1} c$ for every k. Therefore,

$$\dim_H F_m = \log_m \sum_{k=0}^{m-1} \alpha_k c - \frac{1}{\log m} \sum_{k=0}^{m-1} \log(\alpha_k c) \alpha_k$$

$$= -\frac{1}{\log m} \sum_{k=0}^{m-1} \alpha_k \log \alpha_k,$$

which is the desired result.   □

To establish (10.19) we need the following result.

Proposition 10.7.4. *For each* $k \neq \ell$ *and* $\beta \geq 0$ *we have*

$$\dim_H \{x \in [0,1] : \tau_k(x) = \beta \tau_\ell(x)\} = \frac{\log(m - 2 + (\beta+1)/\beta^{\beta/(\beta+1)})}{\log m}.$$

Proof. Set $d = 1$ and

$$\varphi_1 = \chi_{[k/m,(k+1)/m)} - \beta \chi_{[\ell/m,(\ell+1)/m)}.$$

Since $0 \in \text{int } \mathfrak{Q}(\mathcal{M}) = (-\beta, 1)$, it follows from Theorem 10.7.2 that

$$\dim_H\{x \in [0,1] : \tau_k(x) = \beta \tau_\ell(x)\} = \inf_{q \in \mathbb{R}} \frac{\log(e^q + e^{-\beta q} + m - 2)}{\log m}.$$

It is easy to verify that the infimum is attained at $q = \log \beta/(1 + \beta)$, and a straightforward computation yields the desired result. $\qquad\square$

By Propositions 10.7.3 and 10.7.4, for each $k \neq \ell$ and $\beta \geq 0$ we have

$$\dim_H\left\{x \in [0,1] : \frac{\tau_k(x)}{\tau_\ell(x)} = \beta\right\} = \max\left\{-\sum_{j=0}^{m-1} \alpha_j \log_m \alpha_j : \frac{\alpha_k}{\alpha_\ell} = \beta\right\}.$$

Therefore, there exists a set

$$F_m(\alpha_0, \ldots, \alpha_{m-1}) \subset \left\{x \in [0,1] : \frac{\tau_k(x)}{\tau_\ell(x)} = \beta\right\}$$

with

$$\dim_H F_m(\alpha_0, \ldots, \alpha_{m-1}) = \dim_H \left\{x \in [0,1] : \frac{\tau_k(x)}{\tau_\ell(x)} = \beta\right\}.$$

In particular, letting $m = 3$, $k = 1$, $\ell = 0$, and $\beta = 5$ we conclude from (10.20) that the inequality in (10.21) is in fact an identity, and we establish (10.19).

Chapter 11

Multifractal Rigidity

We consider in this chapter the phenomenon of multifractal rigidity. Roughly speaking, it states that if two dynamical systems are topologically equivalent and some of their multifractal spectra coincide, then the original data must be equivalent (in some sense to be made precise). This leads to a "multifractal classification" of hyperbolic maps (either invertible or noninvertible) in terms of the multifractal spectra. Furthermore, the theory of multifractal analysis has a privileged relation with the experimental study of dynamical systems. In particular, the so-called multifractal spectra, that are obtained from the study of the complexity of the level sets, can be determined experimentally with arbitrary precision. On the other hand, we may be able to recover information about a dynamical system from the information contained in its multifractal spectra. Unfortunately, in general, when we use a single spectrum there is no multifractal rigidity even for topological Markov chains on three symbols.

11.1 Multifractal classification of dynamical systems

We illustrate the multifractal rigidity phenomenon with a model class of repellers. Let f and g be piecewise linear expanding maps of the interval $[0,1]$ with repellers given by

$$J_f = \bigcap_{n=0}^{\infty} f^{-n}(A_f \cup B_f) \quad \text{and} \quad J_g = \bigcap_{n=0}^{\infty} g^{-n}(A_g \cup B_g),$$

where A_f, B_f, A_g and B_g are closed intervals in $[0,1]$ such that

$$f(A_f) = f(B_f) = g(A_g) = g(B_g) = [0,1] \quad \text{and} \quad A_f \cap B_f = A_g \cap B_g = \varnothing.$$

See Figure 11.1. Both repellers can be coded by the Bernoulli shift on two symbols. We also consider Bernoulli measures μ_f and μ_g (on two symbols) that are invariant respectively under f and g.

The following result of Barreira, Pesin and Schmeling in [9] gives a multi-fractal classification in terms of the spectrum $\mathcal{D}$ (see Definition 6.1.1).

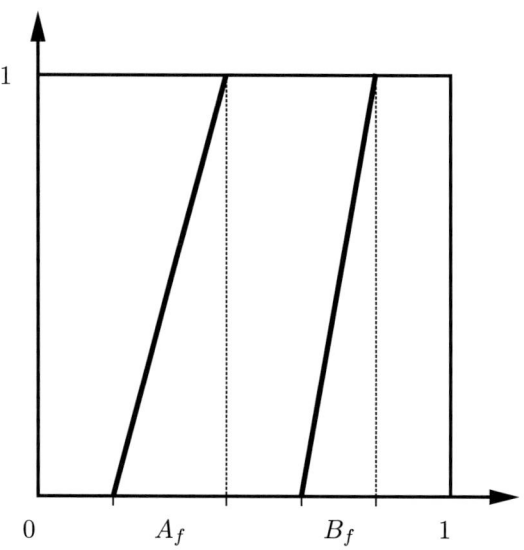

Figure 11.1: Piecewise linear expanding map f

Theorem 11.1.1 (Multifractal rigidity). *If $\mathcal{D}_{\mu_f} = \mathcal{D}_{\mu_g}$ and the spectrum is not a delta function, then there exists a homeomorphism $\chi \colon J_f \to J_g$ such that $f' = g' \circ \chi$ and $\mu_f = \mu_g \circ \chi$.*

Proof. Set $c_1 = f'|A_f$ and $c_2 = f'|B_f$. The Bernoulli measure $\mu_f = \mu(\beta_1, \beta_2)$, $\beta_1 + \beta_2 = 1$ is the equilibrium measure of the function $\psi \colon J_f \to \mathbb{R}$ defined by

$$\psi|A_f = \log \beta_1 \quad \text{and} \quad \psi|B_f = \log \beta_2.$$

As in (6.3), we define a function $T \colon \mathbb{R} \to \mathbb{R}$ by requiring that

$$P(-T(q) \log \|df\| + q\psi) = 0 \tag{11.1}$$

for every $q \in \mathbb{R}$. Proceeding as in (3.16) we show that identity (11.1) can be written in the form

$$c_1^{-T(q)} \beta_1{}^q + c_2^{-T(q)} \beta_2{}^q = 1. \tag{11.2}$$

By statement 4 in Theorem 6.1.2, the spectrum $\mathcal{D}_{\mu_f}$ is the Legendre transform of T. By the uniqueness of the Legendre transform, the spectrum $\mathcal{D}_{\mu_f}$ uniquely determines $T(q)$ for every $q \in \mathbb{R}$, and hence, it is sufficient to show that equation (11.2) uniquely determines the numbers β_1, β_2, c_1, and c_2 up to a permutation of the indices 1 and 2.

Set $\alpha(q) = -T'(q)$ and $\alpha_\pm = \lim_{q\to\pm\infty}\alpha(q)$. We can easily verify that

$$\alpha_\pm = -\lim_{q\to\pm\infty}(T(q)/q).$$

Since the spectrum $\mathcal{D}_{\mu_f}$ is not a delta function, the function T is strictly convex, and $\alpha_+ < \dim_H J_f < \alpha_-$. Therefore, raising both sides of (11.2) to the power $1/q$ and letting $q \to \pm\infty$, we obtain

$$\max\{\beta_1 c_1^{\alpha_+}, \beta_2 c_2^{\alpha_+}\} = \min\{\beta_1 c_1^{\alpha_-}, \beta_2 c_2^{\alpha_-}\} = 1.$$

We assume that $\beta_1 c_1^{\alpha_+} = 1$ (the case when $\beta_2 c_2^{\alpha_+} = 1$ can be treated in a similar manner). Since $\alpha_+ < \alpha_-$, we must have $\beta_2 c_2^{\alpha_-} = 1$.

Setting $q = 0$ in equation (11.2) we obtain $c_1^{-s} + c_2^{-s} = 1$, where $s = \dim_H J_f$. Furthermore, setting $x = c_1^{-s}$, $a = \alpha_+/s < 1$, and $b = \alpha_-/s > 1$, we can show that $x^a + (1-x)^b = 1$. We can easily verify that this equation has a unique solution $x \in (0,1)$, which uniquely determines the numbers c_1 and c_2, and hence, also the numbers β_1 and β_2. $\qquad\square$

See [10] for a version of Theorem 11.1.1 in the case of hyperbolic sets.

11.2 Entropy spectrum and topological Markov chains

We consider in this section the entropy spectrum $\mathcal{E} = \mathcal{E}_E$ of an invariant measure (see Section 7.1), and we study in a systematic manner its multifractal rigidity properties in the case of topological Markov chains. We follow closely Barreira and Saraiva in [13]. In a certain sense, the entropy spectrum is the simplest possible spectrum. In particular, it contains no information about distances, in strong contrast with the dimension spectrum $\mathcal{D}$.

11.2.1 Equivalence classes of functions

Given a Hölder continuous function φ, we would like to recover as much information as possible about the function from the entropy spectrum of its equilibrium measure. Clearly, one cannot expect to recover completely the function φ. For example, for each $c \in \mathbb{R}$ the functions φ and $\varphi + c$ have the same equilibrium measure. We consider instead certain equivalence classes of functions.

Let $\sigma|\Sigma_A^+$ be a topological Markov chain. We denote by $\mathrm{Aut}(\Sigma_A^+)$ the family of *automorphisms* of Σ_A^+, that is, the homeomorphisms $\tau\colon \Sigma_A^+ \to \Sigma_A^+$ such that $\tau \circ \sigma = \sigma \circ \tau$. We show that if two functions are related by an automorphism, then the entropy spectra of their equilibrium measures coincide. Let $\mathcal{M}$ be the family of σ-invariant probability Borel measures in Σ_A^+.

Proposition 11.2.1. *Let $\sigma|\Sigma_A^+$ be a topologically mixing topological Markov chain, and let $\varphi_1, \varphi_2\colon \Sigma_A^+ \to \mathbb{R}$ be Hölder continuous functions. If $\varphi_2 = \varphi_1 \circ \tau$ for some* *automorphism $\tau \in \text{Aut}(\Sigma_A^+)$, then*

$$P(q\varphi_1) = P(q\varphi_2) \quad \text{for every} \quad q \in \mathbb{R}, \tag{11.3}$$

and thus the entropy spectra of the equilibrium measures of φ_1 and φ_2 coincide.

Proof. Given $\nu \in \mathcal{M}$, we define a new measure in Σ_A^+ by $\nu_\tau(C) = \nu(\tau(C))$ for each measurable set $C \in \Sigma_A^+$. Clearly, $\nu_\tau \in \mathcal{M}$. We can easily verify that

$$h_\nu(\sigma, \xi) = h_{\nu_\tau}(\sigma, \tau^{-1}\xi),$$

and thus $h_\nu(\sigma) = h_{\nu_\tau}(\sigma)$. We also have

$$\int_{\Sigma_A^+} \varphi_2 \, d\nu_\tau = \int_{\Sigma_A^+} (\varphi_1 \circ \tau) \, d(\nu \circ \tau) = \int_{\Sigma_A^+} \varphi_1 \, d\nu,$$

and hence, for every $q \in \mathbb{R}$,

$$h_\nu(\sigma) + q \int_{\Sigma_A^+} \varphi_1 \, d\nu = h_{\nu_\tau}(\sigma) + q \int_{\Sigma_A^+} \varphi_2 \, d\nu_\tau.$$

By the variational principle of the topological pressure, we conclude that identity (11.3) holds for every $q \in \mathbb{R}$.

In order to show that the entropy spectra coincide we notice that

$$T_1(q) := P(q\varphi_1) - qP(\varphi_1) = P(q\varphi_2) - qP(\varphi_2) =: T_2(q). \tag{11.4}$$

By statement 6 in Theorem 7.3.2, the entropy spectra $\mathcal{E} = \mathcal{D}_1$ of the equilibrium measures of φ_1 and φ_2 are determined respectively by the functions T_1 and T_2. Thus, it follows from (11.4) that the spectra are equal. $\square$

By Proposition 11.2.1, we can only discuss the multifractal rigidity problem of recovering φ from its spectrum $\mathcal{E}$ up to an equivalence relation.

Definition 11.2.2. We say that two functions $\varphi_1, \varphi_2 \colon \Sigma_A^+ \to \mathbb{R}$ are *equivalent* if $\varphi_1 - \varphi_2 \circ \tau$ is cohomologous to a constant for some $\tau \in \text{Aut}(\Sigma_A^+)$.

We note that in each equivalence class any two functions have the same entropy spectrum.

11.2.2 Locally constant functions

A function $\varphi \colon \Sigma_A^+ \to \mathbb{R}$ is said to be *k-locally constant* if it is constant in cylinder sets of length k. We denote by L_k the family of k-locally constant functions. We say that an equivalence class of functions (see Definition 11.2.2) is an *equivalence class of L_k* if it contains a k-locally constant function.

In fact, it is sufficient to consider 2-locally constant functions.

Proposition 11.2.3. *If $\varphi\colon \Sigma_A^+ \to \mathbb{R}$ is a k-locally constant function, then there exist a topological Markov chain Σ_B^+ and a homeomorphism $\pi\colon \Sigma_A^+ \to \Sigma_B^+$ such that*

$$\pi \circ \sigma = \sigma \circ \pi \quad and \quad \varphi \circ \pi^{-1} \in L_2. \tag{11.5}$$

Proof. Fix a bijection γ between $\{1,\ldots,p\}^{k-1}$ and $\{1,\ldots,p^{k-1}\}$. Given a $p \times p$ matrix $A = (a_{ij})$, we define a $p^{k-1} \times p^{k-1}$ matrix $B = (b_{ij})$ with entries

$$b_{ij} = \begin{cases} 1, & \text{if } a_{\gamma^{-1}(i)_l \gamma^{-1}(i)_{l+1}} = 1 \text{ for } l = 1,\ldots,k-2, \\ & a_{\gamma^{-1}(j)_l \gamma^{-1}(j)_{l+1}} = 1 \text{ for } l = 1,\ldots,k-2, \\ & \text{and } \gamma^{-1}(i)_m = \gamma^{-1}(j)_{m-1} \text{ for } m = 2,\ldots,k-1, \\ 0, & \text{otherwise.} \end{cases}$$

We also define a map π in Σ_A^+ by

$$\pi(\alpha_0 \alpha_1 \cdots) = (\gamma(\alpha_0 \cdots \alpha_{k-2})\gamma(\alpha_1 \cdots \alpha_{k-1})\cdots).$$

We can easily show that $\pi(\Sigma_A^+) = \Sigma_B^+$. Indeed, for each $\alpha = (\alpha_1 \alpha_2 \cdots) \in \Sigma_A^+$ we have $b_{\pi(\alpha)_i \pi(\alpha)_{i+1}} = 1$, since

$$a_{\gamma^{-1}(\pi(\alpha)_i)_l \gamma^{-1}(\pi(\alpha)_i)_{l+1}} = a_{\alpha_{i+l-1}\alpha_{i+l}} = 1 \quad \text{for} \quad l = 1,\ldots,k-2,$$

$$a_{\gamma^{-1}(\pi(\alpha)_{i+1})_l \gamma^{-1}(\pi(\alpha)_{i+1})_{l+1}} = a_{\alpha_{i+l}\alpha_{i+l+1}} = 1 \quad \text{for} \quad l = 1,\ldots,k-2,$$

and

$$\gamma^{-1}(\pi(\alpha)_i)_m = \alpha_{i+m-1} = \gamma^{-1}(\pi(\alpha)_{i+1})_{m-1} \quad \text{for} \quad m = 2,\ldots,k-1.$$

Furthermore, $\pi\colon \Sigma_A^+ \to \Sigma_B^+$ is a homeomorphism since it maps cylinder sets in Σ_A^+ onto cylinder sets in Σ_B^+, and we can easily verify that the identity in (11.5) holds. Since π maps cylinder sets of length k onto cylinder sets of length 2, and $\varphi \in L_k$, we conclude that $\varphi \circ \pi^{-1} \in L_2$. $\square$

For functions in L_2 there is an explicit expression for the topological pressure. Given $\varphi\colon \Sigma_A^+ \to \mathbb{R}$ in L_2, we define the $p \times p$ matrix

$$A(\varphi) = (a_{ij} \exp(\varphi|C_{ij}))_{i,j=1}^p. \tag{11.6}$$

The following property is well known.

Proposition 11.2.4. *If $\varphi \in L_2$, then $P(\varphi) = \log \rho_{A(\varphi)}$, where $\rho_{A(\varphi)}$ is the spectral radius of the matrix $A(\varphi)$.*

By Proposition 11.2.4, if $\varphi \in L_2$ and $P(\varphi) = 0$, then the matrix $A(\varphi)$ has spectral radius equal to 1.

Now we study the relations between the matrices $A(\varphi)$ and $A(\psi)$ of two equivalent functions in L_2. For a permutation γ of $\{1,\ldots,p\}$ such that $a_{\gamma(i)\gamma(j)} = 1$ whenever $a_{ij} = 1$, we define an automorphism $\tau\colon \Sigma_A^+ \to \Sigma_A^+$ by

$$\tau((\alpha_i)_{i\in\mathbb{N}}) = (\gamma(\alpha_i))_{i\in\mathbb{N}}, \tag{11.7}$$

and we call it a *permutation automorphism*.

Theorem 11.2.5 ([13]). *Let $\varphi, \psi \colon \Sigma_A^+ \to \mathbb{R}$ be functions in L_2. Then:*

1. *$\varphi - \psi$ is cohomologous to a constant if and only if*

$$A(\varphi) = e^c D^{-1} A(\psi) D \tag{11.8}$$

for some positive diagonal matrix D and some constant $c \in \mathbb{R}$;

2. *$\varphi = \psi \circ \tau$ for some permutation automorphism τ if and only if*

$$A(\varphi) = P^{-1} A(\psi) P \tag{11.9}$$

for some permutation matrix P such that $A = P^{-1} A P$.

Proof. We begin with an auxiliary result.

Lemma 11.2.6. *If $\varphi, \psi \in L_2$, and*

$$\varphi - \psi = u \circ \sigma - u + c \tag{11.10}$$

for some continuous function u and some constant $c \in \mathbb{R}$, then $u \in L_1$.

Proof of the lemma. We proceed by contradiction. If u was not in L_1, then there would exist sequences $\alpha = (\alpha_1 \alpha_2 \cdots)$ and $\alpha' = (\alpha_1 \alpha_2' \cdots)$ with $\alpha_2 \neq \alpha_2'$ such that $u(\alpha) \neq u(\alpha')$. Now take β_0 with $a_{\beta_0 \alpha_1} = 1$, and consider the sequences

$$\beta = (\beta_0 \alpha_1 \alpha_2 \cdots) \quad \text{and} \quad \beta' = (\beta_0 \alpha_1 \alpha_2' \cdots)$$

in $C_{\beta_0 \alpha_1}$. Since $\varphi, \psi \in L_2$ we have

$$\varphi(\beta) - \psi(\beta) = \varphi(\beta') - \psi(\beta'),$$

which is equivalent to

$$u(\alpha) - u(\alpha') = u(\beta) - u(\beta').$$

Similarly, for each $n \in \mathbb{N}$ we obtain sequences $\beta^{(n)}, \beta'^{(n)} \in C_{\beta_{n-1} \cdots \beta_0 \alpha_1}$ such that

$$u(\beta^{(n)}) - u(\beta'^{(n)}) = u(\alpha) - u(\alpha') \neq 0.$$

Since $d(\beta^{(n)}, \beta'^{(n)}) \to 0$ when $n \to \infty$, we obtain a contradiction (recall that u is continuous). This shows that $u \in L_1$. $\qquad\square$

We proceed with the proof of the theorem. Assume that $\varphi, \psi \in L_2$ are such that $\varphi - \psi$ is cohomologous to a constant, say $c \in \mathbb{R}$. By Lemma 11.2.6, there is a continuous function $u \in L_1$ satisfying (11.10), and we obtain

$$\varphi | C_{ij} - \psi | C_{ij} = u | C_j - u | C_i + c.$$

This implies that (11.8) holds for the diagonal matrix

$$D = \operatorname{diag}(\exp(u|C_i))_{i=1}^p. \tag{11.11}$$

On the other hand, we can easily verify that if (11.8) is satisfied for some positive diagonal matrix D and some constant c, then (11.10) holds with u specified by (11.11).

Now we assume that $\varphi = \psi \circ \tau$ for some permutation automorphism τ. Then $\varphi|C_{ij} = \psi|C_{\gamma(i)\gamma(j)}$, and the matrices $A(\varphi)$ and $A(\psi)$ are conjugated by the permutation matrix

$$P = (\delta_{i\gamma(j)})_{i,j=1}^p, \tag{11.12}$$

where $\delta_{\alpha\beta} = 1$ if $\alpha = \beta$, and $\delta_{\alpha\beta} = 0$ otherwise. We show that $A = P^{-1}AP$. Set $B = P^{-1}AP$. Since P^{-1} coincides with the transpose of P, we have

$$b_{ij} = \sum_{\alpha=1}^p \sum_{\beta=1}^p (P^{-1})_{i\alpha} a_{\alpha\beta} P_{\beta j} = \sum_{\alpha=1}^p \sum_{\beta=1}^p \delta_{\alpha\gamma(i)} a_{\alpha\beta} \delta_{\beta\gamma(j)} = a_{\gamma(i)\gamma(j)},$$

where a_{ij} and b_{ij} are respectively the entries of A and B. If $a_{ij} = 1$, then there exists $\omega = (ij\cdots) \in \Sigma_A^+$. The sequence $\tau(\omega) = (\gamma(i)\gamma(j)\cdots)$ is in Σ_A^+, and thus $a_{\gamma(i)\gamma(j)} = 1$. On the other hand, if $a_{\gamma(i)\gamma(j)} = 1$, then there exists

$$\omega' = (\gamma(i)\gamma(j)\cdots) \in \Sigma_A^+.$$

Since τ is a bijection, there exists $\omega \in \Sigma_A^+$ with $\tau(\omega) = \omega'$. Clearly, $\omega = (ij\cdots)$, and thus $a_{ij} = 1$. Therefore, $a_{\gamma(i)\gamma(j)} = 1$ if and only if $a_{ij} = 1$. This shows that $b_{ij} = a_{ij}$ for each i and j, and hence $B = A$.

On the other hand, we can easily verify that if (11.9) is satisfied for some permutation matrix P such that $A = P^{-1}AP$, then $\varphi = \psi \circ \tau$ for some permutation automorphism τ specified by the entries of P as in (11.12). $\qquad\square$

11.2.3 Multifractal rigidity for locally constant functions

The following result of Barreira and Saraiva in [13] shows that for a full topological Markov chain $\sigma|\Sigma_n^+$, the entropy spectrum of an equivalence class of L_1 completely determines this class. This is the strongest possible multifractal rigidity.

Theorem 11.2.7. *Let $\sigma|\Sigma_n^+$ be the full topological Markov chain on n symbols, and let $\mathcal{E}$ be the entropy spectrum of an equivalence class of L_1. Then $\mathcal{E}$ completely determines the equivalence class.*

Proof. Let $\varphi \in L_1$, and write

$$\varphi|C_i = \log \alpha_i \quad \text{for} \quad i = 1, \ldots, n. \tag{11.13}$$

Without loss of generality, we assume that $\alpha_i \geq \alpha_{i+1}$ for $i = 1, \ldots, n-1$, and that $P(\varphi) = 0$.

We show how to determine the equivalence class of φ from the function $T(q) = P(q\varphi)$. By statement 6 in Theorem 7.3.2, this is equivalent to determining the equivalence class of φ from $\mathcal{E}$. We have

$$P(q\varphi) = \lim_{m\to\infty} \frac{1}{m} \log \sum_{i_1\cdots i_m} \exp\left[q \sup_{C_{i_1\cdots i_m}} S_m\varphi(x)\right],$$

where

$$S_m\varphi(x) = \sum_{k=0}^{m-1} \varphi(\sigma^k x).$$

Since

$$\sup_{C_{i_1\cdots i_m}} S_m\varphi(x) = \log(\alpha_{i_1}\cdots\alpha_{i_m}),$$

we obtain

$$T(q) = \lim_{m\to\infty} \frac{1}{m} \log \sum_{i_1\cdots i_m} (\alpha_{i_1}\cdots\alpha_{i_m})^q$$

$$= \lim_{m\to\infty} \frac{1}{m} \log(\alpha_1^q + \cdots + \alpha_n^q)^m = \log(\alpha_1^q + \cdots + \alpha_n^q). \tag{11.14}$$

In order to determine the constants $\alpha_1,\ldots,\alpha_n$ from the function T, we use the following result.

Lemma 11.2.8 (see, for example, [158]). *Assume that the polynomial*

$$p(x) = x^n + a_{n-1}x^{n-1} + \cdots + a_0 \tag{11.15}$$

has roots α_j, $j = 1,\ldots,n$. Setting $\beta_k = \sum_{j=1}^n \alpha_j^k$ for each $k \in \mathbb{N}$, we have

$$\beta_k + a_{n-1}\beta_{k-1} + \cdots + a_0\beta_{k-n} = 0 \quad \text{for} \quad k > n,$$

and

$$\beta_k + a_{n-1}\beta_{k-1} + \cdots + a_{n-k+1}\beta_1 = -ka_{n-k} \quad \text{for} \quad 1 \le k \le n.$$

By (11.14), we have $\beta_k = \exp T(k)$ for $k = 1,\ldots,n$, with $\alpha_1,\ldots,\alpha_n$ as in (11.13). It follows from Lemma 11.2.8 that

$$-a_{n-1} = \beta_1,$$
$$-2a_{n-2} = \beta_2 + a_{n-1}\beta_1,$$
$$-3a_{n-3} = \beta_3 + a_{n-2}\beta_2 + a_{n-3}\beta_1,$$
$$\cdots$$
$$-na_0 = \beta_n + a_{n-1}\beta_{n-1} + \cdots + a_1\beta_1.$$

These identities allow us to determine the coefficients $a_0,\ldots,a_{n-1}$ of the polynomial p in (11.15) from the function T. To determine φ and hence its equivalence class we only need to compute the roots $\alpha_1,\ldots,\alpha_n$ of the polynomial. $\square$

Now we consider 2-locally constant functions in topological Markov chains on two symbols, and we show that there is also a strong multifractal rigidity, although not as strong as for 1-locally constant functions in Theorem 11.2.7. We start with the full topological Markov chain.

Theorem 11.2.9 ([13]). *Let $\sigma|\Sigma_2^+$ be the full topological Markov chain on two symbols, and let $\mathcal{E}$ be the entropy spectrum of an equivalence class of L_2. Then the equivalence class is completely determined by $\mathcal{E}$, except when*

$$T(q) = \log(\alpha^q + (1-\alpha)^q)$$

for some $\alpha \in (0, 1/2)$, in which case there exist three equivalence classes represented by functions φ_i, $i = 1, 2, 3$ with matrices $A(\varphi_i)$ given by

$$\begin{pmatrix} 1-\alpha & 1-\alpha \\ \alpha & \alpha \end{pmatrix}, \quad \begin{pmatrix} 1-\alpha & \alpha \\ \alpha & 1-\alpha \end{pmatrix}, \quad \begin{pmatrix} \alpha & 1-\alpha \\ 1-\alpha & \alpha \end{pmatrix}. \tag{11.16}$$

Proof. We first notice that there exists $\varphi \in L_2$ in the equivalence class of functions determining the spectrum $\mathcal{E}$ such that $P(\varphi) = 0$, and

$$\varphi|C_{11} = \log \alpha_{11}, \quad \varphi|C_{12} = \log \alpha_{12}, \quad \varphi|C_{21} = 0, \quad \varphi|C_{22} = \log \alpha_{22},$$

with $\alpha_{11} \geq \alpha_{22}$. Indeed, consider a function $\varphi \in L_2$ with $P(\varphi) = 0$. We define $u \in L_1$ by $u|C_1 = -\varphi|C_{21}$ and $u|C_2 = 0$, and we set

$$\psi = \varphi + u \circ \sigma - u.$$

Clearly, $\psi \in L_2$, $P(\psi) = 0$, and

$$\psi|C_{21} = \varphi|C_{21} + u|C_1 - u|C_2 = 0.$$

We have (see (11.6))

$$A(\varphi) = \begin{pmatrix} \alpha_{11} & \alpha_{12} \\ 1 & \alpha_{22} \end{pmatrix}.$$

We continue with an auxiliary result.

Lemma 11.2.10. *Let $\psi \in L_2$ be a function with $\psi|C_{ij} = \beta_{ij}$ for $i, j \in \{1, 2\}$. Then*

$$\lim_{q \to +\infty} \int_{\Sigma_2^+} \psi \, d\mu_q = \max_{\nu \in \mathcal{M}} \int_{\Sigma_2^+} \psi \, d\nu = \max\{\beta_{11}, \tfrac{1}{2}(\beta_{12} + \beta_{21}), \beta_{22}\}, \tag{11.17}$$

where μ_q is the equilibrium measure of $q\psi$.

Proof of the lemma. We first show that

$$\lim_{q \to +\infty} \int_{\Sigma_2^+} \psi \, d\mu_q = \sup_{\nu \in \mathcal{M}} \int_{\Sigma_2^+} \psi \, d\nu. \tag{11.18}$$

Let $\nu \in \mathcal{M}$. By the variational principle of the topological pressure,

$$h_{\mu_q}(\sigma) + \int_{\Sigma_2^+} q\psi\, d\mu_q \geq h_\nu(\sigma) + \int_{\Sigma_2^+} q\psi\, d\nu.$$

Since $h_\nu(\sigma) \leq \log 2$ for every $\nu \in \mathcal{M}$, dividing by q and letting $q \to +\infty$ we obtain

$$\lim_{q \to +\infty} \int_{\Sigma_2^+} \psi\, d\mu_q \geq \int_{\Sigma_2^+} \psi\, d\nu.$$

Since $\mu_q \in \mathcal{M}$ for every $q \in \mathbb{R}$, we obtain (11.18).

Now we show that the supremum in (11.18) is in fact a maximum, and that it is equal to the right-hand side of (11.17). Let $\nu \in \mathcal{M}$. We have $\nu(C_1) = \nu(C_{11}) + \nu(C_{12})$ as well as

$$\nu(C_1) = \nu(\sigma^{-1}C_1) = \nu(C_{11}) + \nu(C_{21}).$$

Therefore, $\nu(C_{12}) = \nu(C_{21})$, and

$$\int_{\Sigma_2^+} \psi\, d\nu = \nu(C_{11})\beta_{11} + \nu(C_{12})(\beta_{12} + \beta_{21}) + \nu(C_{22})\beta_{22}.$$

We consider the function $\rho\colon \mathbb{R}^3 \to \mathbb{R}$ defined by

$$\rho(x, y, z) = x\beta_{11} + y(\beta_{12} + \beta_{21}) + z\beta_{22},$$

and the compact set

$$B = \{(x, y, z) \in \mathbb{R}^3 : x + 2y + z = 1 \text{ and } x, y, z \geq 0\}.$$

Clearly,

$$\sup_{\nu \in \mathcal{M}} \int_{\Sigma_2^+} \psi\, d\nu = \max\{\rho(x, y, z) : (x, y, z) \in B\}. \tag{11.19}$$

Furthermore, the maximum in (11.19) is attained at one of the vertices of B, that is, at $(1, 0, 0)$, $(0, 1/2, 0)$ or $(0, 0, 1)$. Thus,

$$\max\{\rho(x, y, z) : (x, y, z) \in B\} = \max\{\beta_{11}, \tfrac{1}{2}(\beta_{12} + \beta_{21}), \beta_{22}\}.$$

Now let $\delta(i_1 \cdots i_k) \in \mathcal{M}$ be the delta measure supported on the periodic point $(i_1 \cdots i_k i_1 \cdots i_k \cdots)$. Setting

$$\nu_1 = \delta(1), \quad \nu_2 = \tfrac{1}{2}(\delta(12) + \delta(21)), \quad \nu_3 = \delta(2),$$

we have

$$\int_{\Sigma_2^+} \psi\, d\nu_1 = \beta_{11}, \quad \int_{\Sigma_2^+} \psi\, d\nu_2 = \frac{1}{2}(\beta_{12} + \beta_{21}), \quad \int_{\Sigma_2^+} \psi\, d\nu_3 = \beta_2.$$

This shows that the supremum in (11.19) is in fact a maximum. $\square$

Denoting by $\rho(q)$ the spectral radius of the matrix

$$A(q\varphi) = \begin{pmatrix} \alpha_{11}^q & \alpha_{12}^q \\ 1 & \alpha_{22}^q \end{pmatrix}, \tag{11.20}$$

it follows from Proposition 11.2.4 that $T(q) = \log \rho(q)$. By (7.15) we have

$$T'(q) = \int_{\Sigma_2^+} \varphi \, d\mu_{q\varphi}.$$

Set

$$\bar{c} = \lim_{q \to +\infty} T'(q) \quad \text{and} \quad \underline{c} = \lim_{q \to -\infty} T'(q). \tag{11.21}$$

It follows from Lemma 11.2.10 that

$$\bar{c} = \max\{\log \alpha_{11}, \tfrac{1}{2}\log \alpha_{12}\} \quad \text{and} \quad \underline{c} = \min\{\tfrac{1}{2}\log \alpha_{12}, \log \alpha_{22}\}.$$

Furthermore, since μ_0 is the Bernoulli measure $(1/2, 1/2)$, we have

$$T'(0) = \frac{1}{4}(\log \alpha_{11} + \log \alpha_{12} + \log \alpha_{22}). \tag{11.22}$$

Now we consider three cases:

1.

$$\log \alpha_{22} \leq \frac{1}{2}\log \alpha_{12} \leq \log \alpha_{11}, \tag{11.23}$$

in which case $\bar{c} = \log \alpha_{11}$ and $\underline{c} = \log \alpha_{22}$;

2.

$$\frac{1}{2}\log \alpha_{12} \leq \log \alpha_{22} \leq \log \alpha_{11}, \tag{11.24}$$

in which case $\bar{c} = \log \alpha_{11}$ and $2\underline{c} = \log \alpha_{12}$;

3.

$$\log \alpha_{22} \leq \log \alpha_{11} \leq \frac{1}{2}\log \alpha_{12}, \tag{11.25}$$

in which case $2\bar{c} = \log \alpha_{12}$ and $\underline{c} = \log \alpha_{22}$.

In the three cases, two of the numbers α_{11}, α_{12}, and α_{22} are determined by $\bar{c}$ and $\underline{c}$, although we are not able to say which ones. In order to determine the third number we note that the matrix $A(\varphi)$ is positive, and thus it has a positive real eigenvalue which is greater than the absolute value of the other one. Since the spectral radius of $A(\varphi)$ is $\exp P(\varphi) = 1$ (by Proposition 11.2.4), this maximal eigenvalue is exactly 1, and we can easily verify that the matrix $A(\varphi)$ must be, in each case,

$$\begin{pmatrix} e^{\bar{c}} & (1-e^{\bar{c}})(1-e^{\underline{c}}) \\ 1 & e^{\underline{c}} \end{pmatrix}, \quad \begin{pmatrix} e^{\bar{c}} & e^{2\underline{c}} \\ 1 & 1 - \frac{e^{2\underline{c}}}{1-e^{\bar{c}}} \end{pmatrix}, \quad \begin{pmatrix} 1 - \frac{e^{2\bar{c}}}{1-e^{\underline{c}}} & e^{2\bar{c}} \\ 1 & e^{\underline{c}} \end{pmatrix}.$$

In Case 1 we have $e^{\underline{c}} < 1$. Indeed, if $e^{\underline{c}} > 1$, then the trace of $A(\varphi)$ is greater than 2, and thus its spectral radius is greater than 1. Furthermore, if $e^{\underline{c}} = 1$, then $\alpha_{12} = 0$, which is impossible. We also have $e^{\overline{c}} < 1$, since otherwise $\alpha_{12} \leq 0$ which again is impossible. It follows from (11.23) that

$$e^{2\underline{c}} \leq (1 - e^{\overline{c}})(1 - e^{\underline{c}}) \leq e^{2\overline{c}},$$

which is equivalent to

$$\frac{e^{\underline{c}} - 1 + \sqrt{(1 - e^{\underline{c}})^2 + 4(1 - e^{\underline{c}})}}{2} \leq e^{\overline{c}} \leq 1 - \frac{e^{2\underline{c}}}{1 - e^{\underline{c}}}. \qquad (11.26)$$

In Case 2 we have $e^{\overline{c}} < 1$. Indeed, similarly, if $e^{\overline{c}} > 1$, then the trace of $A(\varphi)$ is greater than 2, and thus its spectral radius is greater than 1. Furthermore, if $e^{\overline{c}} = 1$, then α_{22} would be undefined. Since $\underline{c} \leq \overline{c}$ we also have $e^{\underline{c}} < 1$. It follows from (11.24) that

$$e^{\underline{c}} \leq 1 - \frac{e^{2\underline{c}}}{1 - e^{\overline{c}}} \leq e^{\overline{c}},$$

which is equivalent to

$$1 - e^{\underline{c}} \leq e^{\overline{c}} \leq 1 - \frac{e^{2\underline{c}}}{1 - e^{\underline{c}}}. \qquad (11.27)$$

In Case 3, proceeding as in Case 2, we show that $e^{\underline{c}} < 1$. Furthermore, we also have $e^{\overline{c}} < 1$, since otherwise α_{11} would be negative, which is impossible. It follows from (11.25) that

$$e^{\underline{c}} \leq 1 - \frac{e^{2\overline{c}}}{1 - e^{\underline{c}}} \leq e^{\overline{c}},$$

which is equivalent to

$$\frac{e^{\underline{c}} - 1 + \sqrt{(1 - e^{\underline{c}})^2 + 4(1 - e^{\underline{c}})}}{2} \leq e^{\overline{c}} \leq 1 - e^{\underline{c}}. \qquad (11.28)$$

We can easily verify that for $e^{\underline{c}} > 1/2$ the numbers

$$\frac{e^{\underline{c}} - 1 + \sqrt{(1 - e^{\underline{c}})^2 + 4(1 - e^{\underline{c}})}}{2}, \quad 1 - \frac{e^{2\underline{c}}}{1 - e^{\underline{c}}}, \quad \text{and} \quad 1 - e^{\underline{c}}$$

are smaller than $e^{\underline{c}}$. Since $e^{\overline{c}}$ is smaller than or equal to some of these numbers, we conclude that for $e^{\underline{c}} > 1/2$ we would have $e^{\overline{c}} < e^{\underline{c}}$, which is impossible. Therefore, we must have $e^{\underline{c}} \leq 1/2$. Note that for $e^{\underline{c}} \in [0, 1/2]$,

$$\frac{e^{\underline{c}} - 1 + \sqrt{(1 - e^{\underline{c}})^2 + 4(1 - e^{\underline{c}})}}{2} \leq 1 - e^{\underline{c}} \leq 1 - \frac{e^{2\underline{c}}}{1 - e^{\underline{c}}},$$

with equalities when $e^{\underline{c}} = 1/2$.

The three inequalities in (11.26), (11.27), and (11.28) can be satisfied simultaneously only if

$$e^{\bar{c}} = 1 - e^{\underline{c}}. \tag{11.29}$$

When this happens, we are in a situation in which it is impossible to completely recover the equivalence class. Moreover, we have three equivalence classes represented by the functions φ_i, $i = 1, 2, 3$ with matrices $A(\varphi_i)$ given by

$$\begin{pmatrix} 1 - e^{\underline{c}} & e^{\underline{c}}(1 - e^{\underline{c}}) \\ 1 & e^{\underline{c}} \end{pmatrix}, \quad \begin{pmatrix} 1 - e^{\underline{c}} & e^{2\underline{c}} \\ 1 & 1 - e^{\underline{c}} \end{pmatrix}, \quad \begin{pmatrix} e^{\underline{c}} & (1 - e^{\underline{c}})^2 \\ 1 & e^{\underline{c}} \end{pmatrix}.$$

When $e^{\underline{c}} = 1/2$ the three matrices coincide, and thus we obtain a single equivalence class. When $e^{\underline{c}} \neq 1/2$ the functions are cohomologous to the ones represented by the matrices in (11.16) with $\alpha = e^{\underline{c}}$. Now we show that the three functions are not equivalent although they have the same entropy spectrum.

Lemma 11.2.11. *Let $\sigma|\Sigma_2^+$ be the full topological Markov chain on two symbols, and let $\varphi_1, \varphi_2, \varphi_3 \in L_2$ be the functions with matrices as in (11.16) for some $\alpha \in (0, 1/2)$. Then the equilibrium measures of φ_1, φ_2, and φ_3 have the same entropy spectrum, but the functions are not equivalent.*

Proof of the lemma. Computing the spectral radius of the matrices $A(q\varphi_i)$ we obtain

$$P(q\varphi_i) = \log(\alpha^q + (1 - \alpha)^q), \quad i = 1, 2, 3.$$

This shows that the entropy spectra of the three functions coincide.

Now we show that the functions are not equivalent. Note that any automorphism $\tau \in \mathrm{Aut}(\Sigma_2^+)$ transforms fixed points of σ^n into fixed points of σ^n. In particular, if $\gamma_1 = (11 \cdots)$ and $\gamma_2 = (22 \cdots)$, then $\{\tau(\gamma_1), \tau(\gamma_2)\} = \{\gamma_1, \gamma_2\}$. Therefore, if two functions $\psi_1, \psi_2 \in L_2$ with $P(\psi_1) = P(\psi_2)$ are equivalent, then

$$\{\psi_1(\gamma_1), \psi_1(\gamma_2)\} = \{\psi_2(\gamma_1), \psi_2(\gamma_2)\}.$$

Since the functions φ_i, $i = 1, 2, 3$ have the same topological pressure $P(q\varphi_i)$ for each $q \in \mathbb{R}$, and the sets $\{\varphi_i(\gamma_1), \varphi_i(\gamma_2)\}$, $i = 1, 2, 3$ are distinct, we conclude that φ_1, φ_2, and φ_3 are not equivalent. $\quad\square$

Now we assume that (11.29) does not occur, i.e., that $e^{\bar{c}} \neq 1 - e^{\underline{c}}$. There are two possibilities: when

$$1 - e^{\underline{c}} < e^{\bar{c}} \leq 1 - \frac{e^{2\underline{c}}}{1 - e^{\underline{c}}} \tag{11.30}$$

we are in Case 1 or Case 2 (see (11.23) and (11.24)), and when

$$\frac{e^{\underline{c}} - 1 + \sqrt{(1 - e^{\underline{c}})^2 + 4(1 - e^{\underline{c}})}}{2} \leq e^{\bar{c}} < 1 - e^{\underline{c}} \tag{11.31}$$

we are in Case 1 or Case 3 (see (11.23) and (11.25)). Using the parameter $T'(0)$ we determine in which case we are, and thus we identify a single equivalence class.

Assume first that (11.30) holds. By (11.22), in Case 1 we have

$$e^{4T'(0)} = e^{\bar{c}}e^{\underline{c}}(1 - e^{\bar{c}})(1 - e^{\underline{c}}), \tag{11.32}$$

and in Case 2 we have

$$e^{4T'(0)} = e^{\bar{c}}e^{2\underline{c}}\left(1 - \frac{e^{2\underline{c}}}{1 - e^{\bar{c}}}\right). \tag{11.33}$$

If only one of these identities holds, then we identify in which case we are, and thus also the equivalence class. Otherwise, when both identities in (11.32) and (11.33) are satisfied we must have

$$e^{\bar{c}} = 1 - e^{\underline{c}} \quad \text{or} \quad e^{\bar{c}} = 1 - \frac{e^{2\underline{c}}}{1 - e^{\underline{c}}}. \tag{11.34}$$

The first identity was already analyzed and corresponds to the situation when we obtain three equivalence classes. When the second identity in (11.34) holds, the equivalence classes in Case 1 and Case 2 coincide, and contain the function with matrix

$$\begin{pmatrix} 1 - \frac{e^{2\underline{c}}}{1-e^{\underline{c}}} & e^{2\underline{c}} \\ 1 & e^{\underline{c}} \end{pmatrix}.$$

Therefore, when (11.30) holds, since $e^{\bar{c}} \neq 1 - e^{\underline{c}}$ we identify a single equivalence class of L_2 that generates the spectrum.

Similarly, when (11.31) holds, in Case 1 we have

$$e^{4T'(0)} = e^{\bar{c}}e^{\underline{c}}(1 - e^{\bar{c}})(1 - e^{\underline{c}}), \tag{11.35}$$

and in Case 3 we have

$$e^{4T'(0)} = e^{2\bar{c}}e^{\underline{c}}\left(1 - \frac{e^{2\bar{c}}}{1 - e^{\underline{c}}}\right). \tag{11.36}$$

If only one of these identities holds, then we identify a single equivalence class. Otherwise, when both identities in (11.35) and (11.36) are satisfied we must have

$$e^{\bar{c}} = 1 - e^{\underline{c}} \quad \text{or} \quad e^{\underline{c}} = 1 - \frac{e^{2\bar{c}}}{1 - e^{\bar{c}}}.$$

When the second identity holds, the equivalence classes in Case 1 and Case 3 coincide, and contain the function with matrix

$$\begin{pmatrix} e^{\bar{c}} & e^{2\bar{c}} \\ 1 & 1 - \frac{e^{2\underline{c}}}{1-e^{\underline{c}}} \end{pmatrix}.$$

Therefore, when (11.31) holds, since $e^{\bar{c}} \neq 1 - e^{\underline{c}}$ we identify a single equivalence class of L_2 that generates the spectrum. This concludes the proof of the theorem.

$\square$

Now we consider the topological Markov chains $\sigma|\Sigma_A^+$ with transition matrix $A = \left(\begin{smallmatrix} 1 & 1 \\ 1 & 0 \end{smallmatrix}\right)$ or $A = \left(\begin{smallmatrix} 0 & 1 \\ 1 & 1 \end{smallmatrix}\right)$.

Theorem 11.2.12 ([13]). *Let $\sigma|\Sigma_A^+$ be the topological Markov chain with transition matrix $A = \left(\begin{smallmatrix} 1 & 1 \\ 1 & 0 \end{smallmatrix}\right)$ or $A = \left(\begin{smallmatrix} 0 & 1 \\ 1 & 1 \end{smallmatrix}\right)$, and let $\mathcal{E}$ be the entropy spectrum of an equivalence class of L_2. Then $\mathcal{E}$ completely determines the equivalence class.*

Proof. We only consider the case when $A = \left(\begin{smallmatrix} 1 & 1 \\ 1 & 0 \end{smallmatrix}\right)$ since the other one is entirely similar. Let $\varphi \in L_2$ be a function in the equivalence class generating the spectrum $\mathcal{E}$ such that $P(\varphi) = 0$, and

$$\varphi|C_{11} = \log \alpha_{11}, \quad \varphi|C_{12} = \log \alpha_{12}, \quad \varphi|C_{21} = 0$$

(proceeding as in the proof of Theorem 11.2.9 we can easily verify that there exists a function φ with these properties). Let $\bar{c}$ and $\underline{c}$ be as in (11.21). Proceeding as in the proof of Lema 11.2.10, we show that

$$\bar{c} = \max\{\log \alpha_{11}, \tfrac{1}{2} \log \alpha_{12}\} \quad \text{and} \quad \underline{c} = \min\{\log \alpha_{11}, \tfrac{1}{2} \log \alpha_{12}\}.$$

We can verify that there are two possibilities for the matrix $A(\varphi)$, namely

$$\begin{pmatrix} e^{\bar{c}} & e^{2\underline{c}} \\ 1 & 0 \end{pmatrix} \quad \text{and} \quad \begin{pmatrix} e^{\underline{c}} & e^{2\bar{c}} \\ 1 & 0 \end{pmatrix}. \tag{11.37}$$

Since $P(\varphi) = 0$, the function φ is represented by a matrix with spectral radius 1, and thus 1 is an eigenvalue of the matrices in (11.37). Therefore, in Case 1 we have

$$e^{\bar{c}} + e^{2\underline{c}} = 1, \tag{11.38}$$

and in Case 2 we have

$$e^{2\bar{c}} + e^{\underline{c}} = 1. \tag{11.39}$$

In order that the identities (11.38) and (11.39) are satisfied simultaneously, the number $e^{\underline{c}}$ must be 0, $(-1 + \sqrt{5})/2$ or 1. But it cannot be 0, and also cannot be 1 since $e^{\bar{c}}$ would then be 0. Thus, we must have $e^{\underline{c}} = (-1 + \sqrt{5})/2$. We can easily verify that for this value the two matrices in (11.37) are equal.

To determine the equivalence class we only need to verify which of the identities in (11.38) and (11.39) is satisfied. When (11.38) holds, the equivalence class is represented by the first matrix in (11.37). Otherwise, when (11.39) holds, the equivalence class is represented by the second matrix in (11.37). When both identities hold, as we observed, the two matrices in (11.37) are equal, and thus we obtain a single equivalence class. Therefore, each entropy spectrum determined by an equivalence class of L_2 is determined by a single equivalence class. $\square$

11.2.4   Failure of multifractal rigidity

Following [13] we give an explicit example of a topological Markov chain on three symbols for which there is no multifractal rigidity, even more in some generic sense.

Let $\varphi \in L_2$ be a function with $P(\varphi) = 0$. By Proposition 11.2.4 we have

$$T(q) = P(q\varphi) = \log \rho_{A(q\varphi)}.$$

On the other hand, given a square matrix $A = (a_{ij})$ we consider its *characteristic polynomial*

$$p_A(z, q) = \det(z\,\mathrm{Id} - A^{(q)}),$$

where $A^{(q)} = (a_{ij}^q)$. We note that T can be computed from $p_{A(\varphi)}$. This follows from the identity

$$p_{A(\varphi)}(z, q) = \det(z\,\mathrm{Id} - A(\varphi)^{(q)}) = \det(z\,\mathrm{Id} - A(q\varphi))$$

(see (11.20)). Therefore, the entropy spectrum $\mathcal{E}$ can also be computed from the characteristic polynomial $p_{A(\varphi)}$. However, we show that in general the knowledge of $p_{A(\varphi)}$ is not sufficient to determine the equivalence class of φ. More precisely, we exhibit functions in distinct equivalence classes that have the same characteristic polynomial, and thus the same entropy spectrum.

Theorem 11.2.13 ([13]). *Let $\sigma|\Sigma_A^+$ be the topological Markov chain with transition matrix*

$$A = \begin{pmatrix} 0 & 1 & 1 \\ 1 & 0 & 1 \\ 1 & 1 & 0 \end{pmatrix},$$

and let $\varphi, \psi \in L_2$ be functions satisfying $P(\varphi) = P(\psi) = 0$, with matrices

$$A(\varphi) = \begin{pmatrix} 0 & \alpha_{12} & \alpha_{13} \\ 1 & 0 & \alpha_{23} \\ 1 & \alpha_{32} & 0 \end{pmatrix} \quad and \quad A(\psi) = \begin{pmatrix} 0 & \alpha_{12} & \alpha_{13} \\ 1 & 0 & \frac{\alpha_{13}}{\alpha_{12}}\alpha_{32} \\ 1 & \frac{\alpha_{12}}{\alpha_{13}}\alpha_{23} & 0 \end{pmatrix},$$

where $\alpha_{12} > \alpha_{13} > \alpha_{23}\alpha_{32}$ and $\alpha_{13}\alpha_{32} \neq \alpha_{12}\alpha_{23}$. Then φ and ψ have the same characteristic polynomial, and thus the same entropy spectrum, but are not equivalent.

Proof. We can easily verify that $p_{A(\varphi)} = p_{A(\psi)} = p$, where

$$p(z, q) = z^3 - (\alpha_{12}^q + \alpha_{13}^q + (\alpha_{23}\alpha_{32})^q)z + (\alpha_{12}\alpha_{23})^q + (\alpha_{13}\alpha_{32})^q.$$

In order to show that the functions φ and ψ are not equivalent we first observe that $\mathrm{Aut}(\Sigma_A^+) \approx S_3$, where S_3 is the permutation group of three elements (see Example 2.19 in [41]). Furthermore, to each permutation γ of $\{1, 2, 3\}$ there corresponds a permutation automorphism in $\mathrm{Aut}(\Sigma_A^+)$ (see (11.7)). Therefore, the automorphisms of Σ_A^+ are precisely the permutation automorphisms.

We proceed by contradiction. Assume that φ and ψ are equivalent, that is, there exist $\tau \in \mathrm{Aut}(\Sigma_A^+)$ obtained from a permutation γ as in (11.7), and a continuous function $u\colon \Sigma_A^+ \to \mathbb{R}$ such that

$$\varphi \circ \tau = \psi + u \circ \sigma - u. \tag{11.40}$$

Since $\varphi, \psi \in L_2$, the function $\varphi \circ \tau$ is also in L_2. By Lemma 11.2.6 we have $u \in L_1$, and thus $u|C_i = \log d_i$ for some constants $d_i > 0$, $i = 1, 2, 3$. We can easily verify that identity (11.40) can be written in matrix form as

$$\begin{pmatrix} 0 & \alpha_{\gamma(1)\gamma(2)} & \alpha_{\gamma(1)\gamma(3)} \\ \alpha_{\gamma(2)\gamma(1)} & 0 & \alpha_{\gamma(2)\gamma(3)} \\ \alpha_{\gamma(3)\gamma(1)} & \alpha_{\gamma(3)\gamma(2)} & 0 \end{pmatrix} = \begin{pmatrix} 0 & \frac{d_1}{d_2}\alpha_{12} & \frac{d_1}{d_3}\alpha_{13} \\ \frac{d_2}{d_1} & 0 & \frac{d_2}{d_3}\frac{\alpha_{13}}{\alpha_{12}}\alpha_{32} \\ \frac{d_3}{d_1} & \frac{d_3}{d_1}\frac{\alpha_{12}}{\alpha_{13}}\alpha_{23} & 0 \end{pmatrix}. \tag{11.41}$$

We note that

$$\begin{aligned} \alpha_{\gamma(1)\gamma(2)}\alpha_{\gamma(2)\gamma(1)} &= \alpha_{12}, \\ \alpha_{\gamma(1)\gamma(3)}\alpha_{\gamma(3)\gamma(1)} &= \alpha_{13}, \\ \alpha_{\gamma(2)\gamma(3)}\alpha_{\gamma(3)\gamma(2)} &= \alpha_{23}\alpha_{32}. \end{aligned} \tag{11.42}$$

The matrix in the left-hand side of (11.41) can also be written in the form $P^{-1}A(\varphi)P$ where P is the permutation matrix in (11.12). Since the set of entries of $PA(\varphi)P^{-1}$ is the same as the set of entries of $A(\varphi)$, we can easily verify that the identities in (11.42) together with the hypothesis $\alpha_{12} > \alpha_{13} > \alpha_{23}\alpha_{32}$ imply that P is the identity matrix (and thus τ is the identity automorphism). Indeed, if $\gamma(1) = 2$, then $\alpha_{2\gamma(2)}\alpha_{\gamma(2)2} = \alpha_{12}$. Since $\alpha_{23}\alpha_{32} < \alpha_{12}$ it cannot be $\gamma(2) = 3$. Therefore, $\gamma(2) = 1$ and $\gamma(3) = 3$. But then, from the second identity in (11.42), we should have $\alpha_{23}\alpha_{32} = \alpha_{13}$, which contradicts the hypotheses in the theorem. We can show in a similar manner that $\gamma(1) = 3$ yields a contradiction. Indeed, if $\gamma(1) = 3$, then from the first identity in (11.42) we would have $\alpha_{3\gamma(2)}\alpha_{\gamma(2)3} = \alpha_{12}$. It follows from the hypotheses in the theorem that it cannot be $\gamma(2) = 2$, and thus we must have $\gamma(2) = 1$ and $\gamma(3) = 2$. But then the second identity in (11.42) gives $\alpha_{32}\alpha_{23} = \alpha_{13}$, which again contradicts the hypotheses in the theorem. Therefore, we must have $\gamma(1) = 1$, and the first identity in (11.42) gives $\alpha_{1\gamma(2)}\alpha_{\gamma(2)1} = \alpha_{12}$. If $\gamma(2) = 3$, then we obtain $\alpha_{13} = \alpha_{12}$, which is forbidden by the hypotheses in the theorem. Therefore, we must have $\gamma(2) = 2$ and $\gamma(3) = 3$. This shows that $\gamma = \mathrm{Id}$, and thus τ is the identity automorphism.

Hence, equation (11.41) reduces to

$$\begin{pmatrix} 0 & \alpha_{12} & \alpha_{13} \\ 1 & 0 & \alpha_{23} \\ 1 & \alpha_{32} & 0 \end{pmatrix} = \begin{pmatrix} 0 & \frac{d_1}{d_2}\alpha_{12} & \frac{d_1}{d_3}\alpha_{13} \\ \frac{d_2}{d_1} & 0 & \frac{d_2}{d_3}\frac{\alpha_{13}}{\alpha_{12}}\alpha_{32} \\ \frac{d_3}{d_1} & \frac{d_3}{d_1}\frac{\alpha_{12}}{\alpha_{13}}\alpha_{23} & 0 \end{pmatrix}.$$

This implies that $d_1 = d_2 = d_3$, and hence u is constant. Therefore, $\psi = \varphi$. On the other hand, since $\alpha_{13}\alpha_{32} \neq \alpha_{12}\alpha_{23}$ we must have $\psi \neq \varphi$. This contradiction shows that the functions ψ and φ cannot be equivalent. $\square$

We note that it could happen that using only the information given by the function T we could obtain more equivalence classes than the two in Theorem 11.2.13. However, it is shown in [13] that T contains sufficient information to obtain exactly the two equivalence classes.

Chapter 12

Hyperbolic Sets: Past and Future

We give a complete description of the dimension spectra of Birkhoff averages in a hyperbolic set of a conformal diffeomorphism, considering *simultaneously* Birkhoff averages into the past and into the future, i.e., both for negative and positive time. We emphasize that the description of these spectra is not a consequence of the results in Chapter 6. The main difficulty is that although the local product structure provided by the intersection of local stable and unstable manifolds is a Lipschitz homeomorphism with Lipschitz inverse, the level sets of Birkhoff averages are never compact. This causes their box dimension to be strictly larger than their Hausdorff dimension, and thus a product of level sets may have a Hausdorff dimension that a priori need not be the sum of the dimensions of the level sets. Instead, we construct explicitly *noninvariant* measures concentrated on each product of level sets having the appropriate pointwise dimension.

12.1 A model case: the Smale horseshoe

We briefly explain in this section why the results are nontrivial, even in the particular case of the linear Smale horseshoe. This is the product $\Lambda = C \times C$ of two standard middle-third Cantor sets. Let $f\colon \Lambda \to \Lambda$ be the dynamics in the horseshoe, here assumed to be expanding in the vertical direction and contracting in the horizontal direction. Given continuous functions $\varphi, \psi\colon \Lambda \to \mathbb{R}$ we consider the level sets of Birkhoff averages given for each $\alpha, \beta \in \mathbb{R}$ by

$$K_{\alpha\beta} = \left\{ x \in \Lambda : \lim_{n\to\infty} \frac{1}{n} \sum_{k=0}^{n-1} \varphi(f^k x) = \alpha \text{ and } \lim_{n\to\infty} \frac{1}{n} \sum_{k=0}^{n-1} \psi(f^{-k} x) = \beta \right\}.$$

We define the *dimension spectrum* by

$$\mathcal{D}(\alpha, \beta) = \dim_H K_{\alpha\beta}.$$

We want to explain why we cannot obtain a description of the spectrum $\mathcal{D}$ from the results in Chapter 6. Let P and Q be the orthogonal projections respectively onto the horizontal and vertical axes. It follows from the exponential behavior of f along the stable and unstable manifolds that (see Theorem 12.2.2)

$$P(K_{\alpha\beta}) \times C = \left\{ x \in \Lambda : \lim_{n \to \infty} \frac{1}{n} \sum_{k=0}^{n-1} \psi(f^{-k}x) = \beta \right\},$$

and

$$C \times Q(K_{\alpha\beta}) = \left\{ x \in \Lambda : \lim_{n \to \infty} \frac{1}{n} \sum_{k=0}^{n-1} \varphi(f^{k}x) = \alpha \right\}.$$

Therefore,

$$K_{\alpha\beta} = [P(K_{\alpha\beta}) \times C] \cap [C \times Q(K_{\alpha\beta})] = P(K_{\alpha\beta}) \times Q(K_{\alpha\beta}), \qquad (12.1)$$

and each level set $K_{\alpha\beta}$ is a product of level sets of Birkhoff averages. We could try to obtain a description of the spectrum $\mathcal{D}$ from the results in Chapter 6 applied to $P(K_{\alpha\beta})$ and $Q(K_{\alpha\beta})$. The problem is that in general the Hausdorff dimension of a product $A \times B$ need not be the sum of the Hausdorff dimensions of A and B, unless, for example, $\dim_H A = \overline{\dim}_B A$ or $\dim_H B = \overline{\dim}_B B$. Even more, we can show that if the functions φ and ψ are not cohomologous to constants, then

$$\dim_H P(K_{\alpha\beta}) < \overline{\dim}_B P(K_{\alpha\beta}) \quad \text{and} \quad \dim_H Q(K_{\alpha\beta}) < \overline{\dim}_B Q(K_{\alpha\beta})$$

for all α, β, except for one value of α and one value of β. So, even though it follows immediately from (12.1) that

$$\mathcal{D}(\alpha, \beta) \geq \dim_H P(K_{\alpha\beta}) + \dim_H Q(K_{\alpha\beta}), \qquad (12.2)$$

a priori this inequality could be strict. The main objective of this chapter is to show that (12.2) is in fact an identity for every α and β (see Section 12.4). We follow closely Barreira and Valls in [22].

12.2   Dimension spectra

We consider in this section the dimension spectra of Birkhoff averages on a locally maximal hyperbolic set Λ of a diffeomorphism f. We assume that f is conformal on Λ.

We continue to denote by $C^\varepsilon(\Lambda)$ the space of Hölder continuous functions $\varphi \colon \Lambda \to \mathbb{R}$ with a given Hölder exponent $\varepsilon \in (0, 1]$. Fix $\kappa \in \mathbb{N}$. We consider two pairs of functions (Φ^+, Ψ^+) and (Φ^-, Ψ^-) in $H(\Lambda) := C^\varepsilon(\Lambda)^\kappa \times C^\varepsilon(\Lambda)^\kappa$, and we write

$$\Phi^\pm = (\varphi_1^\pm, \ldots, \varphi_\kappa^\pm) \quad \text{and} \quad \Psi^\pm = (\psi_1^\pm, \ldots, \psi_\kappa^\pm).$$

We always assume that $\psi_i^\pm > 0$ for $i = 1, \ldots, \kappa$ (for simplicity we write $\Psi^\pm > 0$). For each $\alpha = (\alpha_1, \ldots, \alpha_\kappa) \in \mathbb{R}^\kappa$ we consider the level sets

$$K_\alpha^+ = \bigcap_{i=1}^\kappa \left\{ x \in \Lambda : \lim_{n\to\infty} \frac{\sum_{k=0}^n \varphi_i^+(f^k x)}{\sum_{k=0}^n \psi_i^+(f^k x)} = \alpha_i \right\},$$

and

$$K_\alpha^- = \bigcap_{i=1}^\kappa \left\{ x \in \Lambda : \lim_{n\to\infty} \frac{\sum_{k=0}^n \varphi_i^-(f^{-k} x)}{\sum_{k=0}^n \psi_i^-(f^{-k} x)} = \alpha_i \right\}.$$

Definition 12.2.1. We define the *dimension spectrum* $\mathcal{D} \colon \mathbb{R}^\kappa \times \mathbb{R}^\kappa \to \mathbb{R}$ by

$$\mathcal{D}(\alpha, \beta) = \dim_H(K_\alpha^+ \cap K_\beta^-).$$

We first consider separately the level sets K_α^+ and K_α^-. Given $\rho > 0$, for each $x \in \Lambda$ we consider the local stable and unstable manifolds $V^s(x) = V_\rho^s(x)$ and $V^u(x) = V_\rho^u(x)$ (see (4.21) and (4.22)). We also consider the *global stable* and *unstable* manifolds at $x \in \Lambda$ defined respectively by

$$W^s(x) = \bigcup_{n\in\mathbb{N}} f^{-n} V^s(f^n x) \quad \text{and} \quad W^u(x) = \bigcup_{n\in\mathbb{N}} f^n V^u(f^{-n} x).$$

We recall the numbers t_s and t_u defined by (4.42).

Theorem 12.2.2. *Let Λ be a locally maximal hyperbolic set of a $C^{1+\varepsilon}$ diffeomorphism f, for some $\varepsilon > 0$, such that f is conformal on Λ. Given pairs of functions $(\Phi^\pm, \Psi^\pm) \in H(\Lambda)$ with $\Psi^\pm > 0$, for each $\alpha \in \mathbb{R}^\kappa$ and $x^\pm \in K_\alpha^\pm$ we have*

$$\Lambda \cap W^s(x^+) \subset K_\alpha^+, \quad \Lambda \cap W^u(x^-) \subset K_\alpha^-, \tag{12.3}$$

and

$$\dim_H K_\alpha^+ = \dim_H(K_\alpha^+ \cap V^u(x^+)) + t_s,$$
$$\dim_H K_\alpha^- = \dim_H(K_\alpha^- \cap V^s(x^-)) + t_u.$$

Proof. Let $a, b \colon \Lambda \to \mathbb{R}$ be continuous functions with $b > 0$. It follows from the exponential behavior of f along the unstable manifolds and the uniform continuity of a and b in Λ that for each $x \in \Lambda$ and $\delta > 0$, given $n \in \mathbb{N}$ sufficiently large we have

$$|a(f^m y) - a(f^m x)| < \delta \quad \text{and} \quad |b(f^m y) - b(f^m x)| < \delta$$

for every $y \in V^s(x)$ and $m > n$. Therefore,

$$
\begin{aligned}
&\left| \frac{\sum_{k=0}^{m} a(f^k y)}{\sum_{k=0}^{m} b(f^k y)} - \frac{\sum_{k=0}^{m} a(f^k x)}{\sum_{k=0}^{m} b(f^k x)} \right| \\
&\leq \frac{\sum_{k=0}^{m} |a(f^k y) - a(f^k x)|}{\sum_{k=0}^{m} b(f^k y)} + \left| \frac{\sum_{k=0}^{m} a(f^k x)}{\sum_{k=0}^{m} b(f^k y)} - \frac{\sum_{k=0}^{m} a(f^k x)}{\sum_{k=0}^{m} b(f^k x)} \right| \\
&\leq \frac{n \sup |b| + (m - n + 1)\delta}{(m + 1) \inf b} + (m + 1) \sup |a| \frac{n \sup b + (m - n + 1)\delta}{(m + 1)^2 (\inf b)^2} \\
&\rightarrow \frac{\delta}{\inf b} + \frac{\delta \sup |a|}{(\inf b)^2}
\end{aligned}
\tag{12.4}
$$

as $m \to \infty$. Now assume that there exists $\beta \in \mathbb{R}$ such that

$$
\lim_{m \to \infty} \frac{\sum_{k=0}^{m} a(f^k x)}{\sum_{k=0}^{m} b(f^k x)} = \beta.
$$

Letting $\delta \to 0$ in (12.4) we obtain

$$
\lim_{m \to \infty} \frac{\sum_{k=0}^{m} a(f^k y)}{\sum_{k=0}^{m} b(f^k y)} = \beta \quad \text{for every} \quad y \in V^s(x).
$$

This implies that $\Lambda \cap V^s(x) \subset K_\alpha^+$ for every $x \in K_\alpha^+$. Furthermore, since the set K_α^+ is f-invariant we conclude that

$$
\Lambda \cap f^{-n} V^s(f^n x) \subset K_\alpha^+
$$

for every $x \in K_\alpha^+$ and $n \in \mathbb{N}$. Therefore, $\Lambda \cap W^s(x) \subset K_\alpha^+$. Similar arguments establish the second inclusion in (12.3).

Since f is conformal on Λ, by results of Hasselblatt in [71] the local product structure (see Definition 4.2.5) is a Lipschitz homeomorphism with Lipschitz inverse. This implies that for each $x \in K_\alpha^+$ the set

$$
\Lambda \cap \bigcup_{y \in K_\alpha^+ \cap V^u(x)} V^s(y)
$$

is taken onto the product $(K_\alpha^+ \cap V^u(x)) \times (\Lambda \cap V^s(x))$ by a Lipschitz map with Lipschitz inverse. In view of (4.40) and (4.41) we obtain

$$
\dim_H K_\alpha^+ = \dim_H(K_\alpha^+ \cap V^u(x)) + t_s.
$$

Similar arguments establish the corresponding identity for K_α^-. This completes the proof of the theorem. $\square$

12.3   Existence of full measures

We denote by $\mathcal{M}$ the family of f-invariant probability Borel measures in Λ, and we define functions $\mathcal{P}^{\pm} \colon \mathcal{M} \to \mathbb{R}^{\kappa}$ by

$$\mathcal{P}^{\pm}(\mu) = \left(\frac{\int_{\Lambda} \varphi_1^{\pm} \, d\mu}{\int_{\Lambda} \psi_1^{\pm} \, d\mu}, \dots, \frac{\int_{\Lambda} \varphi_{\kappa}^{\pm} \, d\mu}{\int_{\Lambda} \psi_{\kappa}^{\pm} \, d\mu} \right).$$

Theorem 12.3.1 ([22]). *Let Λ be a locally maximal hyperbolic set of a $C^{1+\varepsilon}$ diffeomorphism f, for some $\varepsilon > 0$, such that f is conformal and topologically mixing on Λ. Given pairs of functions $(\Phi^{\pm}, \Psi^{\pm}) \in H(\Lambda)$ with $\Psi^{\pm} > 0$, if $\alpha \in \operatorname{int} \mathcal{P}^{+}(\mathcal{M})$ and $\beta \in \operatorname{int} \mathcal{P}^{-}(\mathcal{M})$, then there exists a probability measure ν in Λ with $\nu(K_{\alpha}^{+} \cap K_{\beta}^{-}) = 1$, such that*

$$\lim_{r \to 0} \frac{\log \nu(B(x,r))}{\log r} = \dim_H K_{\alpha}^{+} + \dim_H K_{\beta}^{-} - \dim_H \Lambda \tag{12.5}$$

for ν-almost every $x \in \Lambda$, and

$$\limsup_{r \to 0} \frac{\log \nu(B(x,r))}{\log r} \le \dim_H K_{\alpha}^{+} + \dim_H K_{\beta}^{-} - \dim_H \Lambda \tag{12.6}$$

for every $x \in K_{\alpha}^{+} \cap K_{\beta}^{-}$.

Proof. Consider a Markov partition of Λ, and let $\sigma | \Sigma_A$ be the associated two-sided topological Markov chain. We also consider the coding map $\chi \colon \Sigma_A \to \Lambda$ obtained from the Markov partition, and the maps $\sigma^{+}, \sigma^{-}, \pi^{+}$, and π^{-} defined by (4.31) and (4.32). The following statement is an immediate consequence of Proposition 4.2.11.

Lemma 12.3.2. *For each $i = 1, \dots, \kappa$ there exist Hölder continuous functions φ_i^{u}, ψ_i^{u}, $d^{u} \colon \Sigma_A^{+} \to \mathbb{R}$ and φ_i^{s}, ψ_i^{s}, $d^{s} \colon \Sigma_A^{-} \to \mathbb{R}$, and continuous functions $g_i^{\pm}$, $h_i^{\pm}$, $\rho^{\pm} \colon \Sigma_A \to \mathbb{R}$ such that*

$$\varphi_i^{+} \circ \chi = \varphi_i^{u} \circ \pi^{+} + g_i^{+} - g_i^{+} \circ \sigma,$$
$$\psi_i^{+} \circ \chi = \psi_i^{u} \circ \pi^{+} + h_i^{+} - h_i^{+} \circ \sigma,$$
$$\log \|df|E^{u}\| \circ \chi = d^{u} \circ \pi^{+} + \rho^{+} - \rho^{+} \circ \sigma,$$

and

$$\varphi_i^{-} \circ \chi = \varphi_i^{s} \circ \pi^{-} + g_i^{-} - g_i^{-} \circ \sigma,$$
$$\psi_i^{-} \circ \chi = \psi_i^{s} \circ \pi^{-} + h_i^{-} - h_i^{-} \circ \sigma,$$
$$\log \|df^{-1}|E^{s}\| \circ \chi = d^{s} \circ \pi^{-} + \rho^{-} - \rho^{-} \circ \sigma.$$

Now we initiate the process of construction of the measure ν. Set

$$d^+ = \dim_H K_\alpha^+ - t_s \quad \text{and} \quad d^- = \dim_H K_\beta^- - t_u.$$

By (4.43) we have

$$d^+ + d^- = \dim_H K_\alpha^+ + \dim_H K_\beta^- - \dim_H \Lambda. \tag{12.7}$$

We also write

$$\Phi^u = (\varphi_1^u, \ldots, \varphi_\kappa^u), \quad \Psi^u = (\psi_1^u, \ldots, \psi_\kappa^u),$$

$$\Phi^s = (\varphi_1^s, \ldots, \varphi_\kappa^s), \quad \Psi^s = (\psi_1^s, \ldots, \psi_\kappa^s).$$

Given vectors $q^\pm \in \mathbb{R}^\kappa$, we define Hölder continuous functions $a^u \colon \Sigma_A^+ \to \mathbb{R}$ and $b^s \colon \Sigma_A^- \to \mathbb{R}$ by

$$\begin{aligned} a^u &= \langle q^+, \Phi^u - \alpha * \Psi^u \rangle - d^+ d^u, \\ b^s &= \langle q^-, \Phi^s - \beta * \Psi^s \rangle - d^- d^s, \end{aligned} \tag{12.8}$$

where $\langle \cdot, \cdot \rangle$ is the standard inner product in $\mathbb{R}^\kappa$, and where

$$\alpha * (\varphi_1, \ldots, \varphi_\kappa) = (\alpha_1 \varphi_1, \ldots, \alpha_\kappa \varphi_\kappa).$$

Since f is topologically mixing on Λ (and hence the same happens with f^{-1}), there exist a unique equilibrium measure μ^u of a^u in Σ_A^+ (with respect to σ^+), and a unique equilibrium measure μ^s of b^s in Σ_A^- (with respect to σ^-). We note that μ^u and μ^s are Gibbs measures. Since $\alpha \in \text{int } \mathcal{P}^+(\mathcal{M})$ and $\beta \in \text{int } \mathcal{P}^-(\mathcal{M})$, the following statement is an immediate consequence of Theorem 10.1.4.

Lemma 12.3.3. *There exist vectors $q^\pm \in \mathbb{R}^\kappa$ such that the corresponding measures μ^u and μ^s satisfy*

$$P_{\sigma^+}(a^u) = P_{\sigma^-}(b^s) = 0, \tag{12.9}$$

and

$$\int_{\Sigma_A^+} \Phi^u \, d\mu^u = \alpha * \int_{\Sigma_A^+} \Psi^u \, d\mu^u, \quad \int_{\Sigma_A^-} \Phi^s \, d\mu^s = \beta * \int_{\Sigma_A^-} \Psi^s \, d\mu^s.$$

Take $x \in \Lambda$, and let $R(x) \subset \Lambda$ be a rectangle of the Markov partition containing x. We define measures ν^u and ν^s in $R(x)$ by

$$\nu^u = \mu^u \circ \pi^+ \circ \chi^{-1} \quad \text{and} \quad \nu^s = \mu^s \circ \pi^- \circ \chi^{-1},$$

using in (12.8) the vectors $q^\pm$ in Lemma 12.3.3. Finally, we consider the measure $\nu = \nu^u \times \nu^s$ in $R(x)$. Since μ^u and μ^s are Gibbs measures, we have (see (4.36) and (4.37))

$$\nu(R(x)) = \mu^u(C_{i_0}^+) \mu^s(C_{i_0}^-) > 0.$$

Lemma 12.3.4. *Given $\gamma > 1$, there exists $K > 0$ such that for every $x \in \Lambda$ and every sufficiently small $r > 0$ we have*

$$\nu(B(x, \gamma r)) \leq K\nu(B(x, r)).$$

Proof of the lemma. Consider the Hölder continuous functions a, $b\colon \Lambda \to \mathbb{R}$ defined by

$$a = \langle q^+, \Phi^+ - \alpha * \Psi^+ \rangle - d^+ \log \|df|E^u\|,$$
$$b = \langle q^-, \Phi^- - \beta * \Psi^- \rangle + d^- \log \|df|E^s\|.$$

We note that ν^u is the equilibrium measure of a with respect to f, and that ν^s is the equilibrium measure of b with respect to f^{-1}. Repeating arguments in the proof of Lemma 6.1.5 we obtain the desired statement. $\qquad\square$

We proceed with the proof of the theorem.

Lemma 12.3.5. *We have*

$$\liminf_{r \to 0} \frac{\log \nu(B(x, r))}{\log r} \geq \dim_H K_\alpha^+ + \dim_H K_\beta^- - \dim_H \Lambda$$

for ν-almost every $x \in \Lambda$.

Proof of the lemma. Using the variational principle of the topological pressure for the functions in (12.8) it follows from Lemma 12.3.3 that

$$\frac{h_{\mu^u}(\sigma^+)}{\int_{\Sigma_A^+} d^u \, d\mu^u} = d^+ \quad \text{and} \quad \frac{h_{\mu^s}(\sigma^-)}{\int_{\Sigma_A^-} d^s \, d\mu^s} = d^-. \tag{12.10}$$

By Shannon–McMillan–Breiman's theorem and Birkhoff's ergodic theorem, it follows from (12.10) that given $\delta > 0$, for μ^u-almost every $\omega^+ \in C_{i_0}^+$ and μ^s-almost every $\omega^- \in C_{i_0}^-$ there exists $l(\omega) \in \mathbb{N}$ such that for every n, $m > l(\omega)$ we have

$$d^+ - \delta < -\frac{\log \mu^u(C_{i_0 \cdots i_n}^+)}{\sum_{k=0}^n d^u((\sigma^+)^k \omega^+)} < d^+ + \delta,$$

and

$$d^- - \delta < -\frac{\log \mu^s(C_{i_{-m} \cdots i_0}^-)}{\sum_{k=0}^m d^s((\sigma^-)^k \omega^-)} < d^- + \delta.$$

For each $\omega \in \Sigma_A$ and each sufficiently small $r > 0$, let $n = n(\omega, r)$ and $m = m(\omega, r)$ be the unique integers such that

$$-\sum_{k=0}^n d^u((\sigma^+)^k \omega^+) > \log r, \quad -\sum_{k=0}^{n+1} d^u((\sigma^+)^k \omega^+) \leq \log r, \tag{12.11}$$

and

$$-\sum_{k=0}^{m} d^s((\sigma^-)^k \omega^-) > \log r, \quad -\sum_{k=0}^{m+1} d^s((\sigma^-)^k \omega^-) \le \log r, \tag{12.12}$$

where $\omega^+ = \pi^+ \omega$ and $\omega^- = \pi^- \omega$. In a similar manner to that in the proof of Theorem 6.2.1, there exists a constant $\rho > 1$ (independent of $x = \chi(\omega) \in \Lambda$ and $r > 0$) such that

$$B(y, r/\rho) \cap \Lambda \subset \chi(C_{i_{-m} \cdots i_n}) \subset B(x, \rho r) \tag{12.13}$$

for some point $y \in \chi(C_{i_{-m} \cdots i_n})$, where $\omega = (\cdots i_{-1} i_0 i_1 \cdots)$. Furthermore, by Lemma 12.3.4 there exists a constant $c > 0$ (independent of x and r) such that

$$\nu(B(y, 2\rho r)) \le c\nu(B(y, r/\rho)).$$

Since $B(x, r) \subset B(y, 2\rho r)$, it follows from (12.13) that

$$\nu(B(x,r)) \le \nu(B(y, 2\rho r)) \le c\nu(B(y, r/\rho))$$
$$\le c\nu(\chi(C_{i_{-m} \cdots i_n})) = c\mu^u(C^+_{i_0 \cdots i_n}) \mu^s(C^-_{i_{-m} \cdots i_0})$$
$$< c \exp\left[(-d^+ + \delta) \sum_{k=0}^{n} d^u((\sigma^+)^k \omega^+) \right]$$
$$\times \exp\left[(-d^- + \delta) \sum_{k=0}^{m} d^s((\sigma^-)^k \omega^-) \right]$$
$$\le c \exp[(\log r + \sup |d^u|)(d^+ - \delta) + (\log r + \sup |d^s|)(d^- - \delta)],$$

and hence

$$\liminf_{r \to 0} \frac{\log \nu(B(x,r))}{\log r} \ge d^+ + d^- - 2\delta,$$

for ν-almost every point $x \in \Lambda$. In view of (12.7), the arbitrariness of δ implies the desired result. $\square$

Now let $\Lambda_{\alpha\beta} \subset \Sigma_A$ be the set of points $\omega \in \Sigma_A$ such that for $i = 1, \ldots, \kappa$ we have

$$\lim_{n \to \infty} \frac{\sum_{k=0}^{n} \varphi_i^u((\sigma^+)^k \omega^+)}{\sum_{k=0}^{n} \psi_i^u((\sigma^+)^k \omega^+)} = \alpha_i, \quad \lim_{n \to \infty} \frac{\sum_{k=0}^{n} \varphi_i^s((\sigma^-)^k \omega^-)}{\sum_{k=0}^{n} \psi_i^s((\sigma^-)^k \omega^-)} = \beta_i.$$

Lemma 12.3.6. *The inequality in (12.6) holds for every $x \in \chi(\Lambda_{\alpha\beta})$.*

Proof of the lemma. Given $\delta > 0$ and $\omega \in \Lambda_{\alpha\beta}$, there exists $r(\omega) \in \mathbb{N}$ such that for every $n > r(\omega)$ we have

$$\left\| \left\langle q^+, \sum_{k=0}^{n} (\Phi^u - \alpha * \Psi^u)((\sigma^+)^k \omega^+) \right\rangle \right\| < \delta n \sup |\langle q^+, \Psi^u \rangle|, \tag{12.14}$$

and

$$\left\| \left\langle q^-, \sum_{k=0}^n (\Phi^s - \beta * \Psi^s)((\sigma^-)^k \omega^-) \right\rangle \right\| < \delta n \sup \left| \langle q^-, \Psi^s \rangle \right|. \tag{12.15}$$

Since μ^u and μ^s are Gibbs measures, in view of (12.9) there exists a constant $D > 0$ such that for every $\omega^+ \in C_{i_0}^+$, $\omega^- \in C_{i_0}^-$, and $n, m \in \mathbb{N}$ we have

$$D^{-1} < \frac{\mu^u(C_{i_0 \cdots i_n}^+)}{\exp \sum_{k=0}^n a^u((\sigma^+)^k \omega^+)} < D,$$

and

$$D^{-1} < \frac{\mu^s(C_{i_{-m} \cdots i_0}^-)}{\exp \sum_{k=0}^m b^s((\sigma^-)^k \omega^-)} < D.$$

Combining these inequalities with (12.14)–(12.15) we obtain

$$\mu^u(C_{i_0 \cdots i_n}^+) > D^{-1} \exp\left[-d^+ \sum_{k=0}^n d^u((\sigma^+)^k \omega^+) - \delta n \sup \left| \langle q^+, \Psi^u \rangle \right| \right], \tag{12.16}$$

and

$$\mu^s(C_{i_{-m} \cdots i_0}^-) > D^{-1} \exp\left[-d^- \sum_{k=0}^m d^s((\sigma^-)^k \omega^-) - \delta m \sup \left| \langle q^-, \Psi^s \rangle \right| \right]. \tag{12.17}$$

Given $\omega \in \Lambda_{\alpha\beta}$, we take $r > 0$ sufficiently small such that $n(\omega, r) > r(\omega)$ and $m(\omega, r) > r(\omega)$ (the hyperbolicity of f on Λ guarantees that this is always possible). Combining (12.16)–(12.17) with (12.11)–(12.12) we obtain

$$\nu(B(x, \rho r)) \geq \nu(\chi(C_{i_{-m} \cdots i_n})) = \mu^u(C_{i_0 \cdots i_n}^+) \mu^s(C_{i_{-m} \cdots i_0}^-)$$
$$\geq D^{-2} r^{d^+ + d^-} \exp(-\delta n \sup \left| \langle q^+, \Psi^u \rangle \right| - \delta m \sup \left| \langle q^-, \Psi^s \rangle \right|)$$

for all sufficiently small $r > 0$. Note that by (12.11)–(12.12) we have

$$-n \inf d^u > \log r \quad \text{and} \quad - m \inf d^s > \log r.$$

Therefore, for every $x = \chi(\omega)$ with $\omega \in \Lambda_{\alpha\beta}$ we obtain

$$\limsup_{r \to 0} \frac{\log \nu(B(x, r))}{\log r} \leq d^+ + d^- + \delta \left(\frac{\sup \left| \langle q^+, \Psi^u \rangle \right|}{\inf d^u} + \frac{\sup \left| \langle q^-, \Psi^s \rangle \right|}{\inf d^s} \right).$$

Since δ is arbitrary, for every $x \in \chi(\Lambda_{\alpha\beta})$ we have

$$\limsup_{r \to 0} \frac{\log \nu(B(x, r))}{\log r} \leq d^+ + d^-.$$

In view of (12.7) this establishes the desired statement. $\qquad\square$

Lemma 12.3.7. *We have $\chi(\Lambda_{\alpha\beta}) = K_\alpha^+ \cap K_\beta^-$.*

Proof of the lemma. It follows from Lemma 12.3.2 that

$$
\begin{aligned}
\varphi_i^+(f^k(\chi(\omega))) &= \psi_i^+(\chi(\sigma^k\omega)) \\
&= \varphi_i^u(\pi^+(\sigma^k\omega)) + g_i^+(\sigma^k\omega) - g_i^+(\sigma^{k+1}\omega) \\
&= \varphi_i^u((\sigma^+)^k\omega^+) + g_i^+(\sigma^k\omega) - g_i^+(\sigma^{k+1}\omega),
\end{aligned}
$$

with similar identities for the functions ψ_i^+, φ_i^-, and ψ_i^-. Therefore,

$$
\frac{\sum_{k=0}^{n-1} \varphi_i^+(f^k(\chi(\omega)))}{\sum_{k=0}^{n-1} \psi_i^+(f^k(\chi(\omega)))} = \frac{\sum_{k=0}^{n-1} \varphi_i^u((\sigma^+)^k\omega^+) + g_i^+(\omega) - g_i^+(\sigma^n\omega)}{\sum_{k=0}^{n-1} \psi_i^u((\sigma^+)^k\omega^+) + h_i^+(\omega) - h_i^+(\sigma^n\omega)}, \tag{12.18}
$$

and

$$
\frac{\sum_{k=0}^{n-1} \varphi_i^-(f^{-k}(\chi(\omega)))}{\sum_{k=0}^{n-1} \psi_i^-(f^{-k}(\chi(\omega)))} = \frac{\sum_{k=0}^{n-1} \varphi_i^s((\sigma^-)^k\omega^-) + g_i^-(\omega) - g_i^-(\sigma^n\omega)}{\sum_{k=0}^{n-1} \psi_i^s((\sigma^-)^k\omega^-) + h_i^-(\omega) - h_i^-(\sigma^n\omega)}. \tag{12.19}
$$

Now we observe that

$$
\sum_{k=0}^{n-1} \psi_i^u((\sigma^+)^k\omega^+) \geq n \inf \psi_i^+ - 2\sup|h_i^+|,
$$

and

$$
\sum_{k=0}^{n-1} \psi_i^s((\sigma^-)^k\omega^-) \geq n \inf \psi_i^- - 2\sup|h_i^-|.
$$

Since $\psi_i^\pm > 0$ for $i = 1,\ldots,\kappa$, it follows from these inequalities that the limits

$$
\lim_{n\to\infty} \frac{\sum_{k=0}^{n-1} \varphi_i^+(f^k(\chi(\omega)))}{\sum_{k=0}^{n-1} \psi_i^+(f^k(\chi(\omega)))} \quad \text{and} \quad \lim_{n\to\infty} \frac{\sum_{k=0}^{n-1} \varphi_i^-(f^{-k}(\chi(\omega)))}{\sum_{k=0}^{n-1} \psi_i^-(f^{-k}(\chi(\omega)))}
$$

exist if and only if the limits

$$
\lim_{n\to\infty} \frac{\sum_{k=0}^{n-1} \varphi_i^u((\sigma^+)^k\omega^+)}{\sum_{k=0}^{n-1} \psi_i^u((\sigma^+)^k\omega^+)} \quad \text{and} \quad \lim_{n\to\infty} \frac{\sum_{k=0}^{n-1} \varphi_i^s((\sigma^-)^k\omega^-)}{\sum_{k=0}^{n-1} \psi_i^s((\sigma^-)^k\omega^-)}
$$

exist. In this case we have

$$
\lim_{n\to\infty} \frac{\sum_{k=0}^{n-1} \varphi_i^+(f^k(\chi(\omega)))}{\sum_{k=0}^{n-1} \psi_i^+(f^k(\chi(\omega)))} = \lim_{n\to\infty} \frac{\sum_{k=0}^{n-1} \varphi_i^u((\sigma^+)^k\omega^+)}{\sum_{k=0}^{n-1} \psi_i^u((\sigma^+)^k\omega^+)},
$$

and

$$
\lim_{n\to\infty} \frac{\sum_{k=0}^{n-1} \varphi_i^-(f^{-k}(\chi(\omega)))}{\sum_{k=0}^{n-1} \psi_i^-(f^{-k}(\chi(\omega)))} = \lim_{n\to\infty} \frac{\sum_{k=0}^{n-1} \varphi_i^s((\sigma^-)^k\omega^-)}{\sum_{k=0}^{n-1} \psi_i^s((\sigma^-)^k\omega^-)}.
$$

In particular, $\omega \in \Lambda_{\alpha\beta}$ if and only if $\chi(\omega) \in K_\alpha^+ \cap K_\beta^-$. This shows that $\chi(\Lambda_{\alpha\beta}) = K_\alpha^+ \cap K_\beta^-$. $\qquad\square$

Combining the above lemmas we readily obtain the statement in the theorem.
□

We note that the measure ν constructed in the proof of Theorem 12.3.1 is not invariant.

12.4 Formula for the spectrum

We can use the former results to obtain a formula for the spectrum $\mathcal{D}$.

Theorem 12.4.1 ([22]). *Let Λ be a locally maximal hyperbolic set of a $C^{1+\varepsilon}$ diffeomorphism f, for some $\varepsilon > 0$, such that f is conformal and topologically mixing on Λ. Given pairs of functions $(\Phi^{\pm}, \Psi^{\pm}) \in H(\Lambda)$ with $\Psi^{\pm} > 0$, if $\alpha \in$ int $\mathcal{P}^{+}(\mathcal{M})$ and $\beta \in$ int $\mathcal{P}^{-}(\mathcal{M})$, then the set $K_{\alpha}^{+} \cap K_{\beta}^{-}$ is dense in Λ, and*

$$\mathcal{D}(\alpha, \beta) = \dim_H K_{\alpha}^{+} + \dim_H K_{\beta}^{-} - \dim_H \Lambda. \tag{12.20}$$

Proof. It follows easily from the construction of the functions Φ^u, Ψ^u, Φ^s, and Ψ^s that the sets K_{α}^{+} and K_{β}^{-} are dense in Λ (we note that by Theorem 12.3.1 they are nonempty). Namely, by Lemma 12.3.2 (see also (12.18) and (12.19)) the ratios of Birkhoff averages of these functions only depend on the symbolic past (in the case of K_{α}^{+}) or on the symbolic future (in the case of K_{β}^{-}). The fact that the set $K_{\alpha}^{+} \cap K_{\beta}^{-}$ is dense in Λ is thus an immediate consequence of the identities

$$\overline{\bigcup_{k \in \mathbb{N}} (\sigma^{+})^{-k} \omega^{+}} = \Sigma_A^{+} \quad \text{and} \quad \overline{\bigcup_{k \in \mathbb{N}} (\sigma^{-})^{-k} \omega^{-}} = \Sigma_A^{-},$$

valid for every $\omega^{+} \in \Sigma_A^{+}$ and $\omega^{-} \in \Sigma_A^{-}$.

Now let ν be the measure constructed in Theorem 12.3.1. By Theorem 2.1.5, it follows from (12.5) that

$$\dim_H \nu = \dim_H K_{\alpha}^{+} + \dim_H K_{\beta}^{-} - \dim_H \Lambda.$$

Since $\nu(K_{\alpha}^{+} \cap K_{\beta}^{-}) = 1$, we obtain

$$\dim_H(K_{\alpha}^{+} \cap K_{\beta}^{-}) \geq \dim_H \nu = \dim_H K_{\alpha}^{+} + \dim_H K_{\beta}^{-} - \dim_H \Lambda.$$

For the reverse inequality, we note that by Theorem 2.1.5 it follows from (12.6) that

$$\dim_H(K_{\alpha}^{+} \cap K_{\beta}^{-}) \leq \dim_H K_{\alpha}^{+} + \dim_H K_{\beta}^{-} - \dim_H \Lambda.$$

This completes the proof of the theorem.
□

12.5   Conditional variational principle

We also obtain a conditional variational principle for the spectrum $\mathcal{D}$.

Theorem 12.5.1 ([22]). *Let Λ be a locally maximal hyperbolic set of a $C^{1+\varepsilon}$ diffeomorphism f, for some $\varepsilon > 0$, such that f is conformal and topologically mixing on Λ. Given pairs of functions $(\Phi^\pm, \Psi^\pm) \in H(\Lambda)$ with $\Psi^\pm > 0$, the following properties hold:*

  1. *if $\alpha \in \operatorname{int} \mathcal{P}^+(\mathcal{M})$ and $\beta \in \operatorname{int} \mathcal{P}^-(\mathcal{M})$, then*

$$
\mathcal{D}(\alpha, \beta) = \max\left\{ \frac{h_\mu(f)}{-\int_\Lambda \log \|df|E^s\|\, d\mu} : \mu \in \mathcal{M} \text{ and } \mathcal{P}^+(\mu) = \alpha \right\}
$$
$$
+ \max\left\{ \frac{h_\mu(f)}{\int_\Lambda \log \|df|E^u\|\, d\mu} : \mu \in \mathcal{M} \text{ and } \mathcal{P}^-(\mu) = \beta \right\};
$$

  2. *the spectrum $\mathcal{D}$ is analytic in $\operatorname{int} \mathcal{P}^+(\mathcal{M}) \times \operatorname{int} \mathcal{P}^-(\mathcal{M})$.*

Proof. In view of Theorem 12.4.1 (see (12.20)), the statements are immediate consequences of Theorems 10.1.4 and 10.3.1. $\qquad\square$

Part IV

Hyperbolicity and Recurrence

Chapter 13

Pointwise Dimension for Hyperbolic Dynamics

Sometimes a given *global* invariant can be built with the help of a *local* quantity. For example, the Kolmogorov–Sinai entropy and the Hausdorff dimension, which are quantities of global nature, can be built (in a rigorous mathematical sense) respectively with the help of the local entropy and the pointwise dimension. In the case of the entropy this is due to Shannon–McMillan–Breiman's theorem: the Kolmogorov–Sinai entropy is obtained integrating the local entropy. In this chapter we are mostly interested in the Hausdorff dimension of an invariant measure. In particular, for repellers and hyperbolic sets of conformal maps we establish explicit formulas for the pointwise dimension of an arbitrary invariant measure in terms of the local entropy and the Lyapunov exponents. This allows us to show that the Hausdorff dimension of a (nonergodic) invariant measure is equal to the essential supremum of the Hausdorff dimensions of the measures in an ergodic decomposition.

13.1 Repellers of conformal maps

13.1.1 Formula for the pointwise dimension

Let J be a repeller of a $C^{1+\varepsilon}$ transformation f, for some $\varepsilon > 0$, such that f is conformal on J. Let also μ be an f-invariant probability measure in J.

By Birkhoff's ergodic theorem, for μ-almost every $x \in J$ there exists the limit

$$\lambda(x) = \lim_{n \to \infty} \frac{1}{n} \log \|d_x f^n\| = \lim_{n \to \infty} \frac{1}{n} \sum_{k=0}^{n-1} \varphi(f^k x), \tag{13.1}$$

where $\varphi(x) = \log \|d_x f\|$. We note that the function $x \mapsto \lambda(x)$ is f-invariant μ-almost everywhere.

For μ-almost every $x \in J$, let $h_\mu(x)$ be the local entropy of μ at the point x (see Lemma 7.2.8). Let also $\mathcal{R} = \{R_1, \ldots, R_k\}$ be any partition of J. Given $i_0, \ldots, i_n \in \{1, \ldots, k\}$ we define the *rectangle*

$$R_{i_0 \cdots i_n} = \{x \in J : f^j x \in R_{i_j} \text{ for } j = 0, \ldots, n\}. \tag{13.2}$$

By Shannon–McMillan–Breiman's theorem, for μ-almost every $x \in J$ we have

$$h_\mu(x) = \lim_{n \to \infty} -\frac{1}{n} \log \mu(R_n(x)), \tag{13.3}$$

where $R_n(x)$ is any rectangle $R_{i_0 \cdots i_n}$ containing x. For definiteness, we assume from the beginning that for each x we make a particular choice of rectangles $R_n(x)$ for all $n \in \mathbb{N}$. Let X be a fixed full μ-measure f-invariant set of points $x \in J$ such that:

1. the number $\lambda(x)$ in (13.1) is well-defined;

2. the number $h_\mu(x)$ in (7.9) is well-defined and satisfies (13.3).

The following is a local formula for the pointwise dimension of invariant measures that are not necessarily ergodic. It was obtained by Barreira and Wolf in [24].

Theorem 13.1.1. *Let J be a repeller of a $C^{1+\varepsilon}$ transformation f, for some $\varepsilon > 0$, such that f is conformal on J, and let μ be an f-invariant probability measure in J. Then for μ-almost every $x \in J$ we have*

$$\underline{d}_\mu(x) = \overline{d}_\mu(x) = \frac{h_\mu(x)}{\lambda(x)}.$$

Proof. Fix $\delta > 0$. For each $x \in X$ there exists $p(x) \in \mathbb{N}$ such that if $n \geq p(x)$, then

$$\lambda(x) - \delta < \frac{1}{n} \log \|d_x f^n\| < \lambda(x) + \delta, \tag{13.4}$$

and

$$-h_\mu(x) - \delta < \frac{1}{n} \log \mu(R_n(x)) < -h_\mu(x) + \delta. \tag{13.5}$$

For each $\ell \in \mathbb{N}$, let

$$Q_\ell = \{x \in X : p(x) \leq \ell\}.$$

We note that $\bigcup_{\ell \in \mathbb{N}} Q_\ell = X$. Furthermore, for each $x \in X$ there exists $r(x) > 0$ such that for every $r \in (0, r(x))$ there exists a unique integer $n = n(x, r) \geq p(x)$ such that

$$\|d_x f^n\|^{-1} \geq r \quad \text{and} \quad \|d_x f^{n+1}\|^{-1} < r. \tag{13.6}$$

We write $R(x, r) = R_{n(x,r)}(x)$ for each $x \in X$ and $r \in (0, r(x))$.

We first obtain an upper bound for the pointwise dimension. Since f is conformal on J, there exists $\kappa > 0$ (independent of x and r) such that $B(x, \kappa r) \supset R(x, r)$ for each $x \in X$ and $r \in (0, r(x))$. Hence,

$$\mu(B(x, \kappa r)) \geq \mu(R(x, r)) \geq \exp[(-h_\mu(x) - \delta)n].$$

By (13.4) and (13.6) we obtain

$$\mu(B(x, \kappa r)) \geq \exp\left[(h_\mu(x) + \delta)\frac{\log r}{\lambda(x) - \delta}\right],$$

and thus,

$$\overline{d}_\mu(x) \leq \frac{h_\mu(x) + \delta}{\lambda(x) - \delta}.$$

The arbitrariness of δ implies that $\overline{d}_\mu(x) \leq h_\mu(x)/\lambda(x)$ for every $x \in X$, and hence for μ-almost every $x \in J$.

Now we obtain a lower bound for the pointwise dimension. Given $x \in X$ we define

$$\Gamma(x) = \{y \in X : |h_\mu(y) - h_\mu(x)| < \delta \text{ and } |\lambda(y) - \lambda(x)| < \delta\}. \tag{13.7}$$

Note that $\Gamma(x)$ is f-invariant. The sets $\Gamma(x)$ cover X, and we can choose points $y_i \in X$, $i \in \mathbb{N}$ such that setting $\Gamma_i = \Gamma(y_i)$ we have $\mu(\Gamma_i) > 0$ for each i, and $\mu(\bigcup_{i \in \mathbb{N}} \Gamma_i) = 1$.

Fix $i, \ell \in \mathbb{N}$. We construct a Moran cover of $\Gamma_i \cap Q_\ell$ by sets of the form $R(x, r) \cap \Gamma_i \cap Q_\ell$. For each $x \in \Gamma_i \cap Q_\ell$ and $r > 0$, we denote by $R'(x, r)$ the largest rectangle containing x (among those in (13.2)) with the property that

$$R'(x, r) = R(y, r) \quad \text{for some} \quad y \in R'(x, r) \cap \Gamma_i \cap Q_\ell,$$

and that

$$R(z, r) \subset R'(x, r) \quad \text{for every} \quad z \in R'(x, r) \cap \Gamma_i \cap Q_\ell.$$

We note that two sets $R'(x, r)$ and $R'(y, r)$ either coincide or intersect at most along their boundaries.

The Borel density lemma (see, for example, [59, Theorem 2.9.11]) tells us that for μ-almost every $x \in \Gamma_i \cap Q_\ell$ we have

$$\lim_{r \to 0} \frac{\mu(B(x, r) \cap \Gamma_i \cap Q_\ell)}{\mu(B(x, r))} = 1, \tag{13.8}$$

and thus there exists $\bar{r}(x) > 0$ such that for each $r \in (0, \bar{r}(x))$,

$$\mu(B(x, r)) \leq 2\mu(B(x, r) \cap \Gamma_i \cap Q_\ell). \tag{13.9}$$

Since f is conformal on J, there exist a constant $K > 0$ (independent of x and r) and points $x_1, \ldots, x_k \in \Gamma_i \cap Q_\ell$ with $k \leq K$ such that

$$B(x,r) \cap \Gamma_i \cap Q_\ell \subset \bigcup_{j=1}^{k} R'(x_j, r).$$

By (13.5) and (13.9) we obtain

$$\mu(B(x,r)) \leq 2\mu(B(x,r) \cap \Gamma_i \cap Q_\ell) \leq 2\sum_{j=1}^{k} \mu(R'(x_j,r))$$

$$\leq 2\sum_{j=1}^{k} \exp[(-h_\mu(x_j) + \delta)n(x_j, r)].$$

By the definition of Γ_i and (13.6) we conclude that

$$\mu(B(x,r)) \leq 2\sum_{j=1}^{k} \exp\left[(h_\mu(y_i) - 2\delta)\frac{\log r + \max\varphi}{\lambda(x_j) + \delta}\right]$$

$$\leq 2K \exp\left[(h_\mu(x) - 3\delta)\frac{\log r + \max\varphi}{\lambda(x) + 2\delta}\right],$$

where $\varphi = \log \|df\|$, and thus,

$$\underline{d}_\mu(x) \geq \frac{h_\mu(x) - 3\delta}{\lambda(x) + 2\delta}.$$

The arbitrariness of δ implies that

$$\underline{d}_\mu(x) \geq h_\mu(x)/\lambda(x) \tag{13.10}$$

for μ-almost every $x \in \Gamma_i \cap Q_\ell$. Letting $\ell \to \infty$ we conclude that (13.10) holds for μ-almost every $x \in \Gamma_i$. Finally, since $\bigcup_{\ell \in \mathbb{N}} \Gamma_i$ has full μ-measure, (13.10) holds for μ-almost every $x \in J$. $\qquad\square$

Theorem 13.1.1 can be used to describe how the Hausdorff dimension of an invariant measure behaves under an ergodic decomposition. We first recall the notion of ergodic decomposition. Let $\mathfrak{M}$ be the family of f-invariant probability Borel measures in a compact metric space X, and let $\mathfrak{M}_E \subset \mathfrak{M}$ be the subset of all ergodic measures.

Definition 13.1.2. A probability Borel measure τ in $\mathfrak{M}$ (with the weak* topology) is an *ergodic decomposition* of a measure $\mu \in \mathfrak{M}$ if $\tau(\mathfrak{M}_E) = 1$, and

$$\int_X \varphi \, d\mu = \int_{\mathfrak{M}} \left(\int_X \varphi \, d\nu\right) d\tau(\nu)$$

for every continuous function $\varphi \colon X \to \mathbb{R}$.

We establish a lower bound for the dimension of an invariant measure.

Proposition 13.1.3. *Let $f: X \to X$ be a Borel measurable transformation preserving a probability measure μ in X. For any ergodic decomposition τ of μ we have*

$$\dim_H \mu \geq \operatorname{ess\,sup}\{\dim_H \nu : \nu \in \mathcal{M}_E\}, \tag{13.11}$$

with the essential supremum taken with respect to τ.

Proof. If $\mu(X \setminus Z) = 0$, then $\nu(X \setminus Z) = 0$ for τ-almost every $\nu \in \mathcal{M}$. Hence,

$$\dim_H Z \geq \dim_H \nu \text{ for } \tau\text{-almost every } \nu \in \mathcal{M},$$

and we obtain

$$\dim_H Z \geq \operatorname{ess\,sup}\{\dim_H \nu : \nu \in \mathcal{M}_E\}.$$

Taking the infimum over all sets Z with $\mu(X \setminus Z) = 0$ we obtain the desired statement. $\qquad\square$

Example 13.1.4. We note that inequality (13.11) may be strict. A simple example is given by a rational rotation of the circle. In this case each measure supported on a periodic orbit has zero Hausdorff dimension, and thus when μ is the Lebesgue measure we have a strict inequality in (13.11).

We observe that when the number of ergodic components is finite or even infinite countable it is straightforward to verify that (13.11) becomes an identity, that is,

$$\dim_H \mu = \operatorname{ess\,sup}\{\dim_H \nu : \nu \in \mathcal{M}_E\}. \tag{13.12}$$

Indeed, assume that $X = \bigcup_{n \in \mathbb{N} \cup \{0\}} X_n$ for some pairwise disjoint f-invariant sets X_n, $n \in \mathbb{N} \cup \{0\}$ such that $f|X_n$ is ergodic with respect to μ for each $n \in \mathbb{N}$, and $\mu(X_0) = 0$ (that is, up to a zero measure set the number of ergodic components is countable). Then it is simple to show that

$$\dim_H \mu = \sup\{\dim_H(\mu|X_n) : n \in \mathbb{N}\}.$$

In particular, (13.12) holds in the context of smooth ergodic theory (see [8]). Namely, let μ be a finite hyperbolic measure (see Definition 14.3.1) invariant under a $C^{1+\varepsilon}$ diffeomorphism of a compact manifold. Pesin showed in [113] that if μ is absolutely continuous with respect to the volume, then up to a zero measure set the number of ergodic components is countable, and thus (13.12) holds.

Combining Theorem 13.1.1 with (2.6), and using the μ-almost everywhere f-invariance of the functions h_μ and λ, we obtain the following formulas for the Hausdorff dimension of an invariant measure (that is not necessarily ergodic).

Corollary 13.1.5. *If J is a repeller of a $C^{1+\varepsilon}$ transformation f, for some $\varepsilon > 0$, such that f is conformal on J, and μ is an f-invariant probability measure in J, then*

$$\dim_H \mu = \operatorname{ess\,sup}\left\{\frac{h_\mu(x)}{\lambda(x)} : x \in J\right\}.$$

If, in addition, μ is ergodic, then

$$\underline{d}_\mu(x) = \overline{d}_\mu(x) = \dim_H \mu = \frac{h_\mu(f)}{\int_J \log \|df\| \, d\mu} \qquad (13.13)$$

for μ-almost every $x \in J$.

Identity (13.13) was first established by Pesin in [115]. We emphasize that when μ is not ergodic, in general (13.13) may not hold. Examples can be readily obtained from the fact that given invariant probability measures μ_1, μ_2 and constants c_1, $c_2 > 0$ with $c_1 + c_2 = 1$, the measure $\mu = c_1\mu_1 + c_2\mu_2$ satisfies

$$\dim_H \mu = \max\{\dim_H \mu_1, \dim_H \mu_2\}, \qquad (13.14)$$

and

$$h_\mu(f) = c_1 h_{\mu_1}(f) + c_2 h_{\mu_2}(f). \qquad (13.15)$$

13.1.2　Dimension along ergodic decompositions

The following statement gives a formula for the Hausdorff dimension of an invariant measure in terms of an ergodic decomposition. We recall that $\mathcal{M}_E$ is the set of all ergodic f-invariant probability measures in J.

Theorem 13.1.6 ([24]). *Let J be a repeller of a $C^{1+\varepsilon}$ transformation f, for some $\varepsilon > 0$, such that f is conformal on J, and let μ be an f-invariant probability measure in J. For any ergodic decomposition τ of μ we have*

$$\dim_H \mu = \operatorname{ess\,sup}\{\dim_H \nu : \nu \in \mathcal{M}_E\}, \qquad (13.16)$$

with the essential supremum taken with respect to τ.

Proof. By Proposition 13.1.3 we have

$$\dim_H \mu \geq \operatorname{ess\,sup}\{\dim_H \nu : \nu \in \mathcal{M}_E\}.$$

Now we establish the opposite inequality. By Corollary 13.1.5 we have

$$\dim_H \mu = \operatorname{ess\,sup}\left\{\frac{h_\mu(x)}{\lambda(x)} : x \in X\right\}. \qquad (13.17)$$

Fix $\delta > 0$. As in the proof of Theorem 13.1.1, choose points $y_i \in X$, $i \in \mathbb{N}$ such that the f-invariant sets $\Gamma_i = \Gamma(y_i)$ (see (13.7)) satisfy $\mu(\Gamma_i) > 0$ for each i, and $\mu(\bigcup_{i \in \mathbb{N}} \Gamma_i) = 1$. Given $i \in \mathbb{N}$, we consider the normalized restriction μ_i of μ to Γ_i. It follows from (7.10) and (13.7) that

$$h_{\mu_i}(f|\Gamma_i) = \frac{1}{\mu(\Gamma_i)} \int_{\Gamma_i} h_\mu(x) \, d\mu(x) \geq h_\mu(y_i) - \delta. \qquad (13.18)$$

Set

$$\mathfrak{M}_i = \{\nu \in \mathfrak{M} : \nu(J \setminus \Gamma_i) = 0\},$$

where $\mathfrak{M}$ is the family of all f-invariant probability measures in J. Since Γ_i is f-invariant, there is a one-to-one correspondence between the ergodic f-invariant probability measures in Γ_i and the measures in $\mathfrak{M}_i \cap \mathfrak{M}_E$. Therefore, it is straightforward to verify that $\tau(\mathfrak{M}_i \cap \mathfrak{M}_E) > 0$, and that the normalization τ_i of $\tau|\mathfrak{M}_i$ is an ergodic decomposition of μ_i. Since

$$h_{\mu_i}(f|\Gamma_i) = \int_{\mathfrak{M}_i} h_\nu(f)\, d\tau_i(\nu)$$

(see, for example, [45]), there exists a set $A_i \subset \mathfrak{M}_i \cap \mathfrak{M}_E$ of positive τ_i-measure, and thus also of positive τ-measure, such that

$$h_\nu(f) > h_{\mu_i}(f|\Gamma_i) - \delta \quad \text{for every} \quad \nu \in A_i.$$

By (13.18), for every $\nu \in A_i$ and $x \in \Gamma_i$ we have

$$h_\nu(f) > h_{\mu_i}(f|\Gamma_i) - \delta \geq h_\mu(y_i) - 2\delta > h_\mu(x) - 3\delta,$$

and

$$\lambda(\nu) := \int_{\Gamma_i} \varphi\, d\nu \leq \lambda(y_i) + \delta \leq \lambda(x) + 2\delta.$$

Therefore, for every $\nu \in A_i$ we have

$$\frac{h_\mu(x)}{\lambda(x)} \leq \frac{h_\nu(f) + 3\delta}{\lambda(\nu) - 2\delta} \leq \dim_H \nu + C(\delta),$$

where $\delta \mapsto C(\delta)$ is a function (independent of i and ν) that tends to zero as $\delta \to 0$. Since $\tau(A_i) > 0$, it follows from (13.17) and Corollary 13.1.5 that

$$\dim_H \mu \leq \operatorname{ess\,sup}\{\dim_H \nu : \nu \in \mathfrak{M}_E\} + C(\delta).$$

Letting $\delta \to 0$ we obtain the desired result. $\qquad\qquad\qquad\qquad\qquad$ $\square$

In the case of the entropy it is well known that (see, for example, [45])

$$h_\mu(f) = \int_{\mathfrak{M}_E} h_\nu(f)\, d\tau(\nu) \tag{13.19}$$

for any ergodic decomposition τ of μ. Identities (13.16) and (13.19) are generalizations respectively of (13.14) and (13.15) for an arbitrary number (possibly uncountable) of ergodic invariant probability measures.

13.2   Hyperbolic sets of conformal maps

Now we consider hyperbolic sets of conformal maps, and we derive formulas for the pointwise dimension and the Hausdorff dimension of an invariant measure that is not necessarily ergodic. These formulas are versions of those for repellers in Section 13.1.

13.2.1   Formula for the pointwise dimension

Let Λ be a locally maximal hyperbolic set of a $C^{1+\varepsilon}$ diffeomorphism f, for some $\varepsilon > 0$, such that f is conformal on Λ. Let also μ be an f-invariant probability measure in Λ.

Since f is conformal on Λ, it follows from Birkhoff's ergodic theorem that for μ-almost every $x \in \Lambda$ there exist the limits

$$\lambda_s(x) = \lim_{n\to\infty} \frac{1}{n} \log \|d_x f^n | E^s(x)\| = \lim_{n\to\infty} \frac{1}{n} \sum_{k=0}^{n-1} \varphi_s(f^k x), \tag{13.20}$$

and

$$\lambda_u(x) = \lim_{n\to\infty} \frac{1}{n} \log \|d_x f^n | E^u(x)\| = \lim_{n\to\infty} \frac{1}{n} \sum_{k=0}^{n-1} \varphi_u(f^k x), \tag{13.21}$$

with φ_s and φ_u as in (4.39). The numbers $\lambda_s(x)$ and $\lambda_u(x)$ are respectively the negative and positive values of the Lyapunov exponent at x.

Let $\mathcal{R} = \{R_1, \dots, R_k\}$ be any partition of Λ. Given $i_{-m}, \dots, i_n \in \{1, \dots, k\}$ we define the *rectangle*

$$R_{i_{-m}\cdots i_n} = \{x \in \Lambda : f^j(x) \in R_{i_j} \text{ for } j = -m, \dots, n\}.$$

By Shannon–McMillan–Breiman's theorem, for μ-almost every $x \in \Lambda$ we have

$$h_\mu(x) = \lim_{n,m\to\infty} -\frac{1}{n+m} \log \mu(R_{n,m}(x)), \tag{13.22}$$

for any choice of rectangles $R_{n,m}(x) = R_{i_{-m}\cdots i_n}$ containing x. For definiteness, we assume from the beginning that for each x we make a particular choice of rectangles $R_{n,m}(x)$ for all $n, m \in \mathbb{N}$. Let X be a fixed full μ-measure f-invariant set of points $x \in \Lambda$ such that:

1. the numbers $\lambda_s(x)$ and $\lambda_u(x)$ in (13.20)–(13.21) are well-defined;

2. the number $h_\mu(x)$ in (7.9) is well-defined and satisfies (13.22).

Now we consider invariant measures that are not necessarily ergodic, and following Barreira and Wolf in [24] we establish an explicit formula for the pointwise dimension at a point x in terms of the values $\lambda_s(x)$ and $\lambda_u(x)$ of the Lyapunov exponent, and the local entropy $h_\mu(x)$.

Theorem 13.2.1. *Let Λ be a locally maximal hyperbolic set of a $C^{1+\varepsilon}$ diffeomorphism f, for some $\varepsilon > 0$, such that f is conformal on Λ, and let μ be an f-invariant probability measure in Λ. Then for μ-almost every $x \in \Lambda$ we have*

$$\underline{d}_\mu(x) = \overline{d}_\mu(x) = h_\mu(x)\left(\frac{1}{\lambda_u(x)} - \frac{1}{\lambda_s(x)}\right).$$

Proof. The proof is an elaboration of the proof of Theorem 13.1.1. Fix $\delta > 0$. For each $x \in X$ there exists $p(x) \in \mathbb{N}$ such that if $n, m \geq p(x)$, then

$$\lambda_s(x) - \delta < \frac{1}{n}\log\|d_x f^n|E^s(x)\| < \lambda_s(x) + \delta, \qquad (13.23)$$

$$\lambda_u(x) - \delta < \frac{1}{n}\log\|d_x f^n|E^u(x)\| < \lambda_u(x) + \delta, \qquad (13.24)$$

$$-h_\mu(x) - \delta < \frac{1}{n+m}\log\mu(R_{n,m}(x)) < -h_\mu(x) + \delta. \qquad (13.25)$$

For each $\ell \in \mathbb{N}$, let

$$Q_\ell = \{x \in X : p(x) \leq \ell\}.$$

Clearly, $\bigcup_{\ell \in \mathbb{N}} Q_\ell = X$. Furthermore, for each $x \in X$ there exists $r(x) > 0$ such that for every $r \in (0, r(x))$ there exist unique integers $n = n(x, r) \geq p(x)$ and $m = m(x, r) \geq p(x)$ such that

$$\|d_x f^m|E^s(x)\| \geq r \quad \text{and} \quad \|d_x f^{m+1}|E^s(x)\| < r, \qquad (13.26)$$

$$\|d_x f^n|E^u(x)\|^{-1} \geq r \quad \text{and} \quad \|d_x f^{n+1}|E^u(x)\|^{-1} < r. \qquad (13.27)$$

We write

$$R(x, r) = R_{n(x,r),m(x,r)}(x).$$

Combining (13.23) with (13.26), and (13.24) with (13.27), for all sufficiently small δ we obtain

$$m(\lambda_s(x) - \delta) < \log r - \min\varphi_s \quad \text{and} \quad \log r < m(\lambda_s(x) + \delta), \qquad (13.28)$$

and

$$-\log r + \min\varphi_u < n(\lambda_u(x) + \delta) \quad \text{and} \quad n(\lambda_u(x) - \delta) < -\log r. \qquad (13.29)$$

We first obtain an upper bound for the pointwise dimension. Since f is conformal on Λ, there exists $\kappa > 0$ (independent of x and r) such that $B(x, \kappa r) \supset R(x, r)$. Hence,

$$\mu(B(x, \kappa r)) \geq \mu(R(x, r)) \geq \exp[(-h_\mu(x) - \delta)(n + m)].$$

Therefore,

$$\mu(B(x, \kappa r)) \geq \exp\left[(h_\mu(x) + \delta)\left(\frac{\log r}{\lambda_u(x) - \delta} - \frac{\log r}{\lambda_s(x) + \delta}\right)\right],$$

and hence,

$$\overline{d}_\mu(x) \le (h_\mu(x) + \delta) \left(\frac{1}{\lambda_u(x) - \delta} - \frac{1}{\lambda_s(x) + \delta} \right).$$

The arbitrariness of δ implies that

$$\overline{d}_\mu(x) \le h_\mu(x) \left(\frac{1}{\lambda_u(x)} - \frac{1}{\lambda_s(x)} \right)$$

for every $x \in X$, and hence for μ-almost every $x \in \Lambda$.

Now we obtain a lower bound for the pointwise dimension. Given $x \in X$ we define

$$\Gamma(x) = \big\{ y \in X : |\lambda_s(y) - \lambda_s(x)| < \delta, |\lambda_u(y) - \lambda_u(x)| < \delta,$$
$$\text{and } |h_\mu(y) - h_\mu(x)| < \delta \big\}.$$

Note that $\Gamma(x)$ is f-invariant. The sets $\Gamma(x)$ cover X, and we can choose points $y_i \in X$, $i \in \mathbb{N}$ such that setting $\Gamma_i = \Gamma(y_i)$ we have $\mu(\Gamma_i) > 0$ for each i, and $\mu(\bigcup_{i \in \mathbb{N}} \Gamma_i) = 1$.

Now we proceed in a similar manner to that in the proof of Theorem 13.1.1 to construct a cover of $\Gamma_i \cap Q_\ell$ by sets $R'(x, r)$ of the form $R(x, r) \cap \Gamma_i \cap Q_\ell$. It follows from the Borel density lemma (see (13.8)) that for μ-almost every $x \in \Gamma_i \cap Q_\ell$ there exists a constant $\overline{r}(x) > 0$ such that for each $r \in (0, \overline{r}(x))$,

$$\mu(B(x, r)) \le 2\mu(B(x, r) \cap \Gamma_i \cap Q_\ell).$$

Since f is conformal on Λ, there exist a constant $K > 0$ (independent of x and r) and points $x_1, \ldots, x_k \in \Gamma_i \cap Q_\ell$ with $k \le K$ such that

$$B(x, r) \cap \Gamma_i \cap Q_\ell \subset \bigcup_{j=1}^k R'(x_j, r).$$

Using (13.25) we obtain

$$\mu(B(x, r)) \le 2\mu(B(x, r) \cap \Gamma_i \cap Q_\ell)$$
$$\le 2 \sum_{j=1}^k \mu(R'(x_j, r))$$
$$\le 2 \sum_{j=1}^k \exp[(-h_\mu(x_j) + \delta)(n(x_j, r) + m(x_j, r))].$$

By (13.28)–(13.29) and the definition of Γ_i we conclude that

$$\mu(B(x, r)) \le 2 \sum_{j=1}^k \exp\left[(h_\mu(y_i) - 2\delta) \left(\frac{\log r - \min \varphi_u}{\lambda_u(x_j) + \delta} - \frac{\log r - \min \varphi_s}{\lambda_s(x_j) - \delta} \right) \right]$$
$$\le 2K \exp\left[(h_\mu(x) - 3\delta) \left(\frac{\log r - \min \varphi_u}{\lambda_u(x) + 2\delta} - \frac{\log r - \min \varphi_s}{\lambda_s(x) - 2\delta} \right) \right],$$

and hence,

$$\underline{d}_\mu(x) \geq (h_\mu(x) - 3\delta) \left(\frac{1}{\lambda_u(x) + 2\delta} - \frac{1}{\lambda_s(x) - 2\delta} \right).$$

The arbitrariness of δ implies that

$$\underline{d}_\mu(x) \geq h_\mu(x) \left(\frac{1}{\lambda_u(x)} - \frac{1}{\lambda_s(x)} \right) \tag{13.30}$$

for μ-almost every $x \in \Gamma_i \cap Q_\ell$. Letting $\ell \to \infty$ we conclude that (13.30) holds for μ-almost every $x \in \Gamma_i$. Finally, since $\bigcup_{i \in \mathbb{N}} \Gamma_i$ has full μ-measure, (13.30) holds for μ-almost every $x \in \Lambda$. $\qquad\square$

We note that the existence of the pointwise dimension is known in much greater generality. Namely, for any finite hyperbolic measure μ invariant under a $C^{1+\varepsilon}$ diffeomorphism, it was shown by Barreira, Pesin and Schmeling in [11] that

$$\underline{d}_\mu(x) = \overline{d}_\mu(x) \text{ for } \mu\text{-almost every } x$$

(see Chapter 14 for details). We note that Theorem 13.2.1 can also be obtained from work of Ledrappier and Young in [93] and work of Barreira, Pesin and Schmeling in [11]. However, while these papers require the machinery of smooth ergodic theory, we provide a direct short proof in the case of uniformly hyperbolic dynamics.

Combining Theorem 13.2.1 with (2.6) we obtain a formula for the Hausdorff dimension of an invariant measure.

Corollary 13.2.2. *If Λ is a locally maximal hyperbolic set of a $C^{1+\varepsilon}$ diffeomorphism f, for some $\varepsilon > 0$, such that f is conformal on Λ, and μ is an f-invariant probability measure in Λ, then*

$$\dim_H \mu = \operatorname{ess\,sup} \left\{ h_\mu(x) \left(\frac{1}{\lambda_u(x)} - \frac{1}{\lambda_s(x)} \right) : x \in \Lambda \right\}. \tag{13.31}$$

When μ is *ergodic*, Theorem 13.2.1 (or Corollary 13.2.2) can be used to recover Young's formula in [165] for $\dim_H \mu$ in the uniformly hyperbolic case of diffeomorphisms on surfaces (see also Chapter 14). Let

$$\lambda_s(\mu) = \int_\Lambda \lambda_s(x)\, d\mu(x) \quad \text{and} \quad \lambda_u(\mu) = \int_\Lambda \lambda_u(x)\, d\mu(x).$$

Theorem 13.2.3. *If Λ is a locally maximal hyperbolic set of a $C^{1+\varepsilon}$ diffeomorphism f, for some $\varepsilon > 0$, such that f is conformal on Λ, and μ is an ergodic f-invariant probability measure in Λ, then*

$$\dim_H \mu = h_\mu(f) \left(\frac{1}{\lambda_u(\mu)} - \frac{1}{\lambda_s(\mu)} \right).$$

Proof. Since the functions h_μ, λ_s, and λ_u in the right-hand side of (13.31) are f-invariant μ-almost everywhere and μ is ergodic, we have

$$h_\mu(x) = h_\mu(f), \quad \lambda_s(x) = \lambda_s(\mu), \quad \text{and} \quad \lambda_u(x) = \lambda_u(\mu)$$

for μ-almost every $x \in \Lambda$. Therefore, the desired statement is an immediate consequence of Corollary 13.2.2. $\qquad\qquad\qquad\qquad\qquad\qquad\qquad\qquad\qquad\qquad\qquad\square$

Theorem 13.2.3 was first established by Pesin in [115, Theorem 24.2].

13.2.2   Dimension along ergodic decompositions

Using Theorem 13.2.1 we describe how the dimension of an invariant measure in a hyperbolic set behaves with respect to an ergodic decomposition.

Theorem 13.2.4 ([24]). *Let Λ be a locally maximal hyperbolic set of a $C^{1+\varepsilon}$ diffeomorphism f, for some $\varepsilon > 0$, such that f is conformal on Λ, and let μ be an f-invariant probability measure in Λ. For any ergodic decomposition τ of μ we have*

$$\dim_H \mu = \operatorname{ess\,sup}\{\dim_H \nu : \nu \in \mathcal{M}_E\},$$

with the essential supremum taken with respect to τ.

Proof. The proof is a simple modification of the proof of Theorem 13.1.6. By Corollary 13.2.2 we have

$$\dim_H \mu = \operatorname{ess\,sup}\left\{ h_\mu(x) \left(\frac{1}{\lambda_u(x)} - \frac{1}{\lambda_s(x)} \right) : x \in \Lambda \right\}. \tag{13.32}$$

Fix $\delta > 0$. As in the proof of Theorem 13.2.1, we consider the sets $\Gamma_i = \Gamma(y_i)$, $i \in \mathbb{N}$. Fix i and consider the normalized restriction μ_i of μ to Γ_i. Proceeding as in the proof of Theorem 13.1.6, we show that there exists a set $A_i \subset \mathcal{M}_i \cap \mathcal{M}_E$ of positive τ_i-measure (where τ_i is the normalization of $\tau|\mathcal{M}_i$) such that for every $\nu \in A_i$ and $x \in \Gamma_i$ we have

$$h_\nu(f) > h_{\mu_i}(f|\Gamma_i) - \delta \geq h_\mu(y_i) - 2\delta > h_\mu(x) - 3\delta.$$

Furthermore,

$$\lambda_s(\nu) = \int_{\Gamma_i} \varphi_s \, d\nu \geq \lambda_s(y_i) - \delta \geq \lambda_s(x) - 2\delta,$$

and

$$\lambda_u(\nu) = -\int_{\Gamma_i} \varphi_u \, d\nu \leq \lambda_u(y_i) + \delta \leq \lambda_u(x) + 2\delta.$$

We conclude that

$$h_\mu(x) \left(\frac{1}{\lambda_u(x)} - \frac{1}{\lambda_s(x)} \right) \leq (h_\nu(f) + 3\delta) \left(\frac{1}{\lambda_u(\nu) - 2\delta} - \frac{1}{\lambda_s(\nu) + 2\delta} \right)$$

for every $\nu \in A_i$. Since $\tau(A_i) > 0$, it follows from (13.32) that

$$\dim_H \mu \le \operatorname{ess\,sup}\{\dim_H \nu : \nu \in \mathcal{M}_E\} + C(\delta),$$

where $\delta \mapsto C(\delta)$ is a function (independent of i and ν) that tends to zero as $\delta \to 0$. By Proposition 13.1.3, the arbitrariness of δ implies the desired result. $\square$

The following is an immediate consequence of Theorem 13.2.4.

Corollary 13.2.5. *If Λ is a locally maximal hyperbolic set of a $C^{1+\varepsilon}$ diffeomorphism f, for some $\varepsilon > 0$, such that f is conformal on Λ, then*

$$\sup\{\dim_H \nu : \nu \in \mathcal{M}\} = \sup\{\dim_H \nu : \nu \in \mathcal{M}_E\}.$$

Chapter 14

Product Structure of Hyperbolic Measures

We describe in this chapter the almost product structure of the hyperbolic invariant measures, that is, the invariant measures with nonzero Lyapunov exponents almost everywhere. We note that the existence of a hyperbolic measure ensures the presence of nonuniform hyperbolicity almost everywhere, which together with the nontrivial recurrence given by the invariant measure causes a very complicated behavior of the system. It turns out that, to some extent, the almost product structure of a hyperbolic measure imitates the local product structure defined by the local stable and unstable manifolds, but its study is much more delicate. We also describe the relation between the product structure of hyperbolic invariant measures and the dimension theory of dynamical systems.

14.1 Nonuniform hyperbolicity

The concept of *nonuniform hyperbolicity* originated in seminal work of Pesin in [112, 113, 114] (see also [6, 7, 8, 95, 159] and the references therein).

Let $f \colon M \to M$ be a diffeomorphism.

Definition 14.1.1. The trajectory $\{f^n x : n \in \mathbb{Z}\}$ of a point $x \in M$ is called *nonuniformly hyperbolic* if there exist decompositions

$$T_{f^n x} M = E^s_{f^n x} \oplus E^u_{f^n x}, \quad n \in \mathbb{Z},$$

a constant $\lambda \in (0,1)$, and for each sufficiently small $\rho > 0$ a positive function C_ρ defined in the trajectory of x such that if $k \in \mathbb{Z}$, then:

1. $C_\rho(f^k x) \le e^{\rho|k|} C_\rho(x)$;
2. $d_x f^k E^s_x = E^s_{f^k x}$ and $d_x f^k E^u_x = E^u_{f^k x}$;

3. if $v \in E^s_{f^k x}$ and $m > 0$, then

$$\|d_{f^k x} f^m v\| \le C_\rho(f^k x) \lambda^m e^{\rho m} \|v\|;$$

4. if $v \in E^u_{f^k x}$ and $m < 0$, then

$$\|d_{f^k x} f^m v\| \le C_\rho(f^k x) \lambda^{|m|} e^{\rho|m|} \|v\|;$$

5. $\angle(E^u_{f^k x}, E^s_{f^k x}) \ge C_\rho(f^k x)^{-1}.$

The expression "nonuniform" refers to the estimates in conditions 3 and 4, that may differ from the "uniform" estimate λ^m by a small exponential term. It is immediate that any trajectory in a hyperbolic set is nonuniformly hyperbolic.

Among the most important properties due to nonuniform hyperbolicity is the existence of invariant stable and unstable manifolds (with an appropriate version of Theorem 4.2.2), and their absolute continuity established by Pesin in [112]. The theory also describes the ergodic properties of dynamical systems with an invariant measure absolutely continuous with respect to the volume [113], and the Pesin entropy formula expresses the Kolmogorov–Sinai entropy in terms of the Lyapunov exponents [113] (see also [92]). Combining the nonuniform hyperbolicity with the nontrivial recurrence due to the existence of a finite invariant measure, the work of Katok in [83] revealed a very rich and complicated orbit structure.

Now we state the result concerning the existence of invariant stable and unstable manifolds, established by Pesin in [112].

Theorem 14.1.2 (Existence of invariant manifolds). *If $\{f^n x : n \in \mathbb{Z}\}$ is a nonuniformly hyperbolic trajectory of a $C^{1+\varepsilon}$ diffeomorphism, for some $\varepsilon > 0$, then for each sufficiently small $\rho > 0$ there exist manifolds $V^s(x)$ and $V^u(x)$ containing x, and a function D_ρ defined in the trajectory of x such that:*

1. $T_x V^s(x) = E^s_x$ *and* $T_x V^u(x) = E^u_x;$

2. $D_\rho(f^k x) \le e^{2\rho|k|} D_\rho(x)$ *for every* $k \in \mathbb{Z};$

3. *if* $y \in V^s(x)$, $m > 0$, *and* $k \in \mathbb{Z}$, *then*

$$d(f^{m+k} x, f^{m+k} y) \le D_\rho(f^k x) \lambda^m e^{\rho m} d(f^k x, f^k y); \qquad (14.1)$$

4. *if* $y \in V^u(x)$, $m < 0$, *and* $k \in \mathbb{Z}$, *then*

$$d(f^{m+k} x, f^{m+k} y) \le D_\rho(f^k x) \lambda^{|m|} e^{\rho|m|} d(f^k x, f^k y).$$

Definition 14.1.3. The manifolds $V^s(x)$ and $V^u(x)$ are called respectively *local stable manifold* and *local unstable manifold* at x.

Contrarily to what happens in the case of hyperbolic sets, for nonuniformly hyperbolic trajectories the sizes of these manifolds may not be bounded from below along the orbit (although they decrease at most with a small exponential speed provided that ρ is sufficiently small).

The proof of Theorem 14.1.2 in [112] is an elaboration of the classical work of Perron. In [133], Ruelle obtained a proof of Theorem 14.1.2 based on the study of perturbations of products of matrices in the Multiplicative ergodic theorem (see Theorem 14.2.3 below). Another proof of Theorem 14.1.2 was given by Pugh and Shub in [126] with an elaboration of the classical work of Hadamard. In [125], Pugh constructed a C^1 diffeomorphism in a manifold of dimension 4, that is not of class $C^{1+\varepsilon}$ for any $\varepsilon > 0$ and for which there exists no manifold tangent to E_x^s such that the inequality (14.1) holds in some open neighborhood of x. Therefore, the hypothesis $\varepsilon > 0$ is crucial in Theorem 14.1.2. See [6, 58, 126] for detailed expositions.

14.2 Dynamical systems with nonzero Lyapunov exponents

The concept of nonuniform hyperbolicity is closely related to the study of Lyapunov exponents. These numbers measure the asymptotic exponential rates of contraction and expansion in the neighborhood of a given trajectory.

Let $f\colon M \to M$ be a diffeomorphism.

Definition 14.2.1. Given $x \in M$ and $v \in T_x M$, we define the *(forward) Lyapunov exponent* of (x, v) by

$$\chi(x, v) = \limsup_{n \to +\infty} \frac{1}{n} \log \|d_x f^n v\|,$$

with the convention that $\log 0 = -\infty$.

By the abstract theory of Lyapunov exponents (see [6] for a detailed exposition), for each $x \in M$ there exist a positive integer $p(x) \leq \dim M$, numbers

$$\chi_1(x) < \cdots < \chi_{p(x)}(x), \tag{14.2}$$

and linear subspaces

$$\{0\} = E_0(x) \subset E_1(x) \subset \cdots \subset E_{p(x)}(x) = T_x M$$

such that if $i = 1, \ldots, p(x)$, then

$$E_i(x) = \{v \in T_x M : \chi(x, v) \leq \chi_i(x)\},$$

and $\chi(x, v) = \chi_i(x)$ for every $v \in E_i(x) \setminus E_{i-1}(x)$.

Definition 14.2.2. Given $x \in M$ and $v \in T_x M$, we define the *(backward) Lyapunov exponent* of (x, v) by

$$\chi^-(x, v) = \limsup_{n \to -\infty} \frac{1}{|n|} \log \|d_x f^n v\|.$$

Similarly, by the abstract theory of Lyapunov exponents, for each $x \in M$ there exist a positive integer $p^-(x) \le \dim M$, numbers

$$\chi_1^-(x) > \cdots > \chi_{p^-(x)}^-(x),$$

and linear subspaces

$$T_x M = E_1^-(x) \supset \cdots \supset E_{p^-(x)}^-(x) \supset E_{p^-(x)+1}^-(x) = \{0\}$$

such that if $i = 1, \ldots, p^-(x)$, then

$$E_i^-(x) = \{v \in T_x M : \chi^-(x, v) \le \chi_i^-(x)\},$$

and $\chi^-(x, v) = \chi_i^-(x)$ for every $v \in E_i^-(x) \setminus E_{i+1}^-(x)$.

A priori the structures for positive and negative time could be unrelated. The following result of Oseledets in [108] shows that under mild additional assumptions quite the contrary happens in a set of full measure with respect to any finite invariant measure.

Theorem 14.2.3 (Multiplicative ergodic theorem). *Let $f \colon M \to M$ be a C^1 diffeomorphism, and let μ be an f-invariant finite measure in M such that $\log^+ \|df\|$ and $\log^+ \|df^{-1}\|$ are μ-integrable. Then for μ-almost every $x \in M$ there exist subspaces $H_j(x) \subset T_x M$ for $j = 1, \ldots, p(x)$ such that:*

1. *if $i = 1, \ldots, p(x)$, then $E_i(x) = \bigoplus_{j=1}^i H_j(x)$, and*

$$\lim_{n \to \pm\infty} \frac{1}{n} \log \|d_x f^n v\| = \chi_i(x)$$

for every $v \in E_i(x) \setminus \{0\}$, with uniform convergence in $\{v \in H_i(x) : \|v\| = 1\}$;

2. *for each $i \ne j$ we have*

$$\lim_{n \to \pm\infty} \frac{1}{n} \log |\angle(H_i(f^n x), H_j(f^n x))| = 0.$$

We note that if M is compact, then the functions $\log^+ \|df\|$ and $\log^+ \|df^{-1}\|$ are μ-integrable for any finite measure μ in M. The statement in Theorem 14.2.3 also holds in the more general case of linear cocycles over a measurable transformation. See [6] for a detailed exposition and for a proof of the Multiplicative ergodic theorem.

Let $f\colon M \to M$ be a C^1 diffeomorphism in a compact manifold, and let μ be an f-invariant finite measure in M. We say that f is *nonuniformly hyperbolic* with respect to μ if the set $\Lambda \subset M$ of points whose trajectories are nonuniformly hyperbolic has measure $\mu(\Lambda) > 0$. In this case the constants λ and ρ in the definition of nonuniformly hyperbolic trajectory are replaced by measurable functions $\lambda(x)$ and $\rho(x)$. It follows from Theorem 14.2.3 that the following two conditions are equivalent:

1. f is nonuniformly hyperbolic with respect to μ;

2. $\chi(x, v) \neq 0$ for each $v \in T_x M$ and each x in a set of μ-positive measure.

In other words, the nonuniformly hyperbolic diffeomorphisms with respect to a given measure are precisely those with all Lyapunov exponents nonzero in a set of positive measure.

Let M be a compact manifold. It was shown by Katok in [82] when dim $M = 2$ and by Dolgopyat and Pesin in [47] when dim $M \geq 3$ that there exists a C^∞ diffeomorphism f such that:

1. f preserves the Riemannian volume m in M;

2. f has nonzero Lyapunov exponents m-almost everywhere;

3. f is a Bernoulli diffeomorphism.

When dim $M \geq 5$, Brin constructed in [42] a C^∞ Bernoulli diffeomorphism which preserves the Riemannian volume and has all but one Lyapunov exponent nonzero. In another direction, Bochi showed in [32] that when dim $M = 2$ there exists a residual set D of C^1 area preserving diffeomorphisms such each $f \in D$ is either an Anosov diffeomorphism or has all Lyapunov exponents zero. This result was announced by Mañé but his proof was never published.

14.3   Product structure of hyperbolic measures

Let $f\colon M \to M$ be a diffeomorphism.

Definition 14.3.1. We say that an f-invariant measure μ in M is *hyperbolic* (with respect to f) if all Lyapunov exponents are nonzero μ-almost everywhere, that is, if

$$\limsup_{n \to +\infty} \frac{1}{n} \log \|d_x f^n v\| \neq 0$$

for every $v \in T_x M$ and every x in a full μ-measure set.

Let μ be a hyperbolic f-invariant measure. By Theorem 14.1.2, for μ-almost every $x \in M$ there exist local stable and unstable invariant manifolds $V^s(x)$ and $V^u(x)$ at x. In a certain sense these manifolds reproduce the local product structure that is present in the case of locally maximal hyperbolic sets. But a priori it is unclear whether the hyperbolic measure μ imitates or not the local product structure determined by the invariant manifolds. This problem became known among the specialists as the Eckmann–Ruelle conjecture, claiming that locally a hyperbolic measure indeed imitates the local product structure determined by the local stable and unstable invariant manifolds. For example, one can show that any Gibbs measures (see Definition 3.2.3) has a product structure (see Proposition 4.2.12). Even though Eckmann and Ruelle apparently never formulated the conjecture, their work [49] discusses several related problems and played a fundamental role in the development of the theory.

In order to formulate a rigorous result about the solution of the Eckmann–Ruelle conjecture we need the families of conditional measures μ_x^s and μ_x^u generated by certain measurable partitions constructed by Ledrappier and Young in [93], based on former work of Ledrappier and Strelcyn in [91]. Namely, they constructed two measurable partitions ξ^s and ξ^u of M such that for μ-almost every $x \in M$ we have:

1. $\xi^s(x) \subset V^s(x)$ and $\xi^u(x) \subset V^u(x)$;

2. there exists $\gamma = \gamma(x) > 0$ such that

$$\xi^s(x) \supset V^s(x) \cap B(x, \gamma) \quad \text{and} \quad \xi^u(x) \supset V^s(x) \cap B(x, \gamma).$$

We recall that it was shown by Rohklin in [130] that any measurable partition ξ of M has associated a family of conditional measures. More precisely, for μ-almost every $x \in M$ there is a probability measure μ_x defined in the element $\xi(x)$ of ξ containing x, and the conditional measures are characterized by the following property: if $\mathcal{B}_\xi$ is the σ-subalgebra of the Borel σ-algebra generated by the unions of elements of ξ, then for each Borel set $A \subset M$ the function $x \mapsto \mu_x(A \cap \xi(x))$ is $\mathcal{B}_\xi$-measurable, and

$$\mu(A) = \int_A \mu_x(A \cap \xi(x)) \, d\mu.$$

We denote by μ_x^s and μ_x^u the conditional measures associated respectively with the partitions ξ^s and ξ^u. We represent by

$$B^s(x, r) \subset V^s(x) \quad \text{and} \quad B^u(x, r) \subset V^u(x)$$

the open balls centered at x of radius r with respect to the distances induced respectively in $V^s(x)$ and $V^u(x)$.

We formulate the following result without proof.

Theorem 14.3.2 (Existence of pointwise dimensions). *Let f be a $C^{1+\varepsilon}$ diffeomorphism in a manifold M, for some $\varepsilon > 0$, and let μ be an f-invariant finite measure in M with compact support. If μ is hyperbolic, then the limits*

$$d_\mu^s(x) := \lim_{r \to 0} \frac{\log \mu_x^s(B^s(x, r))}{\log r} \quad \text{and} \quad d_\mu^u(x) := \lim_{r \to 0} \frac{\log \mu_x^u(B^u(x, r))}{\log r} \qquad (14.3)$$

exist for μ-almost every $x \in M$.

We refer to [8] for a detailed proof of Theorem 14.3.2. Due to the need for techniques of smooth ergodic theory the proof is beyond the scope of this book.

Definition 14.3.3. The limits in (14.3), when they exist, are called respectively the *stable* and *unstable pointwise dimensions* of μ at x.

We can easily verify that the functions $x \mapsto d_\mu^s(x)$ and $x \mapsto d_\mu^u(x)$ are f-invariant μ-almost everywhere. Therefore, under the assumptions of Theorem 14.3, if μ is ergodic, then there exist constants d_μ^s and d_μ^u such that

$$d_\mu^s(x) = d_\mu^s \quad \text{and} \quad d_\mu^u(x) = d_\mu^u \tag{14.4}$$

for μ-almost every $x \in M$.

The statement in Theorem 14.3.2 was established by Ledrappier and Young in [93] for C^2 diffeomorphisms. It was also established in [93] that

$$\limsup_{r \to 0} \frac{\log \mu(B(x,r))}{\log r} \leq d_\mu^s(x) + d_\mu^u(x) \tag{14.5}$$

for μ-almost every $x \in M$. We note that Ledrappier and Young consider a more general class of measures for which some Lyapunov exponents may be zero. The only place in [93] where f is required to be of class C^2 concerns the Lipschitz regularity of the holonomies generated by the intermediate foliations (such as any strongly stable foliation inside the stable foliation). In the case of hyperbolic measures a new argument was given in [11] that establishes the Lipschitz regularity of the intermediate foliations in the general case of $C^{1+\varepsilon}$ diffeomorphisms. This ensures that (14.5) holds almost everywhere also when f is of class $C^{1+\varepsilon}$ (see [11] for details).

The following result is due to Barreira, Pesin and Schmeling [11].

Theorem 14.3.4 (Product structure of hyperbolic measures). *Let f be a $C^{1+\varepsilon}$ diffeomorphism in a manifold M, for some $\varepsilon > 0$, and let μ be an f-invariant finite measure in M with compact support. If μ is hyperbolic, then given $\gamma > 0$ there exists a set $\Lambda \subset M$ with $\mu(\Lambda) > \mu(M) - \gamma$ such that for each $x \in \Lambda$ we have*

$$r^\gamma \leq \frac{\mu(B(x,r))}{\mu_x^s(B^s(x,r))\mu_x^u(B^u(x,r))} \leq r^{-\gamma}$$

for all sufficiently small $r > 0$.

See [8] for a detailed proof of Theorem 14.3.4. Again, due to the need for techniques of smooth ergodic theory the proof is beyond the scope of this book. In Section 14.4 we give a complete proof in the case of invariant measures in locally maximal hyperbolic sets, without using techniques of smooth ergodic theory. Ledrappier and Misiurewicz showed in [90] that the hyperbolicity of the measure is essential in Theorem 14.3.4.

The following is an immediate consequence of Theorems 14.3.2 and 14.4.1, using the criterion in Theorem 2.1.6. We recall the numbers d_μ^s and d_μ^u in (14.4) when μ is ergodic.

Theorem 14.3.5. *Let f be a $C^{1+\varepsilon}$ diffeomorphism in a manifold M, for some $\varepsilon > 0$, and let μ be an f-invariant finite measure in M with compact support. If μ is hyperbolic, then*

$$\underline{d}_\mu(x) = \overline{d}_\mu(x) = d^s_\mu(x) + d^u_\mu(x)$$

for μ-almost every $x \in M$. If, in addition, μ is ergodic, then

$$\dim_H \mu = \underline{\dim}_B \mu = \overline{\dim}_B \mu = d^s_\mu + d^u_\mu.$$

The statement in Theorem 14.3.5 was established by Young in [165] when M is a surface and by Ledrappier in [89] when μ is an SRB-measure (after Sinai, Ruelle and Bowen; see, for example, [6] for the definition).

Theorems 14.3.5 and 2.1.5 yield a formula for the Hausdorff dimension of a hyperbolic measure.

Corollary 14.3.6. *If f is a $C^{1+\varepsilon}$ diffeomorphism in a manifold M, for some $\varepsilon > 0$, and μ is an f-invariant hyperbolic finite measure in M with compact support, then*

$$\dim_H \mu = \text{ess} \sup\{d^s_\mu(x) + d^u_\mu(x) : x \in M\}.$$

14.4   Product structure of measures in hyperbolic sets

We show in this section that any finite invariant measure in a locally maximal hyperbolic set possesses an almost product structure. For simplicity we only consider the case of ergodic measures (we briefly describe at the end of the section how to deal with nonergodic measures). We follow closely Barreira, Pesin and Schmeling in [11].

Theorem 14.4.1. *Let f be a $C^{1+\varepsilon}$ diffeomorphism in a manifold M, for some $\varepsilon > 0$, and let Λ be a locally maximal hyperbolic set of f. If μ is an ergodic f-invariant finite measure in Λ, then given $\gamma > 0$ there exists a set $\Lambda' \subset \Lambda$ with $\mu(\Lambda') > \mu(\Lambda) - \gamma$ such that for each $x \in \Lambda'$ we have*

$$r^\gamma \leq \frac{\mu(B(x,r))}{\mu^s_x(B^s(x,r))\mu^u_x(B^u(x,r))} \leq r^{-\gamma} \tag{14.6}$$

for all sufficiently small $r > 0$.

Proof. Since the measure μ is ergodic and the values $\chi_i(x)$ of the Lyapunov exponent (see (14.2)) are f-invariant μ-almost everywhere, they are constant μ-almost everywhere, say equal to $\chi_1 < \chi_2 < \cdots < \chi_p$. Note that $\chi_1 < 0$ and $\chi_p > 0$.

Consider a Markov partition $\mathcal{R}$ of Λ, and define new partitions

$$\mathcal{R}^l_k = \bigvee_{i=-k}^{l} f^i \mathcal{R}$$

for each $k, l \in \mathbb{N}$. For μ-almost every $x \in \Lambda$ there is a unique element $\mathcal{R}_k^l(x)$ of the partition $\mathcal{R}_k^l$ that contains x. Given $\delta \in (0, 1)$, there exists a set $\Gamma \subset \Lambda$ of measure $\mu(\Gamma) > \mu(\Lambda) - \delta/2$, an integer $n_0 \in \mathbb{N}$, and a constant $C > 1$ such that for every $x \in \Gamma$ and $n \geq n_0$ the following properties hold:

(a) setting $h = h_\mu(f)$, for every $k, l \in \mathbb{N}$ we have

$$C^{-1}e^{-(l+k)h-(l+k)\delta} \leq \mu(\mathcal{R}_k^l(x)) \leq Ce^{-(l+k)h+(l+k)\delta}, \tag{14.7}$$

$$C^{-1}e^{-kh-k\delta} \leq \mu_x^s(\mathcal{R}_k^0(x)) \leq Ce^{-kh+k\delta}, \tag{14.8}$$

$$C^{-1}e^{-lh-l\delta} \leq \mu_x^u(\mathcal{R}_0^l(x)) \leq Ce^{-lh+l\delta}; \tag{14.9}$$

(b)

$$\xi^s(x) \cap \bigcap_{n \in \mathbb{N}} \mathcal{R}_0^n(x) \supset B^s(x, e^{-n_0}), \tag{14.10}$$

$$\xi^u(x) \cap \bigcap_{n \in \mathbb{N}} \mathcal{R}_n^0(x) \supset B^u(x, e^{-n_0}); \tag{14.11}$$

(c)

$$e^{-d_\mu^s n - n\delta} \leq \mu_x^s(B^s(x, e^{-n})) \leq e^{-d_\mu^s n + n\delta}, \tag{14.12}$$

$$e^{-d_\mu^u n - n\delta} \leq \mu_x^u(B^u(x, e^{-n})) \leq e^{-d_\mu^u n + n\delta}; \tag{14.13}$$

(d)

$$\mathcal{R}_{an}^{an}(x) \subset B(x, e^{-n}) \subset \mathcal{R}(x), \tag{14.14}$$

$$\mathcal{R}_{an}^0(x) \cap \xi^s(x) \subset B^s(x, e^{-n}) \subset \mathcal{R}(x) \cap \xi^s(x), \tag{14.15}$$

$$\mathcal{R}_0^{an}(x) \cap \xi^u(x) \subset B^u(x, e^{-n}) \subset \mathcal{R}(x) \cap \xi^u(x), \tag{14.16}$$

where a is the integer part of $2(1 + \delta)\max\{\chi_p, -\chi_1, 1\}$;

(e) setting

$$Q_n(x) = \bigcup \mathcal{R}_{an}^{an}(y) \tag{14.17}$$

with the union taken over all $y \in \Gamma$ such that

$$\mathcal{R}_0^{an}(y) \cap B^u(x, 2e^{-n}) \neq \varnothing \quad \text{and} \quad \mathcal{R}_{an}^0(y) \cap B^s(x, 2e^{-n}) \neq \varnothing,$$

we have $\mathcal{R}_{an}^{an}(y) \subset Q_n(x)$ for every $y \in Q_n(x)$, and

$$B(x, e^{-n}) \cap \Gamma \subset Q_n(x) \subset B(x, 4e^{-n}); \tag{14.18}$$

(f)

$$B^s(x, e^{-n}) \cap \Gamma \subset Q_n(x) \cap \xi^s(x) \subset B^s(x, 4e^{-n}), \tag{14.19}$$

$$B^u(x, e^{-n}) \cap \Gamma \subset Q_n(x) \cap \xi^u(x) \subset B^u(x, 4e^{-n}). \tag{14.20}$$

Properties (14.7), (14.8), and (14.9) follow from Shannon–McMillan–Breiman's theorem applied to the partition $\mathcal{R}$. The inequalities in (14.12) and (14.13) are obtained from Theorem 14.3.2. Since the values of the Lyapunov exponent are constant μ-almost everywhere, the properties (14.14), (14.15), and (14.16) follow from (14.10), (14.11), and the choice of a. The inclusions in (14.18) follow from the continuous dependence of the stable and unstable manifolds in the $C^{1+\varepsilon}$ topology on the base point.

By the Borel density lemma (see, for example, [59, Theorem 2.9.11]), there exist an integer $n_1 \geq n_0$ and a set $\Gamma' \subset \Gamma$ of measure $\mu(\Gamma') > 1 - \delta$ such that for every $n \geq n_1$ and $x \in \Gamma'$ we have

$$\mu(B(x, e^{-n}) \cap \Gamma) \geq \frac{1}{2}\mu(B(x, e^{-n})), \tag{14.21}$$

$$\mu_x^s(B^s(x, e^{-n}) \cap \Gamma) \geq \frac{1}{2}\mu_x^s(B^s(x, e^{-n})), \tag{14.22}$$

$$\mu_x^u(B^u(x, e^{-n}) \cap \Gamma) \geq \frac{1}{2}\mu_x^u(B^u(x, e^{-n})). \tag{14.23}$$

We establish two additional properties of the partitions $\mathcal{R}_0^k$ and $\mathcal{R}_k^0$.

Lemma 14.4.2. *There exists a positive constant $D = D(\Gamma') \in (0,1)$ such that for every $k \in \mathbb{N}$ and $x \in \Gamma$ we have*

$$\mu_x^s(\mathcal{R}_0^k(x) \cap \Gamma) \geq D \quad and \quad \mu_x^u(\mathcal{R}_k^0(x) \cap \Gamma) \geq D.$$

Proof of the lemma. By (14.10), for every $k \in \mathbb{N}$ and $x \in \Gamma$ we have

$$\mathcal{R}_0^k(x) \cap \Gamma \supset B^s(x, e^{-n_0}) \cap \Gamma.$$

It follows from (14.22) and (14.12) that

$$\mu_x^s(\mathcal{R}_0^k(x) \cap \Gamma) \geq \frac{1}{2}\mu_x^s(B^s(x, e^{-n_0})) \geq \frac{1}{2}e^{-d_\mu^s n_0 - n_0 \delta} =: D.$$

The second inequality in the lemma can be obtained in a similar manner using the properties (14.11), (14.23), and (14.13). $\qquad\square$

Lemma 14.4.3. *For every $x \in \Gamma$ and $n \geq n_0$ we have*

$$\mathcal{R}_{an}^{an}(x) \cap \xi^s(x) = \mathcal{R}_{an}^0(x) \cap \xi^s(x),$$

$$\mathcal{R}_{an}^{an}(x) \cap \xi^u(x) = \mathcal{R}_0^{an}(x) \cap \xi^u(x).$$

Proof of the lemma. It follows from (14.15) and (14.10) that

$$\mathcal{R}_{an}^0(x) \cap \xi^s(x) \subset \mathcal{R}_{an}^0(x) \cap B^s(x, e^{-n}) \subset \mathcal{R}_{an}^0(x) \cap B^s(x, e^{-n_0})$$
$$\subset \mathcal{R}_{an}^0(x) \cap \mathcal{R}_0^{an}(x) \cap \xi^s(x) = \mathcal{R}_{an}^{an}(x) \cap \xi^s(x).$$

Since $\mathcal{R}_{an}^{an}(x) \subset \mathcal{R}_{an}^0(x)$ this completes the proof of the first identity. The proof of the second identity is entirely analogous. $\qquad\square$

Fix $x \in \Gamma'$ and an integer $n \geq n_1$. We consider two classes $\mathcal{R}(n)$ and $\mathcal{F}(n)$ of elements of the partition $\mathcal{R}_{an}^{an}$ (we call these elements rectangles). Namely, we set

$$\mathcal{R}(n) = \{\mathcal{R}_{an}^{an}(y) \subset \mathcal{R}(x) : \mathcal{R}_{an}^{an}(y) \cap \Gamma \neq \varnothing\},$$

and

$$\mathcal{F}(n) = \{\mathcal{R}_{an}^{an}(y) \subset \mathcal{R}(x) : \mathcal{R}_{an}^{0}(y) \cap \Gamma' \neq \varnothing \text{ and } \mathcal{R}_{0}^{an}(y) \cap \Gamma' \neq \varnothing\}.$$

The rectangles in $\mathcal{R}(n)$ carry the whole measure of the set $\mathcal{R}(x) \cap \Gamma$, that is,

$$\sum_{R \in \mathcal{R}(n)} \mu(R \cap \Gamma) = \mu(\mathcal{R}(x) \cap \Gamma).$$

Clearly, the rectangles in $\mathcal{R}(n)$ that intersect Γ' belong to $\mathcal{F}(n)$.

We want to compare the number of rectangles in $\mathcal{R}(n)$ and $\mathcal{F}(n)$ that intersect a given set. This allows us to evaluate the deviation of the measure μ from the local product structure at the level n. Our main observation is that for "typical" points $y \in \Gamma'$ the number of rectangles in $\mathcal{R}(n)$ that intersect $V^s(y)$ (respectively $V^u(y)$) is "asymptotically" the same up to a subexponential factor. However, in general, the distribution of these rectangles along $V^s(y)$ (respectively $V^u(y)$) may be different for different points y. This causes a deviation from the local product structure.

For each set $A \subset \mathcal{R}(x)$, we define

$$N(n, A) = \operatorname{card}\{R \in \mathcal{R}(n) : R \cap A \neq \varnothing\},$$

$$N^s(n, y, A) = \operatorname{card}\{R \in \mathcal{R}(n) : R \cap \xi^s(y) \cap \Gamma \cap A \neq \varnothing\},$$

$$N^u(n, y, A) = \operatorname{card}\{R \in \mathcal{R}(n) : R \cap \xi^u(y) \cap \Gamma \cap A \neq \varnothing\},$$

$$\widehat{N}^s(n, y, A) = \operatorname{card}\{R \in \mathcal{F}(n) : R \cap \xi^s(y) \cap A \neq \varnothing\},$$

$$\widehat{N}^u(n, y, A) = \operatorname{card}\{R \in \mathcal{F}(n) : R \cap \xi^u(y) \cap A \neq \varnothing\}.$$

We note that $N(n, \mathcal{R}(x))$ is the cardinality of the set $\mathcal{R}(n)$, and $N^s(n, y, \mathcal{R}(x))$ (respectively $N^u(n, y, \mathcal{R}(x))$) is the number of rectangles in $\mathcal{R}(n)$ that intersect Γ and the local stable (respectively unstable) manifold at y. Let $Q_n(x)$ be the set in (14.17).

Lemma 14.4.4. *For every* $y \in \mathcal{R}(x) \cap \Gamma$ *and* $n \geq n_0$ *we have*

$$N^s(n, y, Q_n(y)) \leq \mu_y^s(B^s(y, 4e^{-n})) \cdot Ce^{anh + an\delta},$$

and

$$N^u(n, y, Q_n(y)) \leq \mu_y^u(B^u(y, 4e^{-n})) \cdot Ce^{anh + an\delta}.$$

Proof of the lemma. It follows from (14.19) that

$$\mu_y^s(B^s(y, 4e^{-n})) \geq \mu_y^s(Q_n(y))$$
$$\geq N^s(n, y, Q_n(y))$$
$$\times \min\{\mu_y^s(R) : R \in \mathcal{R}(n) \text{ and } R \cap \xi^s(y) \cap Q_n(y) \cap \Gamma \neq \emptyset\}.$$

We note that the condition $R \cap \xi^s(y) \cap Q_n(y) \neq \emptyset$ implies that $R \in \mathcal{R}(n)$. Let

$$z \in R \cap \xi^s(y) \cap Q_n(y) \cap \Gamma$$

for some $R \in \mathcal{R}(n)$. By Lemma 14.4.3 we obtain

$$\mu_y^s(R) = \mu_y^s(\mathcal{R}_{an}^0(z)) = \mu_z^s(\mathcal{R}_{an}^0(z)).$$

The first inequality in the lemma follows now from (14.8). The second inequality can be obtained with a similar argument using (14.20). □

Lemma 14.4.5. *For every $y \in \mathcal{R}(x) \cap \Gamma'$ and $n \geq n_1$ we have*

$$\mu(B(y, e^{-n})) \leq N(n, Q_n(y)) \cdot 2Ce^{-2anh + 2an\delta}.$$

Proof of the lemma. It follows from (14.21) and (14.18) that

$$\frac{1}{2}\mu(B(y, e^{-n})) \leq \mu(B(y, e^{-n}) \cap \Gamma) \leq \mu(Q_n(y) \cap \Gamma)$$
$$\leq N(n, Q_n(y)) \cdot \max\{\mu(R) : R \in \mathcal{R}(n) \text{ and } R \cap Q_n(y) \neq \emptyset\}.$$

We note that the condition $R \cap Q_n(y) \neq \emptyset$ implies that $R \in \mathcal{R}(n)$. The desired inequality follows now from (14.7). □

Now we estimate the number of rectangles in the classes $\mathcal{R}(n)$ and $\mathcal{F}(n)$.

Lemma 14.4.6. *For μ-almost every $y \in \mathcal{R}(x) \cap \Gamma'$ there exists an integer $n_2(y) \geq n_1$ such that for every $n \geq n_2(y)$ we have*

$$N(n + 2, Q_{n+2}(y)) \leq \widehat{N}^s(n, y, Q_n(y)) \cdot \widehat{N}^u(n, y, Q_n(y)) \cdot 2C^2 e^{4a(h+\varepsilon)} e^{4an\delta}.$$

Proof of the lemma. By the Borel density lemma, for μ-almost every $y \in \Gamma'$ there is an integer $n_2(y) \geq n_1$ such that for every $n \geq n_2(y)$,

$$2\mu(B(y, e^{-n}) \cap \Gamma') \geq \mu(B(y, e^{-n})).$$

Since $\Gamma' \subset \Gamma$, it follows from (14.18) that for every $n \geq n_2(y)$,

$$2\mu(Q_n(y) \cap \Gamma') \geq 2\mu(B(y, e^{-n}) \cap \Gamma') \geq \mu(B(y, e^{-n}))$$
$$\geq \mu(B(y, 4e^{-n-2})) \geq \mu(Q_{n+2}(y)).$$
$$(14.24)$$

For each $m \geq n_2(y)$, by (14.7) and property (e) we have

$$\mu(Q_m(y)) = \sum_{\mathcal{R}_{am}^{am}(z) \subset Q_m(y)} \mu(\mathcal{R}_{am}^{am}(z)) \geq N(m, Q_m(y)) \cdot C^{-1} e^{-2amh - 2am\delta}.$$

Similarly, for each $n \geq n_2(y)$,

$$\mu(Q_n(y) \cap \Gamma') = \sum_{\mathcal{R}_{an}^{an}(z) \subset Q_n(y)} \mu(\mathcal{R}_{an}^{an}(z) \cap \Gamma') \leq N_n \cdot C e^{-2anh + 2an\delta},$$

where N_n is the number of rectangles $\mathcal{R}_{an}^{an}(z) \in \mathcal{R}(n)$ that intersect Γ'. Set $m = n + 2$. The last two inequalities together with (14.24) imply that

$$N(n + 2, Q_{n+2}(y)) \leq N_n \cdot 2C^2 e^{4a(h+\varepsilon) + 4an\delta}. \tag{14.25}$$

On the other hand, since $y \in \Gamma'$ the sets $\mathcal{R}_0^{an}(y) \cap \xi^u(y) \cap \Gamma'$ and $\mathcal{R}_{an}^0(y) \cap \xi^s(y) \cap \Gamma'$ are nonempty.

Consider a rectangle $\mathcal{R}_{an}^{an}(v) \subset Q_n(y)$ intersecting Γ'. Then the rectangles

$$\mathcal{R}_{an}^0(v) \cap \mathcal{R}_0^{an}(y) \quad \text{and} \quad \mathcal{R}_{an}^0(y) \cap \mathcal{R}_0^{an}(v)$$

are in $\mathcal{F}(n)$ and intersect respectively the local stable and unstable manifolds at y. Hence, to each rectangle $\mathcal{R}_{an}^{an}(v) \subset Q_n(y)$ intersecting Γ' we can associate the pair of rectangles

$$(\mathcal{R}_{an}^0(v) \cap \mathcal{R}_0^{an}(y), \mathcal{R}_{an}^0(y) \cap \mathcal{R}_0^{an}(v))$$

in the product

$$\{R \in \mathcal{F}(n) : R \cap \xi^s(y) \cap Q_n(y) \neq \varnothing\} \times \{R \in \mathcal{F}(n) : R \cap \xi^u(y) \cap Q_n(y) \neq \varnothing\}.$$

Clearly, this correspondence is injective. Therefore,

$$\widehat{N}^s(n, y, Q_n(y)) \cdot \widehat{N}^u(n, y, Q_n(y)) \geq N_n,$$

and the desired inequality follows from (14.25).   □

Our next goal is to compare the growth rate in n of the number of rectangles in $\mathcal{F}(n)$ and $\mathcal{R}(n)$. We start with an auxiliary result.

Lemma 14.4.7. *For every $x \in \Gamma'$ and $n \geq n_1$ we have*

$$\widehat{N}^s(n, x, \mathcal{R}(x)) \leq D^{-1} C^2 e^{anh + 3an\delta},$$

and

$$\widehat{N}^u(n, x, \mathcal{R}(x)) \leq D^{-1} C^2 e^{anh + 3an\delta}.$$

Proof of the lemma. Since $\mathcal{R}$ is a finite partition we can find points y_i such that the union of the rectangles $\mathcal{R}_0^{an}(y_i)$ is equal to $\mathcal{R}(x)$, and these rectangles are pairwise disjoint. Without loss of generality we can assume that $y_i \in \Gamma'$ whenever $\mathcal{R}_0^{an}(y_i) \cap \Gamma' \neq \varnothing$. We have

$$N(n, \mathcal{R}(x)) \geq \sum_i N^s(n, y_i, \mathcal{R}_0^{an}(y_i))$$

$$\geq \sum_{i:\mathcal{R}_0^{an}(y_i) \cap \Gamma' \neq \varnothing} N^s(n, y_i, \mathcal{R}_0^{an}(y_i)). \tag{14.26}$$

Now we estimate $N^s(n, y_i, \mathcal{R}_0^{an}(y_i))$ from below for each $y_i \in \Gamma'$. By Lemmas 14.4.2 and 14.4.3, and (14.8) we obtain

$$
\begin{aligned}
N^s(n, y_i, \mathcal{R}_0^{an}(y_i)) &\geq \frac{\mu_{y_i}^s(\mathcal{R}_0^{an}(y_i) \cap \Gamma)}{\max\{\mu_z^s(\mathcal{R}_{an}^{an}(z)) : z \in \xi^s(y_i) \cap \mathcal{R}(x) \cap \Gamma\}} \\
&\geq \frac{D}{\max\{\mu_z^s(\mathcal{R}_{an}^{an}(z)) : z \in \xi^s(y_i) \cap \mathcal{R}(x) \cap \Gamma\}} \\
&= \frac{D}{\max\{\mu_z^s(\mathcal{R}_{an}^0(z)) : z \in \xi^s(y_i) \cap \mathcal{R}(x) \cap \Gamma\}} \\
&\geq DC^{-1}e^{anh-an\delta}.
\end{aligned}
\tag{14.27}
$$

Similarly, it follows from (14.7) that

$$N(n, \mathcal{R}(x)) \leq \frac{\mu(\mathcal{R}(x))}{\min\{\mu(\mathcal{R}_{an}^{an}(z)) : z \in \mathcal{R}(x) \cap \Gamma\}} \leq Ce^{2anh+2an\delta}. \tag{14.28}$$

Now we observe that

$$\widehat{N}^u(n, x, \mathcal{R}(x)) = \operatorname{card}\{i : \mathcal{R}_0^{an}(y_i) \cap \Gamma' \neq \varnothing\}. \tag{14.29}$$

Combining (14.26), (14.27), (14.28), and (14.29) we conclude that

$$
\begin{aligned}
Ce^{2anh+2an\delta} &\geq N(n, \mathcal{R}(x)) \\
&\geq \sum_{i:\mathcal{R}_0^{an}(y_i) \cap \Gamma' \neq \varnothing} N^s(n, y_i, \mathcal{R}_0^{an}(y_i)) \\
&\geq \widehat{N}^u(n, x, \mathcal{R}(x)) \cdot DC^{-1}e^{anh-an\delta}.
\end{aligned}
$$

This yields

$$\widehat{N}^u(n, x, \mathcal{R}(x)) \leq D^{-1}C^2 e^{anh+3an\delta}.$$

The other inequality can be obtained in a similar manner. $\square$

We emphasize that the procedure of filling in rectangles to obtain the class $\mathcal{F}(n)$ may substantially increase the number of rectangles in a neighborhood of some points. However, we show that at almost every point this procedure does not add too many rectangles.

Lemma 14.4.8. *For μ-almost every $y \in \mathcal{R}(x) \cap \Gamma'$ we have*

$$\limsup_{n \to +\infty} \frac{\widehat{N}^s(n, y, Q_n(y))}{N^s(n, y, Q_n(y))} e^{-7an\delta} < 1,$$

and

$$\limsup_{n \to +\infty} \frac{\widehat{N}^u(n, y, Q_n(y))}{N^u(n, y, Q_n(y))} e^{-7an\delta} < 1.$$

Proof of the lemma. By (14.19) and (14.22), for each $n \geq n_1$ and $y \in \Gamma'$,

$$\mu_y^s(Q_n(y)) \geq \mu_y^s(B^s(y, e^{-n}) \cap \Gamma) \geq \frac{1}{2}\mu_y^s(B^s(y, e^{-n})).$$

Since $\mathcal{R}_{an}^{an}(z) \subset \mathcal{R}_{an}^0(z)$ for every z, using (14.8) and (14.12) we obtain

$$
\begin{aligned}
N^s(n, y, Q_n(y)) &\geq \frac{\mu_y^s(Q_n(y))}{\max\{\mu_z^s(\mathcal{R}_{an}^{an}(z)) : z \in \xi^s(y) \cap \mathcal{R}(x) \cap \Gamma\}} \\
&\geq \frac{1}{2} \frac{\mu_y^s(B^s(y, e^{-n}))}{\max\{\mu_z^s(\mathcal{R}_{an}^0(z)) : z \in \xi^s(y) \cap \mathcal{R}(x) \cap \Gamma\}} \\
&\geq \frac{1}{2C} \frac{e^{-d_\mu^s n - n\delta}}{e^{-anh + an\delta}}.
\end{aligned}
\tag{14.30}
$$

Consider the set

$$F = \left\{ y \in \Gamma' : \limsup_{n \to +\infty} \frac{\widehat{N}^s(n, y, Q_n(y))}{N^s(n, y, Q_n(y))} e^{-7an\delta} \geq 1 \right\}.$$

For each $y \in F$ there is an increasing sequence $m_j = m_j(y)$ of positive integers such that

$$
\begin{aligned}
\widehat{N}^s(m_j, y, Q_{m_j}(y)) &\geq \frac{1}{2} N^s(m_j, y, Q_{m_j}(y)) e^{7am_j\delta} \\
&\geq \frac{1}{4C} e^{-d_\mu^s m_j + am_j h + 5am_j\delta}
\end{aligned}
\tag{14.31}
$$

for every $j \in \mathbb{N}$ (note that $a > 1$).

We show that $\mu(F) = 0$. Assume on the contrary that $\mu(F) > 0$. Let $F' \subset F$ be the set of points $y \in F$ for which there exists the limit

$$\lim_{r \to 0} \frac{\log \mu_y^s(B^s(y, r))}{\log r} = d_\mu^s.$$

Clearly, $\mu(F') = \mu(F) > 0$, and we can find $y \in F$ such that

$$\mu_y^s(F) = \mu_y^s(F') = \mu_y^s(F' \cap \mathcal{R}(y) \cap \xi^s(y)) > 0.$$

It follows from Theorem 2.1.5 that

$$\dim_H(F' \cap \xi^s(y)) = d_\mu^s. \tag{14.32}$$

Consider the collection of balls

$$\mathfrak{B} = \{B(z, 4e^{-m_j(z)}) : z \in F' \cap \xi^s(y) \text{ and } j \in \mathbb{N}\}.$$

By the Besicovitch covering lemma (see, for example, [96, Theorem 2.7]), there exists a subcover $\mathcal{C} \subset \mathfrak{B}$ of $F' \cap \xi^s(y)$ with arbitrarily small diameter and finite multiplicity $\rho = \rho(\dim M)$. This means that for any $L > 0$ we can choose a sequence of points $z_i \in F' \cap \xi^s(y)$ and a sequence of integers $t_i \in \{m_j(z_i) : j \in \mathbb{N}\}$ with $t_i > L$ for each i, such that the collection of balls

$$\mathcal{C} = \{B(z_i, 4e^{-t_i}) : i \in \mathbb{N}\}$$

is a cover of $F' \cap \xi^s(y)$ with multiplicity at most ρ. Set $Q(i) = Q_{t_i}(z_i)$. We have

$$\sum_{B \in \mathcal{C}} (\text{diam } B)^{d_\mu^s - \delta} = (8^{d_\mu^s - \delta}) \sum_{i=1}^{\infty} e^{-t_i(d_\mu^s - \delta)},$$

and by (14.31),

$$\sum_{i=1}^{\infty} e^{-t_i(d_\mu^s - \delta)} \leq \sum_{i=1}^{\infty} \widehat{N}^s(t_i, z_i, Q(i)) \cdot 4Ce^{-at_i h - 4at_i \delta}$$

$$\leq 4C \sum_{q=1}^{\infty} e^{-aqh - 4aq\delta} \sum_{i:t_i=q} \widehat{N}^s(q, z_i, Q(i)).$$

Since the multiplicity of the collection $\mathcal{C}$ is at most ρ, each set $Q(i)$ occurs in the sum $\sum_{i:t_i=q} \widehat{N}^s(q, z_i, Q(i))$ at most a number of times equal to ρ. Hence,

$$\sum_{i:t_i=q} \widehat{N}^s(q, z_i, Q(i)) \leq \rho \widehat{N}^s(q, y, \mathcal{R}(y)).$$

It follows from Lemma 14.4.7 that

$$\sum_{B \in \mathcal{C}} (\text{diam } B)^{d_\mu^s - \delta} \leq 4(8^{d_\mu^s - \delta})C \sum_{q=1}^{\infty} e^{-aqh - 4aq\delta} \rho \widehat{N}^s(q, y, \mathcal{R}(y))$$

$$\leq 4(8^{d_\mu^s - \delta})D^{-1}C^3 \rho \sum_{q=1}^{\infty} e^{-aqh - 4aq\delta + aqh + 3aq\delta}$$

$$= 4(8^{d_\mu^s - \delta})D^{-1}C^3 \rho \sum_{q=1}^{\infty} e^{-aq\delta} < \infty.$$

Since L can be chosen arbitrarily large (and thus also the numbers t_i), we obtain

$$\dim_H(F' \cap \xi^s(y)) \le d_\mu^s - \delta < d_\mu^s.$$

This contradicts (14.32). Hence, $\mu(F) = 0$, and this yields the first inequality in the lemma. The proof of the second inequality is similar. $\square$

By Lemma 14.4.8, for μ-almost every $y \in \mathcal{R}(x) \cap \Gamma'$ there exists an integer $n_3(y) \ge n_2(y)$ such that if $n \ge n_3(y)$, then

$$\widehat{N}^s(n, y, Q_n(y)) < N^s(n, y, Q_n(y))e^{7an\delta}, \tag{14.33}$$

and

$$\widehat{N}^u(n, y, Q_n(y)) < N^u(n, y, Q_n(y))e^{7an\delta}. \tag{14.34}$$

By Lemmas 14.4.5 and 14.4.6, for μ-almost every $y \in \mathcal{R}(x) \cap \Gamma'$ and $n \ge n_2(y)$ we have

$$\mu(B(y, e^{-n-2})) \le \widehat{N}^s(n, y, Q_n(y)) \cdot \widehat{N}^u(n, y, Q_n(y)) \cdot 4C^3 e^{4a(h+\delta)} e^{-2anh+6an\delta}.$$

Hence, by (14.33), (14.34), and Lemma 14.4.4 we conclude that

$$\mu(B(y, e^{-n-2})) \le N^s(n, y, Q_n(y)) \cdot N^u(n, y, Q_n(y))$$
$$\times 4C^3 e^{4a(h+\delta)} e^{-2anh+20an\delta} \tag{14.35}$$
$$\le \mu_y^s(B^s(y, 4e^{-n}))\mu_y^u(B^u(y, 4e^{-n})) \cdot 4C^5 e^{4a(h+\delta)} e^{22an\delta}.$$

Moreover, by Lusin's theorem, for each $\delta > 0$ there exists a subset $\Gamma_\delta \subset \Gamma'$ such that

$$\mu(\Gamma_\delta) > \mu(\Gamma') - \delta, \quad n_\delta := \sup\{n_1, n_3(y) : y \in \Gamma_\delta\} < \infty,$$

and the inequalities (14.33) and (14.34) hold for every $n \ge n_\delta$.

Lemma 14.4.9. *For every $\delta > 0$, if $y \in \Gamma_\delta$ and $n \ge n_\delta$, then*

$$\mu_y^s(B^s(y, e^{-n}))\mu_y^u(B^u(y, e^{-n})) \le \mu(B(y, 4e^{-n})) \cdot 4C^3 e^{11an\delta}.$$

Proof of the lemma. Let $z \in \Gamma_\delta \cap Q_n(y)$. By (14.34), if $n \ge n_\delta$ then

$$N^u(n, y, Q_n(y)) \le \widehat{N}^u(n, y, Q_n(y))$$
$$= \widehat{N}^u(n, z, Q_n(y)) < N^u(n, z, Q_n(y))e^{7an\delta},$$

and

$$N^u(n, y, Q_n(y)) \le \inf\{N^u(n, z, Q_n(y)) : z \in \Gamma_\delta \cap Q_n(y)\}e^{7an\delta}. \tag{14.36}$$

Since $N(n, Q_n(y))$ is equal to the number of rectangles R in $Q_n(y)$ we have

$$\widehat{N}^s(n, y, Q_n(y)) \times \inf\{N^u(n, z, Q_n(y)) : z \in Q_n(y)\} \le N(n, Q_n(y)).$$

By (14.36), if $y \in \Gamma_\delta$ and $n \geq n_\delta$, then

$$N^s(n, y, Q_n(y)) \times N^u(n, y, Q_n(y)) \leq N(n, Q_n(y)) e^{7an\delta}.$$

In a similar manner to that in (14.30), if $y \in \Gamma_\delta$ and $n \geq n_\delta$, then

$$N^s(n, y, Q_n(y)) \geq \mu_y^s(B^s(y, e^{-n})) \cdot (2C)^{-1} e^{anh - an\delta},$$

and

$$N^u(n, y, Q_n(y)) \geq \mu_y^u(B^u(y, e^{-n})) \cdot (2C)^{-1} e^{anh - an\delta}.$$

Moreover, by (14.7) and (14.18),

$$N(n, Q_n(y)) \leq \frac{\mu(Q_n(y))}{\min\{\mu(\mathcal{R}_{an}^{an}(z)) : z \in Q_n(y) \cap \Gamma\}}$$
$$\leq \mu(B(y, 4e^{-n})) \cdot Ce^{2anh + 2an\delta}.$$

Combining these inequalities we obtain the desired result.   □

The statement in the theorem follows now immediately from (14.35) and Lemma 14.4.9.   □

The case of nonergodic measures can be treated using a similar procedure to the one in the proofs of Theorems 13.1.6 and 13.2.4. Given $\delta > 0$, we decompose the space into a countable number of invariant subsets

$$\{y \in M : |h_\mu(x) - h_\mu(y)| < \delta, |d_\mu^s(x) - d_\mu^s(y)| < \delta, \text{ and } |d_\mu^u(x) - d_\mu^u(y)| < \delta\}.$$

Repeating arguments in the proof of Theorem 14.4.1, in each set of these sets we obtain lower and upper estimates that deviate from (14.6) by small terms varying uniformly with δ. Letting $\delta \to 0$ yields the desired result. We refer to [11] for details.

Chapter 15

Quantitative Recurrence and Dimension Theory

Poincaré's recurrence theorem (Theorem 2.2.2) is one of the basic but fundamental results of the theory of dynamical systems. Unfortunately it only provides information of a qualitative nature. In particular, it does not consider the following natural problems: with which frequency the orbit of a point visits a given set of positive measure; with which rate the orbit of a point returns to an arbitrarily small neighborhood of the initial point. Birkhoff's ergodic theorem (Theorem 2.2.3) gives a complete answer to the first problem. The second problem of quantitative recurrence has experienced a growing interest during the last decade, also in connection with other fields, including for example compression algorithms. We describe in this chapter several results that provide partial answers to the problem.

15.1 Basic notions

We consider a transformation f in a metric space X. The *(first) return time* of a point $x \in X$ to the ball $B(x, r)$ (with respect to f) is defined by

$$\tau_r(x) = \inf\{n \in \mathbb{N} : d(f^n x, x) < r\},$$

where d is the distance in X.

Definition 15.1.1. For each $x \in X$, the *lower* and *upper recurrence rates* of x (with respect to f) are defined by

$$\underline{R}(x) = \liminf_{r \to 0} \frac{\log \tau_r(x)}{-\log r} \quad \text{and} \quad \overline{R}(x) = \limsup_{r \to 0} \frac{\log \tau_r(x)}{-\log r}. \tag{15.1}$$

When $\underline{R}(x) = \overline{R}(x)$ we denote the common value by $R(x)$, and we call it the *recurrence rate* of x (with respect to f).

We have the following identities.

Proposition 15.1.2. *For each $a > 0$ and $x \in X$ we have*

$$\underline{R}(x) = \liminf_{n \to \infty} \frac{\log \tau_{ae^{-n}}(x)}{n} \quad and \quad \overline{R}(x) = \limsup_{n \to \infty} \frac{\log \tau_{ae^{-n}}(x)}{n}.$$

Proof. For each sufficiently small $r > 0$ there exists a unique integer $n = n(r) \in \mathbb{N}$ such that

$$ae^{-(n+1)} \le r < ae^{-n} < 1.$$

We have

$$\tau_{ae^{-(n+1)}}(x) \ge \tau_r(x) \ge \tau_{ae^{-n}}(x),$$

and thus

$$\frac{\log \tau_{ae^{-n}}(x)}{-\log(ae^{-(n+1)})} < \frac{\log \tau_r(x)}{-\log r} < \frac{\log \tau_{ae^{-(n+1)}}(x)}{-\log(ae^{-n})}.$$

Note that $n(r) \to \infty$ when $r \to 0$, and that all integers are attained by $n(r)$ as $r \to 0$. Therefore,

$$
\begin{aligned}
\limsup_{n \to \infty} \frac{\log \tau_{ae^{-n}}(x)}{-n} &= \limsup_{n \to \infty} \frac{\log \tau_{ae^{-n}}(x)}{-\log(ae^{-(n+1)})} \\
&\le \limsup_{r \to 0} \frac{\log \tau_r(x)}{-\log r} \\
&\le \limsup_{n \to \infty} \frac{\log \tau_{ae^{-(n+1)}}(x)}{-\log(ae^{-n})} \\
&= \limsup_{n \to \infty} \frac{\log \tau_{ae^{-n}}(x)}{-n},
\end{aligned}
\tag{15.2}
$$

and we obtain the first identity in the proposition. The second identity is obtained replacing $\limsup$ by $\liminf$ everywhere in (15.2). $\square$

15.2   Upper bounds for recurrence rates

The following result gives upper bounds for the lower and upper recurrence rates in terms of the lower and upper pointwise dimensions. It was obtained by Barreira and Saussol in [15].

Theorem 15.2.1. *If f preserves a finite measure μ in $X \subset \mathbb{R}^m$, then for μ-almost every $x \in X$ we have*

$$\underline{R}(x) \le \underline{d}_\mu(x) \quad and \quad \overline{R}(x) \le \overline{d}_\mu(x).$$
$$\tag{15.3}$$

Proof. We start with an auxiliary result.

Lemma 15.2.2. *Given a probability measure μ in $\mathbb{R}^m$, there is a constant $\eta > 1$ such that for μ-almost every $x \in \mathbb{R}^m$ and every $\varepsilon > 0$ there exists $\delta = \delta(x, \varepsilon)$ such that*

$$\mu(B(x, \eta r)) \le \mu(B(x, r)) r^{-\varepsilon} \quad \text{for every} \quad r < \delta. \tag{15.4}$$

Proof of the lemma. Clearly, it is sufficient to show that for μ-almost every $x \in \mathbb{R}^m$ we have

$$\mu(B(x, 2^{-n})) \le n^2 \mu(B(x, 2^{-n-1})) \tag{15.5}$$

for all sufficiently large $n \in \mathbb{N}$. For each $n \in \mathbb{N}$ and $\delta > 0$, let

$$K_n(\delta) = \left\{ x \in \operatorname{supp} \mu : \mu(B(x, 2^{-n-1})) < \delta \mu(B(x, 2^{-n})) \right\}.$$

Taking a maximal 2^{-n-2}-separated set $E \subset K_n(\delta)$ (see Section 2.3) we obtain

$$\mu(K_n(\delta)) \le \sum_{x \in E} \mu(B(x, 2^{-n-1})) \le \sum_{x \in E} \delta \mu(B(x, 2^{-n})),$$

and there exists a constant M (depending only on m) such that E can be written in the form $E = \bigcup_{i=1}^{M} E_i$, where each set E_i is 2^{-n}-separated. For each $i = 1, \ldots, M$ the union $\bigcup_{x \in E_i} B(x, 2^{-n})$ is disjoint, and

$$\mu(K_n(\delta)) \le \sum_{i=1}^{M} \sum_{x \in E_i} \delta \mu(B(x, 2^{-n})) \le M\delta.$$

Since

$$\sum_{n>0} \mu(K_n(n^{-2})) \le M \sum_{n>0} n^{-2} < \infty,$$

it follows from Borel–Cantelli's lemma that (15.5) holds for μ-almost every $x \in \mathbb{R}^m$ and all sufficiently large $n \in \mathbb{N}$. This completes the proof of the lemma. $\square$

It follows easily from Lemma 15.2.2 that for each fixed constant $\eta > 1$, for μ-almost every $x \in Z$ and every $\varepsilon > 0$, there exists $\delta = \delta(x, \varepsilon, \eta) > 0$ such that (15.4) holds for every $r < \delta$. Furthermore, the function $\delta(x, \cdot)$ in the lemma can be chosen measurable for each x.

Fix $\varepsilon > 0$, and choose $\delta > 0$ sufficiently small such that the set

$$G = \{x \in X : \delta(x, \varepsilon) > \delta\}$$

has measure $\mu(G) > \mu(X) - \varepsilon$. Given $r, \lambda > 0$ and $x \in X$, we consider the set

$$A_{r,x} = \left\{ y \in B(x, 4r) : \tau_{4r}(y, x) \ge \lambda^{-1} \mu(B(x, 4r))^{-1} \right\},$$

where

$$\tau_{4r}(y, x) = \inf\{k > 0 : d(f^k x, y) < 4r\}.$$

We can easily verify that if $d(x, y) < r$, then

$$\tau_{8r}(y) \leq \tau_{4r}(y, x) \leq \tau_{2r}(y). \tag{15.6}$$

Chebychev's inequality implies that

$$\mu(A_{r,x}) \leq \lambda \mu(B(x, 4r)) \int_{B(x,4r)} \tau_{4r}(y, x) \, d\mu(y).$$

On the other hand, since μ is invariant, Kac's lemma tells us that

$$\int_{B(x,r)} \tau_r(y) \, d\mu(y) = 1, \tag{15.7}$$

and hence,

$$\int_{B(x,4r)} \tau_{4r}(y, x) \, d\mu(y) = \mu(\{y \in X : \tau_{4r}(y, x) < \infty\}) \leq 1.$$

Since $B(x, 2r) \subset B(x, 4r)$, we obtain

$$\mu(\{y \in B(x, 2r) : \tau_{4r}(y, x)\mu(B(x, 4r)) \geq \lambda^{-1}\}) \leq \lambda \mu(B(x, 4r)).$$

Furthermore,

$$\tau_{4r}(y, x)\mu(B(x, 4r)) \geq \tau_{8r}(y)\mu(B(y, 2r))$$

whenever $d(x, y) < 2r$ (see (15.6)), and thus,

$$\mu(\{y \in B(x, 2r) : \tau_{8r}(y)\mu(B(y, 2r)) \geq \lambda^{-1}\}) \leq \lambda \mu(B(x, 4r)). \tag{15.8}$$

We continue with an auxiliary statement.

Lemma 15.2.3. *Let μ be a finite Borel measure in a separable metric space X, and let $G \subset \operatorname{supp}\mu$ be a measurable set. Given $r > 0$, there exists a countable set $E \subset G$ such that:*

1. $B(x, r) \cap B(y, r) = \varnothing$ *for any distinct points x, $y \in E$;*

2. $\mu(G \setminus \bigcup_{x \in E} B(x, 2r)) = 0$.

Proof of the lemma. The existence of the set E can be obtained applying Zorn's lemma to the nonempty family of subsets of G which satisfy the first property, ordered by inclusion. Then the second property holds for any maximal element. Since $\mu(B(x, r)) > 0$ for every $x \in E \subset \operatorname{supp}\mu$, the set E is at most countable. $\square$

By (15.8) with $\lambda = r^{2\varepsilon}$ and Lemma 15.2.2 with $\eta = 4$ (see also the discussion after the lemma), we obtain

$$
\begin{aligned}
D_\varepsilon(r) : &= \mu(\{y \in G : \tau_{8r}(y)\mu(B(y, 2r)) \geq r^{-2\varepsilon}\}) \\
&\leq \sum_{x \in E} \mu(\{y \in B(x, 2r) : \tau_{8r}(y)\mu(B(y, 2r)) \geq r^{-2\varepsilon}\}) \\
&\leq r^{2\varepsilon} \sum_{x \in E} \mu(B(x, 4r)) \\
&\leq r^\varepsilon \sum_{x \in E} \mu(B(x, r)) \leq r^\varepsilon.
\end{aligned}
$$

We conclude that

$$
\sum_{n > -\log \delta} D_\varepsilon(e^{-n}) \leq \sum_{n > -\log \delta} e^{-\varepsilon n} < \infty.
$$

It follows from Borel–Cantelli's lemma that for μ-almost every $x \in G$,

$$
\frac{\log \tau_{8e^{-n}}(x)}{n} \leq 2\varepsilon + \frac{\log \mu(B(x, 2e^{-n}))}{-n}
$$

for all sufficiently large $n \in \mathbb{N}$. The desired result follows now from Propositions 2.1.4 and 15.1.2, together with the arbitrariness of ε. $\qquad \square$

The following example shows that the inequalities in (15.3) can be strict in a set of positive μ-measure.

Example 15.2.4. Consider a rotation of the circle by an irrational number ω that is well-approximated by rational numbers. This means that there exists $\kappa > 1$ such that $|\omega - p/q| < 1/q^{\kappa+1}$ for an infinite number of coprime integers p and q, say p_n and q_n for each $n \in \mathbb{N}$. Since $|q_n\omega - p_n| < 1/q_n^\kappa$, we have

$$
\tau_{1/q_n^\kappa}(x) = \inf\{k > 0 : k\omega(\mathrm{mod}1) < 1/q_n^\kappa\} \leq q_n
$$

for every point x in the circle. Therefore,

$$
\underline{R}(x) \leq \liminf_{n \to \infty} \frac{\log \tau_{1/q_n^\kappa}(x)}{-\log(1/q_n^\kappa)} \leq \frac{1}{\kappa} < 1.
$$

On the other hand, for any irrational rotation the Lebesgue measure m is the unique invariant probability measure, and it satisfies $\underline{d}_m(x) = \overline{d}_m(x) = 1$ for every x. In particular, $\underline{R}(x) < \underline{d}_m(x)$ for every point x (that is, the first inequality in (15.3) is strict everywhere).

Theorem 15.3.1 below shows that under some hyperbolicity assumptions the inequalities in (15.3) become identities in a full μ-measure set.

Boshernitzan proved earlier in [34] that if the α-dimensional Hausdorff measure m_α is σ-finite in X (that is, X can be written as a countable union of sets X_i, $i \in \mathbb{N}$ such that $m_\alpha(X_i) < \infty$ for every i), and f preserves a finite measure μ in X, then

$$\liminf_{n\to\infty}[n^{1/\alpha}d(f^n x, x)] < \infty$$

for μ-almost every $x \in X$. He also proved that if, in addition, $m_\alpha(X) = 0$, then

$$\liminf_{n\to\infty}[n^{1/\alpha}d(f^n x, x)] = 0 \qquad (15.9)$$

for μ-almost every $x \in X$. This result should be compared with the following consequence of Poincaré's recurrence theorem. Namely, we can easily show that for μ-almost every point $x \in X$ we have

$$\liminf_{n\to\infty} d(f^n x, x) = 0. \qquad (15.10)$$

While (15.10) only indicates that some subsequence of the orbit of x converges to x, the identity in (15.9) gives some quantitative information about the speed of convergence to the point x, and thus about the speed of recurrence.

The following result shows that any upper bound for the lower recurrence rate corresponds to precise quantitative information about the speed of recurrence.

Proposition 15.2.5. *For each $x \in X$ and $d \geq 0$, we have $\underline{R}(x) \leq d$ if and only if*

$$\liminf_{n\to\infty}[n^{1/(d+\varepsilon)}d(T^n x, x)] = 0 \qquad (15.11)$$

for every $\varepsilon > 0$.

Proof. We first assume that $\underline{R}(x) \leq d$. For each $\varepsilon > 0$, there exists a sequence of numbers $r_n > 0$ with $r_n \to 0$, such that

$$\tau_{r_n}(x) < r_n^{-(d+\varepsilon)} \quad \text{for every} \quad n \in \mathbb{N}.$$

Let $m_n = \tau_{r_n}(x)$. If the sequence m_n is bounded, then x is periodic, and (15.11) holds. Now assume that m_n is unbounded. We have $d(T^{m_n} x, x) < r_n$, and

$$m_n^{1/(d+2\varepsilon)}d(T^{m_n} x, x) < \tau_{r_n}(x)^{1/(d+2\varepsilon)} r_n$$
$$< r_n^{-(d+\varepsilon)/(d+2\varepsilon)} r_n = r_n^{\varepsilon/(d+\varepsilon)}.$$

Therefore,

$$\liminf_{n\to\infty}[n^{1/(d+2\varepsilon)}d(T^n x, x)] \leq \liminf_{n\to\infty}[m_n^{1/(d+2\varepsilon)}d(T^{m_n} x, x)] = 0.$$

This establishes (15.11) for every $\varepsilon > 0$.

Now we assume that (15.11) holds for every $\varepsilon > 0$. Setting $r_n = 2d(T^n x, x)$, we have $\tau_{r_n}(x) \leq n$, and it follows from (15.11) that

$$\liminf_{n\to\infty}[\tau_{r_n}(x)^{1/(d+\varepsilon)} r_n] = 0.$$

Thus, there exists an unbounded sequence of positive integers k_n such that

$$\tau_{r_{k_n}}(x)^{1/(d+\varepsilon)} r_{k_n} < 1$$

for every $n \in \mathbb{N}$. Therefore,

$$\underline{R}(x) \leq \liminf_{n \to \infty} \frac{\log \tau_{r_n}(x)}{-\log r_n} \leq \liminf_{n \to \infty} \frac{\log(r_{k_n}^{d+\varepsilon})}{-\log r_{k_n}} = d + \varepsilon.$$

The arbitrariness of ε implies that $\underline{R}(x) \leq d$. $\square$

In view of Proposition 15.2.5 the statement in Theorem 15.2.1 for the lower recurrence rate can be reformulated as follows.

Theorem 15.2.6. *If f preserves a finite measure μ in $X \subset \mathbb{R}^m$, then (15.9) holds for μ-almost every $x \in X$ such that $\underline{d}_\mu(x) < \alpha$.*

15.3 Recurrence rate and pointwise dimension

The following result of Barreira and Saussol in [15] gives a complete answer to the quantitative recurrence problem in the introduction to this chapter in the case of equilibrium measures of Hölder continuous functions on locally maximal hyperbolic sets.

Theorem 15.3.1 (Quantitative recurrence). *Let $\Lambda \subset \mathbb{R}^m$ be a locally maximal hyperbolic set of a $C^{1+\varepsilon}$ diffeomorphism, for some $\varepsilon > 0$. If μ is an equilibrium measure of a Hölder continuous function in Λ, then*

$$R(x) = \lim_{r \to 0} \frac{\log \mu(B(x,r))}{\log r} \tag{15.12}$$

for μ-almost every point $x \in \Lambda$.

Proof. We follow closely the alternative proof given by Saussol in [136]. We start with an auxiliary statement. For each $a > 0$, set

$$X_a = \{x \in \Lambda : \underline{d}_\mu(x) > a\}.$$

Lemma 15.3.2. *Given $a > 0$, for each $\rho > 0$ and μ-almost every $x \in X_a$ there exists $r(x) > 0$ such that if $r \in (0, r(x))$ and $n \in [r^{-a}, \mu(B(x,r))^{-1+\rho}] \cap \mathbb{N}$, then $d(f^n x, x) \geq r$.*

Proof of the lemma. Given $r_0 > 0$, we set $G = G_1 \cap G_2 \cap G_3$ where

$$G_1 = \{x \in X_a : \mu(B(x, 2r)) \leq r^a \text{ for } r \leq r_0\},$$
$$G_2 = \{x \in X_a : \mu(B(x, r/2)) \geq r^{m+\rho} \text{ for } r \leq r_0\},$$
$$G_3 = \{x \in X_a : \mu(B(x, r/2)) \geq \mu(B(x, 4r)) r^{\rho a/2} \text{ for } r \leq r_0\}.$$

We note that $\mu(G) \to \mu(X_a)$ as $r_0 \to 0$. Indeed, by the definition of lower pointwise dimension $\mu(G_1) \to \mu(X_a)$ as $r_0 \to 0$. Furthermore, since $\overline{d}_\mu(x) \le m$ for μ-almost every $x \in \Lambda$, we have $\mu(G_2) \to 1$ as $r_0 \to 0$. Finally, by Lemma 15.2.2 we have $\mu(G_3) \to 1$ as $r_0 \to 0$.

For each $r \le r_0$ we define the set

$$A_\rho(r) = \{y \in X : d(f^n y, y) < r \text{ for some } n \in [r^{-a}, \mu(B(y,3r))^{-1+\rho}] \cap \mathbb{N}\}.$$

Take $x \in G$. It follows from the triangle inequality that

$$
\begin{aligned}
&B(x,r) \cap A_\rho(r) \\
&\subset \{y \in B(x,r) : d(f^n y, x) < 2r \text{ for some } n \in [r^{-a}, \mu(B(x,2r))^{-1+\rho}] \cap \mathbb{N}\} \\
&= \bigcup B(x,r) \cap f^{-n} B(x,2r),
\end{aligned}
$$

$$(15.13)$$

where the union is taken over all integers

$$n \in [r^{-a}, \mu(B(x,2r))^{-1+\rho}] \cap \mathbb{N}.$$

Now let $\eta_r \colon [0,\infty) \to \mathbb{R}$ be the $(1/r)$-Lipschitz map such that $\chi_{[0,r]} \le \eta_r \le \chi_{[0,2r]}$, and define the function

$$\varphi_{x,r}(y) = \eta_r(d(x,y)).$$

Clearly, $\varphi_{x,r}$ is $(1/r)$-Lipschitz, and thus it is also Hölder continuous because Λ is compact. Since $f|\Lambda$ has exponential decay of correlations for any equilibrium measure of a Hölder continuous function (see, for example, [38]), we obtain

$$
\begin{aligned}
\mu(B(x,r) \cap f^{-n} B(x,2r)) &\le \int_\Lambda \varphi_{x,r}(\varphi_{x,2r} \circ f^n)\, d\mu \\
&\le r^{-2}\theta_n + \int_\Lambda \varphi_{x,r}\, d\mu \int_\Lambda \varphi_{x,2r}\, d\mu \\
&\le r^{-2}\theta_n + \mu(B(x,2r))\mu(B(x,4r)),
\end{aligned}
$$

for some exponentially decreasing sequence $\theta_n > 0$. Choose $p > 1$ such that $a(p-1) - 2 \ge m + 2\rho$, and r_0 sufficiently small so that

$$\theta_n \le (p-1)(n+1)^{-p} \quad \text{for} \quad n \ge r_0^{-a}.$$

Since

$$\sum_{n \ge q} n^{-p} \le (q-1)^{1-p}/(p-1),$$

it follows from (15.13) that for each $r \in (0, r_0)$,

$$
\begin{aligned}
\mu(B(x,r) \cap A_\rho(r)) &\le r^{a(p-1)-2} + \mu(B(x,2r))^\rho \mu(B(x,4r)) \\
&\le \mu(B(x,r/2))(r^\rho + r^{\rho a/2}).
\end{aligned}
$$

Now let $E \subset G$ be the set constructed in Lemma 15.2.3 with r replaced by $r/2$. Since

$$\mu \left(G| \bigcup_{x \in E} B(x,r) \right) = 0,$$

and the balls $B(x, r/2)$ are disjoint, we have

$$\mu(G \cap A_\rho(r)) \leq \sum_{x \in E} \mu(B(x,r) \cap A_\rho(r))$$

$$\leq \sum_{x \in E} \mu(B(x,r/2))(r^\rho + r^{\rho a/2})$$

$$\leq r^\rho + r^{\rho a/2}.$$

This implies that

$$\sum_{m=1}^{\infty} \mu(G \cap A_\rho(e^{-m})) < \infty.$$

Therefore, by Borel–Cantelli's lemma, for μ-almost every $y \in G$ there exists an integer $m(y) \in \mathbb{N}$ such that

$$y \notin A_\rho(e^{-m}) \quad \text{for every} \quad m \geq m(y).$$

For each $r \leq e^{-m(y)}$, if m is the unique integer such that $e^{-m-1} < r \leq e^{-m}$, then $e^{\delta m} \leq r^{-\delta}$ and $3e^{-m} < 3er$. Hence, there exists no integer

$$n \in [r^{-a}, \mu(B(y, 3er))^{-1+\rho}] \cap \mathbb{N}$$

such that $d(f^n y, y) < r$. The desired result follows now from Lemma 15.2.2. $\quad\square$

We proceed with the proof of the theorem. By Theorem 15.2.1 we have

$$\underline{R}(x) \leq \underline{d}_\mu(x) \quad \text{and} \quad \overline{R}(x) \leq \overline{d}_\mu(x)$$

for μ-almost every $x \in \Lambda$. Furthermore, the first inequality implies that for all $a > 0$ we have

$$\{x \in \Lambda : \underline{R}(x) > a\} \subset X_a \pmod 0.$$

For each $x \in \Lambda$ such that $\underline{R}(x) > a$ we have $\tau_r(x) \geq r^{-a}$ for all sufficiently small r. Given $\rho > 0$, it follows from Lemma 15.3.2 that

$$\tau_r(x) \geq \mu(B(x,r))^{-1+\rho}$$

for all sufficiently small $r > 0$ and μ-almost every $x \in \Lambda$ with $\underline{R}(x) > a$. Therefore,

$$\underline{R}(x) \geq (1-\rho)\underline{d}_\mu(x) \quad \text{and} \quad \overline{R}(x) \geq (1-\rho)\overline{d}_\mu(x)$$

for μ-almost every $x \in \Lambda$ such that $\underline{R}(x) > a$. The arbitrariness of ρ and a implies that

$$\underline{R}(x) = \underline{d}_\mu(x) \quad \text{and} \quad \overline{R}(x) = \overline{d}_\mu(x)$$

for μ-almost every $x \in X$. The desired result is now an immediate consequence of Theorem 14.4.1. $\quad\square$

The study of the quantitative behavior of recurrence started with the work of Ornstein and Weiss [107], closely followed by the work of Boshernitzan [34] (see Section 15.4 for a description of the results). The first paper considers the case of symbolic dynamics (with the distances d in (3.9) and (4.24)), and for an ergodic σ-invariant measure μ it is shown that $R(x) = h_\mu(\sigma)$ for μ-almost every x (see Theorem 15.4.12). Theorem 15.3.1 is a version of this result in the case of hyperbolic sets. See Theorem 15.4.1 for a related result in the case of repellers. For piecewise-monotone maps of the interval and ergodic invariant measures with nonzero entropy, the property in Theorem 15.3.1 (see (15.12)) was established by Saussol, Troubetzkoy and Vaienti in [137], building on results in [107] and results of Hofbauer and Raith in [75, 76].

We note that identity (15.12) relates two quantities of very different nature. In particular, only $R(x)$ depends on the diffeomorphism and only $d_\mu(x)$ depends on the measure. It follows from (15.1) and (15.12) that

$$\lim_{r \to 0} \frac{\log \inf\{n \in \mathbb{N} : d(f^n x, x) < r\}}{-\log r} = \lim_{r \to 0} \frac{\log \mu(B(x,r))}{\log r}$$

for μ-almost every $x \in \Lambda$. This means that asymptotically

$$\inf\{k \in \mathbb{N} : f^k x \in B(x,r)\} \quad \text{is approximately equal to} \quad 1/\mu(B(x,r))$$

for all sufficiently small $r > 0$. This should be compared to Kac's lemma, which tells us that the average value of τ_r in $B(x,r)$ is equal to $1/\mu(B(x,r))$ (see (15.7)). Thus, Theorem 15.3.1 can be thought of as a local version of Kac's lemma.

It was shown by Saussol and Wu in [138] that for a repeller J of a $C^{1+\varepsilon}$ conformal transformation the recurrence rate has a constant dimension spectrum. Moreover, they showed that

$$\dim_H\{x \in J : \underline{R}(x) = \alpha \text{ and } \overline{R}(x) = \beta\} = \dim_H J$$

for every $\alpha \le \beta$ in $[0, +\infty]$. A corresponding result in the case of symbolic dynamics was established by Feng and Wu in [64].

15.4 Product structure and recurrence

We already described the local product structure of hyperbolic sets (see Definition 4.2.5) and the almost product structure of hyperbolic measures (see Sections 14.3 and 14.4). We describe in this section the product structure of the recurrence rate.

15.4.1 Preliminary results

Let $g \colon W \to W$ be a Borel measurable transformation in the metric space W, and let μ be a g-invariant probability measure in W. We recall that the entropy of a finite or countable partition $\mathcal{Z}$ of W by measurable sets is defined by

$$H_\mu(\mathcal{Z}) = -\sum_{Z \in \mathcal{Z}} \mu(Z) \log \mu(Z).$$

For each $n \in \mathbb{N}$ we consider the new partition $\mathcal{Z}_n = \bigvee_{k=0}^{n-1} g^{-k}\mathcal{Z}$. Given $x \in W$ and $n \in \mathbb{N}$ we denote by $\mathcal{Z}_n(x) \in \mathcal{Z}_n$ the unique (mod 0) element of $\mathcal{Z}_n$ that contains x.

The following statement gives general conditions under which the inequalities in (15.3) become identities in a full measure set. It was established by Barreira and Saussol in [17].

Theorem 15.4.1. *Let $g \colon W \to W$ be a Borel measurable transformation in a set $W \subset \mathbb{R}^d$, for some $d \in \mathbb{N}$, let μ be an ergodic g-invariant nonatomic probability measure in W, and let $\mathcal{Z}$ be a finite or countable partition of W by measurable sets with $H_\mu(\mathcal{Z}) < \infty$. Assume that:*

1. *there exists $\kappa > 1$ such that if $n, m \in \mathbb{N}$ and $x \in W$, then*

$$\mu(\mathcal{Z}_{n+m}(x)) \le \kappa \mu(\mathcal{Z}_n(x))\mu(\mathcal{Z}_m(g^n x)); \qquad (15.14)$$

2. *there exists $\lambda > 0$ such that*

$$\sup\{\operatorname{diam} Z : Z \in \mathcal{Z}_n\} < e^{-\lambda n} \qquad (15.15)$$

for all sufficiently large $n \in \mathbb{N}$;

3. *for μ-almost every $x \in X$ there exists $\gamma > 0$ such that $B(x, e^{-\gamma n}) \subset \mathcal{Z}_n(x)$ for all sufficiently large $n \in \mathbb{N}$.*

Then for μ-almost every $x \in W$ we have

$$\underline{R}(x) = \underline{d}_\mu(x) \quad and \quad \overline{R}(x) = \overline{d}_\mu(x). \qquad (15.16)$$

Proof. We first show that the entropy $h_\mu(g)$ is finite and nonzero. Set

$$\sigma_n = \sup\{\mu(Z) : Z \in \mathcal{Z}_n\}.$$

It follows from (15.15) and the fact that μ is not atomic that $\sigma_n \to 0$ as $n \to \infty$. Otherwise there would exist $x \in W$ and $\varepsilon > 0$ such that $\mu(\mathcal{Z}_n(x)) \to \varepsilon$ as $n \to \infty$. But using (15.15) we have $\bigcap_{n \in \mathbb{N}} \mathcal{Z}_n(x) = \{x\}$, and thus $\mu(\{x\}) = \varepsilon > 0$, which contradicts the fact that μ is not atomic. In particular, there exists $p \in \mathbb{N}$ such that $\sigma_p < 1/\kappa$. By (15.14) we have

$$\mu(\mathcal{Z}_{pn}(x)) \le (\kappa \sigma_p)^n$$

for every $x \in W$ and $n \in \mathbb{N}$. By Shannon–McMillan–Breiman's theorem we obtain

$$h_\mu(g) \ge \liminf_{n \to \infty} \frac{\log \mu(\mathcal{Z}_{pn}(x))}{-pn} \ge -\frac{1}{p} \log(\kappa \sigma_p) > 0.$$

Furthermore, it follows from (15.15) that $\mathcal{Z}$ is a generating partition, and hence

$$h_\mu(g) = h_\mu(g, \mathcal{Z}) \le H_\mu(\mathcal{Z}) < \infty.$$

The hypotheses of the theorem ensure that for μ-almost every $x \in W$ there exists $\gamma > 0$ such that

$$B(x, e^{-\gamma n}) \subset \mathcal{Z}_n(x) \subset B(x, e^{-\lambda n})$$

for all sufficiently large $n \in \mathbb{N}$. It follows from Shannon–McMillan–Breiman's theorem that

$$\gamma \underline{d}_\mu(x) \ge h_\mu(g) \ge \lambda \overline{d}_\mu(x)$$

for μ-almost every $x \in W$. Since $0 < h_\mu(g) < \infty$, we conclude that

$$0 < \underline{d}_\mu(x) \le \overline{d}_\mu(x) < \infty \tag{15.17}$$

for μ-almost every $x \in W$.

The return time of the point $y \in B(x, r)$ to $B(x, r)$ is defined by

$$\tau_r(y, x) = \inf\{k \in \mathbb{N} : d(g^k y, x) < r\}.$$

For each $x \in W$ and $r, \varepsilon > 0$, we consider the set

$$A_\varepsilon(x, r) = \left\{ y \in B(x, r) : \tau_r(y, x) \le \mu(B(x, r))^{-1+\varepsilon} \right\}.$$

The following criterion was obtained by Barreira and Saussol in [15].

Lemma 15.4.2. *If $\underline{d}_\mu(x) > 0$ and*

$$\liminf_{r \to 0} \frac{\log \mu(A_\varepsilon(x, r))}{\log \mu(B(x, r))} > 1 \tag{15.18}$$

for μ-almost every $x \in W$ and every sufficiently small $\varepsilon > 0$, then (15.16) holds for μ-almost every $x \in W$.

Proof of the lemma. By Theorem 15.2.1 it remains to prove that

$$\underline{R}(x) \ge \underline{d}_\mu(x) \quad \text{and} \quad \overline{R}(x) \ge \overline{d}_\mu(x)$$

for μ-almost every $x \in W$.

By the hypotheses and Lemma 15.2.2, given $\varepsilon > 0$ sufficiently small there exist numbers $a, \gamma, \rho > 0$ and a set $G \subset W$ with $\mu(G) > 1 - \varepsilon$ such that if $x \in G$ and $r \in (0, \rho)$, then

$$\mu(A_\varepsilon(x, 2r)) \le \mu(B(x, 2r))^{1+\gamma}, \tag{15.19}$$

$$\mu(B(x, 2r)) \le \mu(B(x, r/2))r^{-a\gamma/2}, \tag{15.20}$$

$$\mu(B(x, r)) \le r^a. \tag{15.21}$$

We consider the set

$$A_\varepsilon(r) = \{y \in G : \tau_r(y) \le \mu(B(y, 3r))^{-1+\varepsilon}\}.$$

By (15.6), if $d(x, y) < r$, then $\tau_r(y) \ge \tau_{2r}(y, x)$. Since $B(x, 2r) \subset B(y, 3r)$, if $x \in G$, then using (15.19), (15.20), and (15.21) we obtain

$$\begin{aligned}
\mu(B(x, r) \cap A_\varepsilon(r)) &\le \mu(\{y \in B(x, r) : \tau_{2r}(y, x) \le \mu(B(x, 3r))^{-1+\varepsilon}\}) \\
&\le \mu(A_\varepsilon(x, 2r)) \\
&\le \mu(B(x, 2r))^{1+\gamma} \\
&\le \mu(B(x, r/2))r^{-a\gamma/2}(2r)^{a\gamma}.
\end{aligned}$$

Let $E \subset G$ be the set given by Lemma 15.2.3. Then

$$\begin{aligned}
\mu(A_\varepsilon(r)) &\le \sum_{x \in E} \mu(B(x, r) \cap A_\varepsilon(r)) \\
&\le \sum_{x \in E} \mu(B(x, r/2))r^{-a\gamma/2}(2r)^{a\gamma} \\
&\le 2^{a\gamma} r^{a\gamma/2}.
\end{aligned}$$

We conclude that

$$\sum_{n=1}^\infty \mu(A_\varepsilon(e^{-n})) < \infty.$$

It follows from Borel–Cantelli's lemma that for μ-almost every $x \in G$ we have

$$\tau_{e^{-n}}(x) > \mu(B(x, 3e^{-n}))^{-1+\varepsilon}$$

for all sufficiently large $n \in \mathbb{N}$. By Propositions 2.1.4 and 15.1.2 we obtain

$$\underline{R}(x) \ge (1 - \varepsilon)\underline{d}_\mu(x) \quad \text{and} \quad \overline{R}(x) \ge (1 - \varepsilon)\overline{d}_\mu(x)$$

for μ-almost every $x \in G$. The desired statement follows now from the arbitrariness of ε. $\qquad\square$

Now we show that (15.18) holds, and the theorem follows from Lemma 15.4.2. We define the return time of a set A to itself by

$$\tau(A) = \inf\{n \in \mathbb{N} : g^n A \cap A \ne \varnothing\}.$$

Saussol, Troubetzkoy and Vaienti show in [137] that the return time of an element of the partition $\mathcal{Z}_n = \bigvee_{k=0}^{n-1} g^{-k}\mathcal{Z}$ to itself is typically large, in the following sense.

Lemma 15.4.3. *Let $g\colon W \to W$ be a measurable transformation preserving an ergodic probability measure μ in W. If $\mathcal{Z}$ is a finite or countable measurable partition of W by measurable sets, and $h_\mu(g, \mathcal{Z}) > 0$, then*

$$\liminf_{n \to \infty} \frac{\tau(\mathcal{Z}_n(x))}{n} \ge 1$$

for μ-almost every $x \in W$.

Proof of the lemma. We follow closely the alternative proof given in [1].

We note that it is sufficient to consider the case of finite partitions. Indeed, if $\mathcal{Z} = \{Z_i : i \in \mathbb{N}\}$ is a countable partition, then for some $m \in \mathbb{N}$ the finite partition

$$\mathcal{Z}' = \left\{Z_1, \ldots, Z_m, \bigcup_{l > m} B_l\right\}$$

has positive entropy. Since $\mathcal{Z}$ is finer than $\mathcal{Z}'$, we have $\tau(\mathcal{Z}_n(x)) \geq \tau(\mathcal{Z}'_n(x))$ and the claim follows.

Let $\mathcal{Z}$ be a finite partition of W with $h = h_\mu(g, \mathcal{Z}) > 0$. By Shannon–McMillan–Breiman's theorem, given $\varepsilon > 0$, for μ-almost every $x \in W$ there exists $n(x) \in \mathbb{N}$ such that if $n > n(x)$, then

$$\left| \frac{1}{n} \log \mu(\mathcal{Z}_n(x)) + h \right| < \varepsilon.$$

By Egoroff's theorem, for all sufficiently large $M = M(\varepsilon)$ the set

$$E_M = \{x \in W : n(x) < M\}$$

has measure $\mu(E_M) > 1 - \varepsilon$. Furthermore, there exists a constant $c > 0$ such that if $x \in E_M$ and $n \in \mathbb{N}$, then

$$c^{-1} e^{-nh - n\varepsilon} \leq \mu(\mathcal{Z}_n(x)) \leq c e^{-nh + n\varepsilon}. \tag{15.22}$$

Set $E = E_{M(\varepsilon)}$, $\delta = 1 - 3\varepsilon/h$, and

$$\mathcal{C}_n = \{x \in E : \tau(Z_n(x)) \leq \delta n\}.$$

Clearly, $\mathcal{C}_n = \bigcup_{k=1}^{\delta n} R_n(k)$, where

$$R_n(k) = \{x \in E : \tau(Z_n(x)) = k\}.$$

We want to show that $\sum_{n \in \mathbb{N}} \mu(\mathcal{C}_n) < \infty$.

Let $k \leq n$, and set

$$\mathcal{F} = \{\mathcal{Z}_k(x) : x \in R_n(k)\}.$$

For each $Z \in \mathcal{F}$ there exists a unique set $Z' \in \mathcal{Z}_n$ such that $Z \cap R_n(k) \subset Z' \subset Z$. Therefore,

$$\mu(R_n(k)) = \sum_{Z \in \mathcal{F}} \mu(Z \cap R_n(k)) \leq \sum_{Z \in \mathcal{F}} \mu(Z').$$

Notice that for each $Z \in \mathcal{F}$ we have $Z \cap E \neq \varnothing$ and $Z' \cap E \neq \varnothing$. Hence, there exists $x \in E$ such that $Z = \mathcal{Z}_k(x)$ and $Z' = \xi_n(x)$. It follows from (15.22) that

$$\mu(\mathcal{Z}_n(x)) \leq c e^{-nh + n\varepsilon} \quad \text{and} \quad 1 \leq c\mu(\mathcal{Z}_k(x)) e^{kh + k\varepsilon}.$$

Multiplying these inequalities we obtain

$$\mu(Z') \le c^2 e^{-nh+n\varepsilon} e^{kh+k\varepsilon} \mu(Z),$$

and thus,

$$\mu(R_n(k)) \le c^2 e^{-(n-k)h+2n\varepsilon}.$$

Therefore,

$$\mu(\mathcal{C}_n) = \sum_{k=1}^{\delta n} \mu(R_n(k)) \le c^2 \frac{e^h}{e^h - 1} e^{-n(h-\delta h - 2\varepsilon)}.$$

Since $h - \delta h - 2\varepsilon = \varepsilon > 0$, we obtain $\sum_{n \in \mathbb{N}} \mu(\mathcal{C}_n) < \infty$. By Borel–Cantelli's lemma, for μ-almost every $x \in E$ we have

$$\tau(Z_n(x)) \ge (1 - 3\varepsilon/h)n$$

for all except finitely many $n \in \mathbb{N}$. Since $\mu(E) > 1 - \varepsilon$, the arbitrariness of ε implies the desired result. $\qquad\square$

Since $h_\mu(g, Z) > 0$ we can apply Lemma 15.4.3. By hypothesis 3 in Theorem 15.4.1 we conclude that for μ-almost every $x \in W$ there exists $\gamma > 0$ such that

$$\liminf_{r \to 0} \frac{\tau(B(x,r))}{-\log r} = \liminf_{n \to \infty} \frac{\tau(B(x, e^{-\gamma n}))}{\gamma n}$$

$$\ge \liminf_{n \to \infty} \frac{\tau(Z_n(x))}{\gamma n} \ge \frac{1}{\gamma}. \tag{15.23}$$

The identity in (15.23) follows easily from the fact that given $r > 0$ there exists a unique integer $n = n(r) \in \mathbb{N}$ such that $e^{-\gamma(n+1)} < r \le e^{-\gamma n}$, and thus

$$\frac{\tau(B(x, e^{-\gamma(n+1)}))}{\gamma n} > \frac{\tau(B(x,r))}{-\log r} > \frac{\tau(B(x, e^{-\gamma n}))}{\gamma(n+1)}.$$

It follows from (15.23) that

$$B(x,r) \cap g^{-k} B(x,r) = \varnothing$$

whenever k is a positive integer such that $k < -\log r/(2\gamma)$, and r is sufficiently small.

Let

$$B_k = \bigcup_{y \in B(x,r)} Z_k(y),$$

and write this set as a disjoint union $\bigcup_{j=1}^{N} Z_k(y_j)$. We choose sets $Z_1, Z_2, \ldots \in \bigcup_{n \in \mathbb{N}} Z_n$ such that $Z_i \cap Z_j = \varnothing \pmod 0$ whenever $i \ne j$, and

$$B(x,r) = \bigcup_{\ell \in \mathbb{N}} Z_\ell \pmod 0.$$

It follows from (15.14) that

$$\mu(B_k \cap g^{-k}B(x,r)) = \sum_{j=1}^{N}\sum_{\ell \in \mathbb{N}} \mu(\mathcal{Z}_k(y_j) \cap g^{-k}Z_\ell)$$

$$\leq \sum_{j=1}^{N}\sum_{\ell \in \mathbb{N}} \kappa\mu(\mathcal{Z}_k(y_j))\mu(Z_\ell)$$

$$= \kappa\mu(B_k)\mu(B(x,r)).$$

By hypothesis 2 in Theorem 15.4.1 we have $B_k \subset B(x, r+e^{-\lambda k})$ for all sufficiently small $r > 0$, and hence

$$\frac{\mu(B(x,r) \cap g^{-k}B(x,r))}{\mu(B(x,r))} \leq \kappa\mu(B(x, r + e^{-\lambda k})). \tag{15.24}$$

By Lemma 15.2.2, if $k \geq -\log r/\lambda$, then

$$\mu(B(x, r + e^{-\lambda k})) \leq \mu(B(x, 2r)) \leq \mu(B(x,r))r^{-\varepsilon(r)}, \tag{15.25}$$

where $\varepsilon(r) \to 0$ as $r \to 0$. We note that the function $\varepsilon(r)$ may depend on x. Eventually rechoosing $\varepsilon(r)$, we can assume that if $k \geq -\log r/(2\gamma)$, then

$$\mu(B(x, r + e^{-\lambda k})) \leq \mu(B(x, r^{\lambda/(3\gamma)})) \leq r^{\lambda \underline{d}_\mu(x)/(3\gamma)}r^{-\varepsilon(r)}, \tag{15.26}$$

since $\lambda/\gamma \leq 1$.

Combining the estimates in (15.25) and (15.26) with (15.24), and eventually rechoosing $\varepsilon(r)$, we obtain

$$\frac{\mu(A_\varepsilon(x,r))}{\mu(B(x,r))} \leq \sum_{k=-\log r/(2\gamma)}^{-\log r/\lambda} r^{\lambda \underline{d}_\mu(x)/(3\gamma)}r^{-\varepsilon(r)} + \sum_{k=-\log r/\lambda}^{\mu(B(x,r))^{-1+\varepsilon}} \mu(B(x,r))r^{-\varepsilon(r)}$$

$$\leq \left(-\frac{1}{\lambda} + \frac{1}{2\gamma}\right)\log r\left(\mu(B(x,r))^{1/(\overline{d}_\mu(x)+\varepsilon)}\right)^{\lambda\underline{d}_\mu(x)/(3\gamma)}r^{-\varepsilon(r)}$$

$$+ \left(\mu(B(x,r))^{-1+\varepsilon} + \frac{1}{\lambda}\log r\right)\mu(B(x,r))r^{-\varepsilon(r)}$$

$$\leq \left[\mu(B(x,r))^{\lambda\underline{d}_\mu(x)/[3\gamma(\overline{d}_\mu(x)+\varepsilon)]} + \mu(B(x,r))^\varepsilon\right]r^{-2\varepsilon(r)}$$

for all sufficiently small $r > 0$. By (15.17), we have $\underline{d}_\mu(x) > 0$ for μ-almost every $x \in W$, and thus,

$$\liminf_{r\to 0}\frac{\log\mu(A_\varepsilon(x,r))}{\log\mu(B(x,r))} \geq 1 + \min\left\{\frac{\lambda\underline{d}_\mu(x)}{3\gamma(\overline{d}_\mu(x)+\varepsilon)}, \varepsilon\right\} > 1$$

for μ-almost every point $x \in W$. The theorem follows now immediately from Lemma 15.4.2. $\qquad\square$

The statement in Theorem 15.4.1 was generalized by Urbański in [156] to the so-called loosely Markov systems, and by Saussol in [136] to systems with super-polynomial decay of correlations (see the proof of Theorem 15.3.1).

15.4.2 Stable and unstable recurrence rates

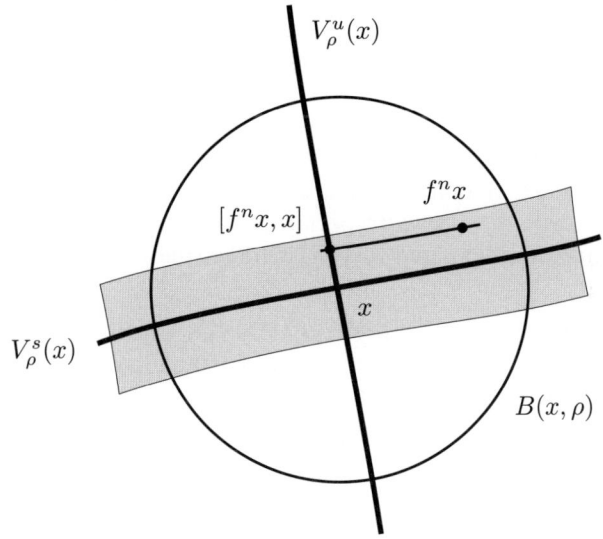

Figure 15.1: Definition of the unstable return time (the grey area is the set of points at a d_u-distance of $V^s_\rho(x)$ at most r)

Let $f: M \to M$ be a $C^{1+\varepsilon}$ diffeomorphism, and let $\Lambda \subset M$ be a locally maximal hyperbolic set of f. Let also d be the distance in M.

We denote by d_s and d_u the distances induced by d respectively in each local stable and unstable manifolds. Given $n \in \mathbb{N}$, when $d(f^n x, x) \leq \delta$ (with δ as in Definition 4.2.5), the distances $d_s([x, f^n x], x)$ and $d_u([f^n x, x], x)$ are well-defined. Thus, for each $\rho \leq \delta$ we can define (see Figure 15.1)

$$\tau^s_r(x, \rho) = \inf\{n \in \mathbb{N} : d(f^{-n} x, x) \leq \rho \text{ and } d_s([x, f^{-n} x], x) < r\},$$

and

$$\tau^u_r(x, \rho) = \inf\{n \in \mathbb{N} : d(f^n x, x) \leq \rho \text{ and } d_u([f^n x, x], x) < r\}.$$

Definition 15.4.4. The integers $\tau^s_r(x, \rho)$ and $\tau^u_r(x, \rho)$ are called respectively the *stable* and *unstable return times*.

We note that the functions $\rho \mapsto \tau^s_r(x, \rho)$ and $\rho \mapsto \tau^u_r(x, \rho)$ are nondecreasing. Set

$$\underline{R}^s(x, \rho) = \liminf_{r \to 0} \frac{\log \tau^s_r(x, \rho)}{-\log r}, \qquad \overline{R}^s(x, \rho) = \limsup_{r \to 0} \frac{\log \tau^s_r(x, \rho)}{-\log r},$$

and

$$\underline{R}^u(x,\rho) = \liminf_{r\to 0} \frac{\log \tau_r^u(x,\rho)}{-\log r}, \quad \overline{R}^u(x,\rho) = \limsup_{r\to 0} \frac{\log \tau_r^u(x,\rho)}{-\log r}.$$

Definition 15.4.5. We define the *lower* and *upper stable recurrence rates* of a point $x \in \Lambda$ (with respect to f) by

$$\underline{R}^s(x) = \lim_{\rho\to 0} \underline{R}^s(x,\rho) \quad \text{and} \quad \overline{R}^s(x) = \lim_{\rho\to 0} \overline{R}^s(x,\rho), \tag{15.27}$$

and the *lower* and *upper unstable recurrence rates* of a point $x \in \Lambda$ (with respect to f) by

$$\underline{R}^u(x) = \lim_{\rho\to 0} \underline{R}^u(x,\rho) \quad \text{and} \quad \overline{R}^u(x) = \lim_{\rho\to 0} \overline{R}^u(x,\rho). \tag{15.28}$$

When $\underline{R}^s(x) = \overline{R}^s(x)$ we denote the common value by $R^s(x)$, and we call it the *stable recurrence rate* of x (with respect to f). When $\underline{R}^u(x) = \overline{R}^u(x)$ we denote the common value by $R^u(x)$, and we call it the *unstable recurrence rate* of x (with respect to f).

It was shown by Barreira and Saussol in [17] that the stable and unstable recurrence rates are related to the stable and unstable pointwise dimensions. The latter were shown to exist by Ledrappier and Young in [93] (see Theorem 14.3.2). We recall the conditional measures μ_x^s and μ_x^u induced by a given measure μ in the measurable partitions of local stable and unstable manifolds (see Section 14.3).

Theorem 15.4.6. *Let Λ be a locally maximal hyperbolic set of a $C^{1+\varepsilon}$ diffeomorphism f, for some $\varepsilon > 0$, such that f is topologically mixing on Λ. If μ is an equilibrium measure of a Hölder continuous function in Λ, then*

$$R^s(x) = \lim_{r\to 0} \frac{\log \mu_x^s(B^s(x,r))}{\log r} \quad \text{and} \quad R^u(x) = \lim_{r\to 0} \frac{\log \mu_x^s(B^u(x,r))}{\log r}$$

for μ-almost every $x \in \Lambda$.

Proof. Let $\mathcal{R} = \{R_1, \ldots, R_\ell\}$ be a Markov partition of Λ. For each $R \in \mathcal{R}$, we denote by R_* the set of points in R that return infinitely often to R. By Poincaré's recurrence theorem we have $\mu(R_*) = \mu(R)$. For each $x \in R_*$ we have

$$T_R(x) := \inf\{k \in \mathbb{N} : f^k x \in R\} < \infty.$$

We define the induced map $f_R \colon R_* \to R_*$ by

$$f_R x = f^{T_R(x)} x. \tag{15.29}$$

Furthermore, given $z \in \operatorname{int} R$ we set

$$W_{R_*}^u(z) := V_\rho^u(z) \cap R_*,$$

and we define the map

$$F = f_{z,u} \colon W_{R_*}^u(z) \to W_{R_*}^u(z)$$

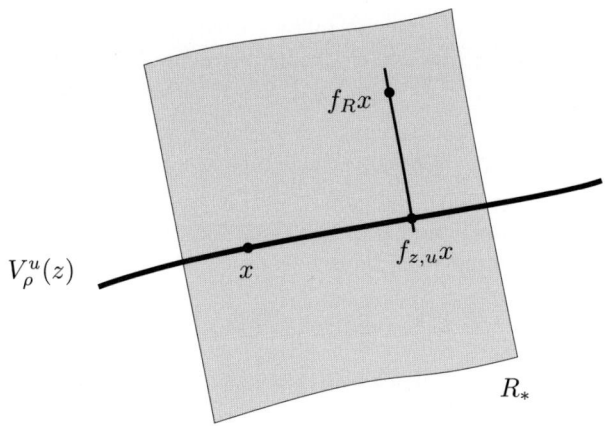

Figure 15.2: Construction of the map $F = f_{z,u}$

by $Fx = [f_R x, x]$ (see the illustration in Figure 15.2). Observe that $Fx = [f_R x, z]$. The return times of the f_R-orbit of x are given by $T_R^0(x) = 0$, and

$$T_R^n(x) = T_R^{n-1}(f_R x) + T_R(x), \quad n \in \mathbb{N}.$$

For each $p \in \mathbb{N}$ we define a new partition of Λ by $\mathcal{R}^p = \bigvee_{k=0}^{p-1} f^{-k}\mathcal{R}$. We also consider the partition $\mathcal{Z}$ of the set $W_{R_*}^u(z)$ defined by

$$\{Z \cap W_{R_*}^u(z) : Z \subset R,\ Z \in \mathcal{R}^p,\ \text{and } T_R|_Z = p,\ \text{for some } p \in \mathbb{N}\}. \qquad (15.30)$$

Finally, for each $n \in \mathbb{N}$ we consider the partition $\mathcal{Z}_n = \bigvee_{k=0}^{n-1} F^{-k}\mathcal{Z}$ of $W_{R_*}^u(z)$. It follows from the construction that

$$T_R(y) = T_R(x) \quad \text{whenever} \quad y \in \mathcal{Z}(x).$$

Therefore, for each $n \in \mathbb{N}$ we have $T_R^n(y) = T_R^n(x)$ whenever $y \in \mathcal{Z}_n(x)$.

We want to apply Theorem 15.4.1 to the map $F|W_{R_*}^u(z)$ and the partition $\mathcal{Z}$. We thus show that the hypotheses in Theorem 15.4.1 are satisfied.

Lemma 15.4.7. *The following properties hold:*

1. *$\mathcal{Z}$ is a countable Markov partition of $W_{R_*}^u(z)$ with respect to the map F, and $F|Z$ is onto for each $Z \in \mathcal{Z}$;*

2. *there exists $\lambda > 0$ such that for all sufficiently large $n \in \mathbb{N}$ we have*

$$\sup\{\mathrm{diam}_{d_u} Z : Z \in \mathcal{Z}_n\} < e^{-\lambda n};$$

3. *there exist $\theta > 0$ and $\alpha \in (0, 1]$ such that if $n \in \mathbb{N}$, $Z \in \mathcal{Z}_n$, and $x, y \in Z$, then*

$$d_u(F^n x, F^n y) \le \exp(\theta T_R^n(x))d_u(x, y)^\alpha.$$

Proof of the lemma. The first statement follows easily from the definitions. For the second statement we first observe that each element $Z \in \mathcal{Z}_n$ is contained in some element of the partition $\mathcal{R}^{T_R^n(x)}$ with $x \in \mathcal{Z}_n$. Choose $\lambda > 0$ such that

$$\limsup_{n \to \infty} \frac{1}{n} \log d_u(f^{-n}x, f^{-n}y) < -2\lambda$$

for all sufficiently close points x and y in the same local unstable manifold. Since $T_R^n \geq n$, for all sufficiently large n we have

$$\sup\{\operatorname{diam}_{d_u} Z : Z \in \mathcal{Z}_n\} < \sup\{e^{-\lambda T_R^n(x)} : Z \in \mathcal{Z}_n\} \leq e^{-\lambda n}$$

whenever $Z \in \mathcal{Z}_n$ and $x \in Z$. It follows from the construction that $F|Z$ is onto for each $Z \in \mathcal{Z}$.

Now we establish the third statement in the lemma. The Markov property of the partition $\mathcal{R}$ implies that

$$F^n x = [f_R[f_R \cdots [f_R x, x], \ldots, x], x] = [f_R{}^n x, x] = [f^{T_R^n(x)}x, x], \tag{15.31}$$

with f_R as in (15.29). Choose $\kappa > 0$ such that e^κ is a Lipschitz constant for f. Let $Z \in \mathcal{Z}_n$ and $x, y \in Z$. Since $T_R^n(x) = T_R^n(y)$, we obtain

$$d(f_R{}^n x, f_R{}^n y) = d(f^{T_R^n(x)}x, f^{T_R^n(x)}y) \leq \exp(\kappa T_R^n(x))d(x, y).$$

Since the local product structure is Hölder continuous, there exist constants $c > 1$ and $\alpha \in (0, 1]$ such that

$$\begin{aligned}
d_u(F^n x, F^n y) &= d_u([f_R{}^n x, z], [f_R{}^n y, z]) \\
&\leq c\, d(f_R{}^n x, f_R{}^n y)^\alpha \\
&\leq c(\exp(\kappa T_R^n(x)))^\alpha d(x, y)^\alpha \\
&\leq c \exp(\kappa \alpha T_R^n(x))d_u(x, y)^\alpha.
\end{aligned}$$

Setting $\theta = \kappa\alpha + \log c$ we obtain the third statement. This completes the proof of the lemma. $\qquad\square$

We notice that the second property in Lemma 15.4.7 corresponds to hypothesis 2 in Theorem 15.4.1.

For each $n \in \mathbb{N}$ we define a new partition of Λ by $\mathcal{R}_n = \bigvee_{p=0}^n f^p \mathcal{R}$. Then the partition

$$\mathcal{R}_\infty = \lim_{n \to \infty} \mathcal{R}_n = \{W_R^u(z) : z \in R \in \mathcal{R}\},$$

where $W_R^u(z) = V_\rho^u(z) \cap R$, is composed of local unstable manifolds. It induces a family of conditional measures μ_z^u for μ-almost every $z \in \Lambda$, given explicitly by

$$\mu_z^u(A) = \lim_{p \to \infty} \frac{\mu(A \cap \mathcal{R}_p(z))}{\mu(\mathcal{R}_p(z))} \tag{15.32}$$

for every measurable subset $A \subset \Lambda$. Each measure μ_z^u can be seen as a measure in $W_R^u(z) = \mathcal{R}_\infty(z)$. However, it may not be invariant under the map F in $W_{R_*}^u(z)$. Because of this we construct a measure m_z^u equivalent to μ_z^u which is F-invariant.

Given a set $A \subset W_R^u(z)$ we write

$$[A, R] := \{[a, y]: a \in A \text{ and } y \in R\}.$$

We define a measure m_z^u in $W_R^u(z)$ by

$$m_z^u(A) := \frac{\mu([A, R])}{\mu(R)} \tag{15.33}$$

for every measurable subset $A \subset W_R^u(z)$. It follows from Theorem 4.2.8 that $\mu(\partial R) = 0$, and thus m_z^u is a well-defined probability measure in $W_R^u(z)$ with the property that $m_z^u(\partial Z) = 0$ for μ-almost every $z \in \Lambda$. Here the boundary ∂Z is computed with respect to the induced topology on $W_R^u(z)$.

Lemma 15.4.8. *There exists a constant $c > 0$ such that for μ-almost every $z \in R$ the following properties hold:*

1. *m_z^u is an ergodic F-invariant measure in $W_R^u(z)$;*

2. *the measures m_z^u and μ_z^u are equivalent, and*

$$c^{-1} < \frac{d\mu_z^u}{dm_z^u} < c.$$

Proof of the lemma. Let $z \in \operatorname{int} R$. Since

$$[F^{-1}A, R] = f_R^{-1}[A, R],$$

the F-invariance of m_z^u follows immediately from the f_R-invariance of the measure $\mu|R$. The ergodicity of m_z^u follows from the ergodicity of μ.

Now we establish the second property. Since the Markov partition is a generating partition, it is sufficient to verify the equivalence of the measures in the elements of the partitions $\mathcal{R}_m \vee \mathcal{R}^n$ for each $n, m \in \mathbb{N}$. We observe that for each $x \in W_R^u(z)$ and $p \in \mathbb{N}$ we have $\mathcal{R}_p(x) = \mathcal{R}_p(z)$. Set $Z = \mathcal{R}_m(x) \cap \mathcal{R}^n(x)$. Since μ is a Gibbs measure, there exists a constant $a > 0$ (independent of m, n, and x) such that

$$a^{-1}\mu(\mathcal{R}_m(x))\mu(\mathcal{R}^n(x)) \le \mu(Z) \le a\mu(\mathcal{R}_m(x))\mu(\mathcal{R}^n(x)).$$

Dividing by $\mu(\mathcal{R}_m(x))$ and letting $m \to \infty$, it follows from (15.32) that

$$a^{-1}\mu(\mathcal{R}^n(x)) \le \mu_z^u(Z) \le a\mu(\mathcal{R}^n(x))$$

for every $n \in \mathbb{N}$. Since $[Z, R] = \mathcal{R}^n(x)$, it follows from (15.33) that $m_z^u(Z) = \mu(\mathcal{R}^n(x))/\mu(R)$, and hence

$$a^{-1}\mu(R) \le \frac{\mu_z^u(Z)}{m_z^u(Z)} \le a\mu(R).$$

This yields the desired result. $\qquad\square$

Now we establish hypothesis 3 in Theorem 15.4.1.

Lemma 15.4.9. *For μ-almost every $z \in \Lambda$ and m_z^u-almost every $x \in W_R^u(z)$, there exists $\gamma > 0$ such that*

$$B(x, e^{-\gamma n}) \subset Z_n(x) \tag{15.34}$$

for all sufficiently large $n \in \mathbb{N}$.

Proof of the lemma. Set $\mu_R = (\mu|R)/\mu(R)$. By Kac's lemma, the induced system (f_R, μ_R) is ergodic, and

$$\int_R T_R(x)\, d\mu_R(x) = \frac{1}{\mu(R)}. \tag{15.35}$$

Observe that

$$T_R^n(x) = \sum_{k=0}^{n-1} T_R(f_R{}^k x)$$

for every $n \in \mathbb{N}$ and $x \in R$. By Birkhoff's ergodic theorem, we have

$$\mu\left(\left\{x \in R: \lim_{n\to\infty} \frac{1}{n} T_R^n(x) = \frac{1}{\mu(R)}\right\}\right) = \mu(R),$$

and thus (see (15.32)),

$$\mu_z^u\left(\left\{x \in W_R^u(z): \lim_{n\to\infty} \frac{1}{n} T_R^n(x) = \frac{1}{\mu(R)}\right\}\right) = 1$$

for μ-almost every $z \in \Lambda$. By Lemma 15.4.8 we conclude that

$$m_z^u\left(\left\{x \in W_R^u(z): \lim_{n\to\infty} \frac{1}{n} T_R^n(x) = \frac{1}{\mu(R)}\right\}\right) = 1$$

for μ-almost every z. Hence, for m_z^u-almost every $x \in W_R^u(z)$ there exists $\delta_x > 1/\mu(R)$ such that $T_R^n(x) < \delta_x n$ for every $n \in \mathbb{N}$. We note that δ_x can be chosen in such a way that $x \mapsto \delta_x$ is measurable.

Let $\varepsilon, \delta > 0$, and set

$$Y_0 = \{x \in W_R^u(z): \delta_x < \delta\}.$$

We have $m_z^u(Y_0) > 1 - \varepsilon$ for all sufficiently large δ. Let $\mathcal{P}_n \subset \mathcal{Z}$ be the collection of elements $Z \in \mathcal{Z}$ such that $T_R|[Z, R] \leq n$. If $x \in Y_0$, then

$$\mathcal{Z}(F^{n-1}x) \in \mathcal{P}_{\delta n} \tag{15.36}$$

for every $n \in \mathbb{N}$, since

$$T_R(F^{n-1}x) \leq T_R^n(x) < \delta_x n < \delta n.$$

Furthermore, by the construction of $\mathcal{Z}$ in (15.30) we have

$$[\mathcal{P}_n, R] \subset \mathcal{R}^1 \cup \cdots \cup \mathcal{R}^n.$$

We show that the orbit of a typical point in Y_0 stays sufficiently far away from $\partial \mathcal{Z}$. Using the Markov property of the partition $\mathcal{R}$, we obtain $[\partial \mathcal{P}_m, R] \subset \partial \mathcal{R}^m$ for every $m \in \mathbb{N}$. It follows from the Hölder continuity of the local product structure that there exists $\alpha > 0$ such that

$$\left[\{x \in W^u_R(z) : d_u(x, \partial \mathcal{P}_m) < r\}, R \right] \subset \{y \in R : d(y, \partial \mathcal{R}^m) < r^\alpha\}. \qquad (15.37)$$

Let

$$\beta_0 = (1 + \log \max\{\|d_x f\| : x \in \Lambda\})\delta/\alpha,$$

and define

$$B_n = \left\{ x \in Y_0 : d_u(F^{n-1}x, \partial \mathcal{Z}(F^{n-1}x)) \le e^{-\beta_0 n} \right\}.$$

Using (15.36) we obtain

$$m^u_z(B_n) \le m^u_z \left(\{x \in Y_0 : d_u(F^{n-1}x, \partial \mathcal{P}_{\delta n}) \le e^{-\beta_0 n}\} \right),$$

and by the F-invariance of m^u_z,

$$m^u_z(B_n) \le m^u_z \left(\{x \in W^u_R(z) : d_u(x, \partial \mathcal{P}_{\delta n}) \le e^{-\beta_0 n}\} \right).$$

Set $L = \max\{\|d_x f\| : x \in \Lambda\}$.

Lemma 15.4.10. *If $\sigma > L$, then there exists $\nu > 0$ such that*

$$\mu(\{x \in \Lambda : d(x, \partial \mathcal{R}^n) < 1/\sigma^n\}) \le c(L/\sigma)^{\nu n} \quad \text{for every} \quad n \in \mathbb{N}.$$

Proof of the lemma. Since L is a Lipschitz constant for f, if $d(x, \partial \mathcal{R}^n) < 1/\sigma^n$, then $d(f^k x, \partial \mathcal{R}) < L^k/\sigma^n$ for some $k < n$. By Theorem 4.2.8 we obtain

$$\mu(\{x \in \Lambda : d(x, \partial \mathcal{R}^n) < 1/\sigma^n\}) \le \sum_{k < n} \mu(\{x \in \Lambda : d(f^k x, \partial \mathcal{R}) < L^k/\sigma^n\})$$

$$\le \sum_{k < n} c(L^k/\sigma^n)^\nu$$

$$\le c(L/\sigma)^{\nu n}.$$

This completes the proof of the lemma. $\qquad \square$

By (15.33), (15.37), and Lemma 15.4.10, there exist constants $c > 0$ and $\nu > 0$ such that

$$m^u_z(B_n) \le \frac{1}{\mu(R)} \mu \left(\{x \in R : d(x, \partial \mathcal{R}^{\delta n}) \le e^{-\alpha \beta_0 n}\} \right)$$

$$\le c e^{-n\alpha\beta_0\nu/\delta} \le c e^{-\nu n}.$$

for every $n \in \mathbb{N}$. This implies that $\sum_{m \in \mathbb{N}} m_z^u(B_m) < \infty$. By Borel–Cantelli's lemma, for m_z^u-almost every $x \in Y_0$ we have $x \notin B_m$ for all sufficiently large m, that is,

$$d_u(F^{m-1}x, \partial \mathcal{Z}(F^{m-1}x)) > e^{-\beta_0 m}$$

for all sufficiently large m. Therefore, for some $\beta > \beta_0$ there exists a set $Y \subset Y_0$ of measure $m_z^u(Y) > 1 - 2\varepsilon$ such that

$$d_u(F^{m-1}x, \partial \mathcal{Z}(F^{m-1}x)) > e^{-\beta m} \tag{15.38}$$

for every $m \in \mathbb{N}$ and $x \in Y$ (recall that by Theorem 4.2.8 the boundary $\partial \mathcal{Z}$ has zero measure).

Fix $\gamma > (\beta + \theta\delta)/\alpha$, with θ as in Lemma 15.4.7. Let $x \in Y$, $n \geq 2$, and $y \in B(x, e^{-\gamma n})$. It is easy to verify that

$$e^{k\theta\delta}e^{-\alpha\gamma n} \leq e^{-\beta n} \tag{15.39}$$

for every $k \leq n$. By (15.38) and (15.39) with $m = 1$ and $k = 0$, we obtain

$$d_u(x, \partial \mathcal{Z}(x)) > e^{-\beta} > e^{-\alpha\gamma n} > d_u(x, y),$$

and hence, $y \in \mathcal{Z}(x)$. By Lemma 15.4.7 we have

$$d_u(Fx, Fy) \leq e^{\theta\delta}d(x, y)^\alpha \leq e^{\theta\delta}e^{-\alpha\gamma n} \leq e^{-\beta n},$$

using (15.39) with $k = 1$. By (15.38) with $m = 2$ we obtain

$$d_u(Fx, \partial \mathcal{Z}(Fx)) > e^{-2\beta} \geq d_u(Fx, Fy),$$

and hence $Fy \in \mathcal{Z}(Fx)$. This shows that $y \in \mathcal{Z}_2(x)$. Again by Lemma 15.4.7, we obtain

$$d_u(F^2x, F^2y) \leq e^{2\theta\delta}e^{-\alpha\gamma n} \leq e^{-\beta n},$$

using (15.39) with $k = 2$. We can repeat the above argument to show that for every $m \leq n$ we have

$$d_u(F^{m-1}x, \partial \mathcal{Z}(F^{m-1}x)) > e^{-m\beta} \geq d_u(F^m x, F^m y),$$

and hence $F^{m-1}x \in \mathcal{Z}(F^{m-1}x)$. This shows that $y \in \mathcal{Z}_m(x)$. Therefore, (15.34) holds for every $x \in Y$ and $n \geq 2$. Since $m_z^u(Y) > 1 - 2\varepsilon$, the arbitrariness of $\varepsilon > 0$ implies the desired statement. $\square$

We proceed with the proof of the theorem. We denote by $\tau_r^u(x, \mathcal{R})$ the return time of x to the ball $B^u(x, r)$ with respect to the map F, that is,

$$\tau_r^u(x, \mathcal{R}) = \inf\{k \in \mathbb{N} : F^k x \in B^u(x, r)\}, \tag{15.40}$$

and we define the corresponding lower and upper recurrence rates by

$$\underline{R}^u(x, \mathcal{R}) = \liminf_{r \to 0} \frac{\log \tau_r^u(x, \mathcal{R})}{-\log r} \quad \text{and} \quad \overline{R}^u(x, \mathcal{R}) = \liminf_{r \to 0} \frac{\log \tau_r^u(x, \mathcal{R})}{-\log r}.$$

We verify that the system (F, m_z^u) satisfies the hypotheses of Theorem 15.4.1.

By Lemma 15.4.8, m_z^u is an ergodic F-invariant measure. We observe that $[Z, R]$ is a partition of R. Furthermore, for each $Z \in \mathcal{Z}$ there exists $p \in \mathbb{N}$ such that $[Z, R] \in \mathcal{R}^p$ and $T_R|[Z, R] = p$. The Gibbs property of the measure μ implies that there exists a constant $b > 0$ (independent of p) such that $\mu([Z, R]) > e^{-bp}$, and hence $m_z^u(Z) > e^{-bp}$ for every $Z \in \mathcal{Z}$. By (15.33), this implies that

$$
-\sum_{Z \in \mathcal{Z}} m_z^u(Z) \log m_z^u(Z) = \sum_{p \in \mathbb{N}} \sum_{Z \in \mathcal{Z}: T_R|[Z,R]=p} m_z^u(Z)(-\log m_z^u(Z))
$$

$$
\leq \sum_{p \in \mathbb{N}} \frac{bp}{\mu(R)} \sum_{Z \in \mathcal{Z}: T_R|[Z,R]=p} \mu([Z, R])
$$

$$
= \frac{b}{\mu(R)} \sum_{p \in \mathbb{N}} p\mu \left(\bigcup_{Z \in \mathcal{Z}: T_R|[Z,R]=p} [Z, R] \right)
$$

$$
= \frac{b}{\mu(R)} \int_R T_R \, d\mu.
$$

It follows from Kac's lemma (see (15.35)) that

$$
H_{m_z^u}(\mathcal{Z}) \leq b/\mu(R) < \infty.
$$

Now we verify the remaining hypotheses of Theorem 15.4.1. We have

$$
[\mathcal{Z}_{n+m}(x), R] = [\mathcal{Z}_n(x), R] \cap f^{-p}[\mathcal{Z}_m(y), R],
$$

with $[\mathcal{Z}_n(x), R] \in \mathcal{R}^p$ and $y = F^n x = f^p x$. It follows from (15.33) and the Gibbs property of μ that there exists a constant $\kappa > 0$ (independent of m, n, and x) such that

$$
m_z^u(\mathcal{Z}_{n+m}(x)) \leq \kappa m_z^u(\mathcal{Z}_n(x)) m_z^u(\mathcal{Z}_m(F^n x)).
$$

This shows that hypothesis 1 in Theorem 15.4.1 is satisfied. Hypothesis 2 is statement 2 in Lemma 15.4.7, and hypothesis 3 is the content of Lemma 15.4.9. Thus, it follows from Theorem 15.4.1 that

$$
\underline{R}^u(x, \mathcal{R}) = \underline{d}_{m_z^u}(x) \quad \text{and} \quad \overline{R}^u(x, \mathcal{R}) = \overline{d}_{m_z^u}(x) \tag{15.41}
$$

for μ-almost every $z \in R$ and m_z^u-almost every $x \in W_R^u(z)$.

On the other hand, by Theorem 14.3.2 (see also (14.4)) there exists a constant d_μ^u such that

$$
\underline{d}_{\mu_x^u}(x) = \overline{d}_{\mu_x^u}(x) = d_\mu^u \tag{15.42}
$$

for μ-almost every $x \in \Lambda$. We recall that for each $x \in W_R^u(z)$ and $p \in \mathbb{N}$ we have $\mathcal{R}_p(x) = \mathcal{R}_p(z)$. Hence, it follows from (15.32) and (15.42) that

$$
\underline{d}_{\mu_z^u}(x) = \overline{d}_{\mu_z^u}(x) = d_\mu^u \tag{15.43}
$$

for μ-almost every $z \in \Lambda$ and μ_z^u-almost every $x \in W_R^u(z)$.

By Lemma 15.4.8 the measures m_z^u and μ_z^u are equivalent for μ-almost every $z \in R$, and hence

$$\underline{d}_{m_z^u}(x) = \underline{d}_{\mu_z^u}(x) \quad \text{and} \quad \overline{d}_{m_z^u}(x) = \overline{d}_{\mu_z^u}(x) \tag{15.44}$$

for μ-almost every $z \in \Lambda$ and μ_z^u-almost every $x \in W_R^u(z)$. Combining (15.41), (15.43), and (15.44) we conclude that

$$\underline{R}^u(x, \mathcal{R}) = \overline{R}^u(x, \mathcal{R}) = \underline{d}_{m_z^u}(x) = \overline{d}_{m_z^u}(x) = d_\mu^u \tag{15.45}$$

for μ-almost every $z \in R$ and μ_z^u-almost every $x \in W_R^u(z)$.

On the other hand, it follows from (15.40) and (15.31) that

$$\tau_r^u(x, \mathcal{R}) = \inf\{k \in \mathbb{N} : d_u([f_R{}^k x, x], x) < r\}.$$

Therefore,

$$T_R^{\tau_r^u(x, \mathcal{R})}(x) = \inf\left\{n \in \mathbb{N} : f^n x \in R \text{ and } d_u([f^n x, x], x) < r\right\}. \tag{15.46}$$

Furthermore, since μ is ergodic it follows from (15.35) that

$$\lim_{n \to \infty} \frac{1}{n} T_{\mathcal{R}(x)}^n(x) = \frac{1}{\mu(\mathcal{R}(x))}$$

for μ-almost every $x \in \Lambda$, where $\mathcal{R}(x)$ is the element of $\mathcal{R}$ that contains x. Therefore,

$$\lim_{n \to \infty} \frac{\log T_{\mathcal{R}(x)}^n(x)}{\log n} = 1 \tag{15.47}$$

for μ-almost every $x \in \Lambda$.

Fix $\rho > 0$, and consider two Markov partitions $\mathcal{R}_+$ and $\mathcal{R}_-$ of Λ. We assume that $\mathcal{R}_-$ has diameter at most ρ (it is well known that there exist Markov partitions of Λ with diameter as small as desired), and we define

$$\Lambda_\rho(\mathcal{R}_+) = \{x \in \Lambda : d(x, \partial \mathcal{R}_+) > \rho\}.$$

Observe that if $x \in \Lambda_\rho(\mathcal{R}_+)$, then

$$\mathcal{R}_-(x) \subset B(x, \rho) \cap \Lambda \subset \mathcal{R}_+(x), \tag{15.48}$$

where $\mathcal{R}_-(x)$ and $\mathcal{R}_+(x)$ are respectively the elements of $\mathcal{R}_-$ and $\mathcal{R}_+$ that contain x. Since Λ is invariant, the orbit of every point $x \in \Lambda$ is contained in Λ. Therefore (even though in general the intersection $B(x, \rho) \cap \Lambda$ in (15.48) cannot be replaced by the ball $B(x, \rho)$), it follows from (15.48) and (15.46) that setting $n_+ = \tau_r^u(x, \mathcal{R}_+)$ and $n_- = \tau_r^u(x, \mathcal{R}_-)$ we have

$$T_{\mathcal{R}_+(x)}^{n_+}(x) \le \tau_r^u(x, \rho) \le T_{\mathcal{R}_-(x)}^{n_-}(x)$$

for μ-almost every $x \in \Lambda_\rho(\mathcal{R}_+)$. We conclude from (15.47) that

$$\underline{R}^u(x, \mathcal{R}_+) \le \underline{R}^u(x, \rho) \le \underline{R}^u(x, \mathcal{R}_-), \tag{15.49}$$

and

$$\overline{R}^u(x, \mathcal{R}_+) \le \overline{R}^u(x, \rho) \le \overline{R}^u(x, \mathcal{R}_-) \tag{15.50}$$

for μ-almost every $x \in \Lambda_\rho(\mathcal{R}_+)$. On the other hand, by (15.45) we have

$$\underline{R}^u(x, \mathcal{R}_-) = \overline{R}^u(x, \mathcal{R}_-) = d_\mu^u \quad \text{and} \quad \underline{R}^u(x, \mathcal{R}_+) = \overline{R}^u(x, \mathcal{R}_+) = d_\mu^u$$

for μ-almost every $x \in \Lambda$. It follows from (15.49) and (15.50) that

$$\underline{R}^u(x, \rho) = \overline{R}^u(x, \rho) = d_\mu^u \tag{15.51}$$

for μ-almost every $x \in \Lambda_\rho(\mathcal{R}_+)$ and all sufficiently small $\rho > 0$. By Theorem 4.2.8, the boundary $\partial\mathcal{R}_+$ has zero measure. Hence, the set $\bigcup_{\rho>0} \Lambda_\rho(\mathcal{R}_+)$ has full μ-measure, and the identities in (13.17) hold for μ-almost every $x \in \Lambda$ and all sufficiently small $\rho > 0$ (possibly depending on x).

Using a version of (15.45) for the stable direction (which can be obtained replacing everywhere the diffeomorphism f by f^{-1}, and the index u by s), we can show that

$$\underline{R}^s(x, \rho) = \overline{R}^s(x, \rho) = d_\mu^s \tag{15.52}$$

for μ-almost every $x \in \Lambda$ and all sufficiently small $\rho > 0$ (possibly depending on x). It follows from (15.51) and (15.52) that the limits in (15.27)–(15.28) are not necessary provided that ρ is sufficiently small. Hence,

$$R^s(x) = \underline{R}^s(x, \rho) = \overline{R}^s(x, \rho) = d_\mu^s, \tag{15.53}$$

and

$$R^u(x) = \underline{R}^u(x, \rho) = \overline{R}^u(x, \rho) = d_\mu^u \tag{15.54}$$

for μ-almost every $x \in \Lambda$ and all sufficiently small $\rho > 0$ (possibly depending on x). This completes the proof of the theorem. □

15.4.3 Product structure of recurrence

The following result of Barreira and Saussol in [17] is now a simple application of Theorems 14.4.1, 15.3.1, and 15.4.6.

Theorem 15.4.11 (Product structure of recurrence). *Let Λ be a locally maximal hyperbolic set of a $C^{1+\varepsilon}$ diffeomorphism f, for some $\varepsilon > 0$, such that f is topologically mixing on Λ, and let μ be an equilibrium measure of a Hölder continuous function in Λ. Then for μ-almost every $x \in \Lambda$ the following properties hold:*

1. *the recurrence rate is equal to the sum of the stable and unstable recurrence rates, that is,*

$$R(x) = R^s(x) + R^u(x);$$

2. *there exists $\rho(x) > 0$ such that for each $\rho \in (0, \rho(x))$ and $\delta > 0$ there is $r(x, \rho, \delta) > 0$ such that if $r < r(x, \rho, \delta)$, then*

$$r^{\delta} < \frac{T_r^s(x, \rho) T_r^u(x, \rho)}{T_r(x)} < r^{-\delta}. \tag{15.55}$$

Proof. By Theorem 14.4.1 we have

$$\underline{d}_{\mu}(x) = \overline{d}_{\mu}(x) = d_{\mu}^s + d_{\mu}^u \tag{15.56}$$

for μ-almost every $x \in \Lambda$. On the other hand, by Theorems 15.3.1 and 15.4.6,

$$R(x) = d_{\mu}(x), \quad R^s(x) = d_{\mu}^s, \quad \text{and} \quad R^u(x) = d_{\mu}^u$$

for μ-almost every $x \in \Lambda$. Together with (15.56) this establishes the first statement in the theorem. In view of the arbitrariness of ρ in (15.53) and (15.54), the second statement is an immediate consequence of the first one. $\qquad\square$

The second statement in Theorem 15.4.11 shows that the return time is approximately equal to the product of the return times in the stable and unstable directions, as if they were independent.

In the case of symbolic dynamics, a version of Theorem 15.4.11 was obtained earlier by Ornstein and Weiss in [107]. We formulate it without proof.

Theorem 15.4.12. *The following properties hold:*

1. *if $\sigma^+|\Sigma^+$ is a one-sided topological Markov chain, and μ^+ is an ergodic σ^+-invariant probability measure in Σ^+, then*

$$\lim_{k \to \infty} \frac{\log \inf\{n \in \mathbb{N} : (i_{n+1} \cdots i_{n+k}) = (i_1 \cdots i_k)\}}{k} = h_{\mu^+}(\sigma^+) \tag{15.57}$$

for μ^+-almost every $(i_1 i_2 \cdots) \in \Sigma^+$;

2. *if $\sigma|\Sigma$ is a two-sided topological Markov chain, and μ is an ergodic σ-invariant probability measure in Σ, then*

$$\lim_{k \to \infty} \frac{\log \inf\{n \in \mathbb{N} : (i_{n-k} \cdots i_{n+k}) = (i_{-k} \cdots i_k)\}}{2k+1} = h_{\mu}(\sigma) \tag{15.58}$$

for μ-almost every $(\cdots i_{-1} i_0 i_1 \cdots) \in \Sigma$.

We recall that $\sigma|\Sigma$ has naturally associated two one-sided topological Markov chains $\sigma^+|\Sigma^+$ and $\sigma^-|\Sigma^-$ (see Section 4.2.3). Furthermore, any σ-invariant measure μ in Σ induces a σ^+-invariant measure μ^+ in Σ^+ and a σ^--invariant measure μ^- in Σ^-, such that

$$h_{\mu^+}(\sigma^+) = h_{\mu^-}(\sigma^-) = h_{\mu}(\sigma). \tag{15.59}$$

For each $w = (\cdots i_{-1}i_0 i_1 \cdots) \in \Sigma$ and $k \in \mathbb{N}$, we set

$$\tau_k^+(w) = \inf\{n \in \mathbb{N} : (i_{n+1} \cdots i_{n+k}) = (i_1 \cdots i_k)\},$$
$$\tau_k^-(w) = \inf\{n \in \mathbb{N} : (i_{-n-k} \cdots i_{-n-1}) = (i_{-k} \cdots i_{-1})\},$$
$$\tau_k(w) = \inf\{n \in \mathbb{N} : (i_{n-k} \cdots i_{n+k}) = (i_{-k} \cdots i_k)\}.$$

The following is an immediate consequence of (15.57), (15.58), and (15.59).

Theorem 15.4.13 ([107]). *Let μ be an ergodic σ-invariant measure in Σ. Given $\varepsilon > 0$, for μ-almost every $w \in \Sigma$ and all sufficiently large k we have*

$$e^{-k\varepsilon} \le \frac{\tau_k^+(w)\tau_k^-(w)}{\tau_k(w)} \le e^{k\varepsilon}.$$

Theorem 15.4.13 is a version of 15.55 in the case of symbolic dynamics.

Bibliography

[1] V. Afraimovich, J. Chazottes and B. Saussol, *Pointwise dimensions for Poincaré recurrences associated with maps and special flows*, Discrete Contin. Dyn. Syst. **9** (2003), 263–280.

[2] D. Anosov, *Geodesic Flows on Closed Riemann Manifolds with Negative Curvature*, Proc. Steklov Inst. Math. 90, Amer. Math. Soc., 1969.

[3] L. Barreira, *A non-additive thermodynamic formalism and applications to dimension theory of hyperbolic dynamical systems*, Ergodic Theory Dynam. Systems **16** (1996), 871–927.

[4] L. Barreira and K. Gelfert, *Multifractal analysis for Lyapunov exponents on nonconformal repellers*, Comm. Math. Phys. **267** (2006), 393–418.

[5] L. Barreira and G. Iommi, *Suspension flows over countable Markov shifts*, J. Stat. Phys. **124** (2006), 207–230.

[6] L. Barreira and Ya. Pesin, *Lyapunov Exponents and Smooth Ergodic Theory*, University Lecture Series 23, Amer. Math. Soc., 2002.

[7] L. Barreira and Ya. Pesin, *Smooth ergodic theory and nonuniformly hyperbolic dynamics*, with appendix by O. Sarig, in Handbook of Dynamical Systems 1B, edited by B. Hasselblatt and A. Katok, Elsevier, 2006, pp. 57–263.

[8] L. Barreira and Ya. Pesin, *Nonuniform Hyperbolicity: Dynamics of Systems with Nonzero Lyapunov Exponents*, Encyclopedia of Mathematics and Its Applications 115, Cambridge University Press, 2007.

[9] L. Barreira, Ya. Pesin and J. Schmeling, *On a general concept of multifractality. Multifractal spectra for dimensions, entropies, and Lyapunov exponents. Multifractal rigidity*, Chaos **7** (1997), 27–38.

[10] L. Barreira, Ya. Pesin and J. Schmeling, *Multifractal spectra and multifractal rigidity for horseshoes*, J. Dynam. Control Systems **3** (1997), 33–49.

[11] L. Barreira, Ya. Pesin and J. Schmeling, *Dimension and product structure of hyperbolic measures*, Ann. of Math. (2) **149** (1999), 755–783.

[12] L. Barreira and L. Radu, *Multifractal analysis of nonconformal repellers: a model case*, Dyn. Syst. **22** (2007), 147–168.

[13] L. Barreira and V. Saraiva, *Multifractal nonrigidity of topological Markov chains*, J. Stat. Phys. **130** (2008), 387–412.

[14] L. Barreira and B. Saussol, *Multifractal analysis of hyperbolic flows*, Comm. Math. Phys. **214** (2000), 339–371.

[15] L. Barreira and B. Saussol, *Hausdorff dimension of measures via Poincaré recurrence*, Comm. Math. Phys. **219** (2001), 443–463.

[16] L. Barreira and B. Saussol, *Variational principles and mixed multifractal spectra*, Trans. Amer. Math. Soc. **353** (2001), 3919–3944.

[17] L. Barreira and B. Saussol, *Product structure of Poincaré recurrence*, Ergodic Theory Dynam. Systems **22** (2002), 33–61.

[18] L. Barreira and B. Saussol, *Variational principles for hyperbolic flows*, in Differential Equations and Dynamical Systems, edited by A. Galves, J. Hale and C. Rocha, Fields Inst. Comm. 31, 2002, pp. 43–63.

[19] L. Barreira, B. Saussol and J. Schmeling, *Distribution of frequencies of digits via multifractal analysis*, J. Number Theory **97** (2002), 410–438.

[20] L. Barreira, B. Saussol and J. Schmeling, *Higher-dimensional multifractal analysis*, J. Math. Pures Appl. **81** (2002), 67–91.

[21] L. Barreira and J. Schmeling, *Sets of "non-typical" points have full topological entropy and full Hausdorff dimension*, Israel J. Math. **116** (2000), 29–70.

[22] L. Barreira and C. Valls, *Multifractal structure of two-dimensional horseshoes*, Comm. Math. Phys. **266** (2006), 455–470.

[23] L. Barreira and C. Wolf, *Measures of maximal dimension for hyperbolic diffeomorphisms*, Comm. Math. Phys. **239** (2003), 93–113.

[24] L. Barreira and C. Wolf, *Pointwise dimension and ergodic decompositions*, Ergodic Theory Dynam. Systems **26** (2006), 653–671.

[25] J. Barrow-Green, *Poincaré and the Three Body Problem*, History of Mathematics 11, Amer. Math. Soc., 1997.

[26] T. Bedford, *The box dimension of self-affine graphs and repellers*, Nonlinearity **2** (1989), 53–71.

[27] T. Bedford and M. Urbański, *The box and Hausdorff dimension of self-affine sets*, Ergodic Theory Dynam. Systems **10** (1990), 627–644.

[28] A. Besicovitch, *On the sum of digits of real numbers represented in the dyadic system*, Math. Ann. **110** (1934), 321–330.

[29] G. Birkhoff, *Proof of a recurrence theorem for strongly transitive systems*, Proc. Nat. Acad. Sci. U.S.A. **17** (1931), 650–655.

[30] G. Birkhoff, *Proof of the ergodic theorem*, Proc. Nat. Acad. Sci. U.S.A. **17** (1931), 656–660.

[31] G. Birkhoff and B. Koopman, *Recent contributions to the ergodic theory*, Proc. Nat. Acad. Sci. U.S.A. **18** (1932), 279–282.

[32] J. Bochi, *Genericity of zero Lyapunov exponents*, Ergodic Theory Dynam. Systems **22** (2002), 1667–1696.

[33] E. Borel, *Sur les probabilités dénombrables et leurs applications arithmétiques*, Rend. Circ. Mat. Palermo **26** (1909), 247–271.

[34] M. Boshernitzan, *Quantitative recurrence results*, Invent. Math. **113** (1993), 617–631.

[35] H. Bothe, *The Hausdorff dimension of certain solenoids*, Ergodic Theory Dynam. Systems **15** (1995), 449–474.

[36] R. Bowen, *Symbolic dynamics for hyperbolic flows*, Amer. J. Math. **95** (1973), 429–460.

[37] R. Bowen, *Topological entropy for noncompact sets*, Trans. Amer. Math. Soc. **184** (1973), 125–136.

[38] R. Bowen, *Equilibrium States and the Ergodic Theory of Anosov Diffeomorphisms*, Lecture Notes in Mathematics 470, Springer, 1975.

[39] R. Bowen, *Markov partitions are not smooth*, Proc. Amer. Math. Soc. **71** (1978), 130–132.

[40] R. Bowen, *Hausdorff dimension of quasi-circles*, Inst. Hautes Études Sci. Publ. Math. **50** (1979), 259–273.

[41] M. Boyle, J. Franks and B. Kitchens, *Automorphisms of one-sided subshifts of finite type*, Ergodic Theory Dynam. Systems **10** (1990), 421–449.

[42] M. Brin, *Bernoulli diffeomorphisms with nonzero exponents*, Ergodic Theory Dynam. Systems **1** (1981), 1–7.

[43] M. Brin and A. Katok, *On local entropy*, in Geometric Dynamics (Rio de Janeiro, 1981), edited by J. Palis, Lecture Notes in Mathematics 1007, Springer, 1983, pp. 30–38.

[44] P. Collet, J. Lebowitz and A. Porzio, *The dimension spectrum of some dynamical systems*, J. Stat. Phys. **47** (1987), 609–644.

[45] M. Denker, C. Grillenberger and K. Sigmund, *Ergodic Theory on Compact Spaces*, Lecture Notes in Mathematics 527, Springer, 1976.

[46] M. Denker and M. Urbański, *On Sullivan's conformal measures for rational maps on the Riemann sphere*, Nonlinearity **4** (1991), 365–384.

[47] D. Dolgopyat and Ya. Pesin, *Every compact manifold carries a completely hyperbolic diffeomorphism*, Ergodic Theory Dynam. Systems **22** (2002), 409–435.

[48] A. Douady and J. Oesterlé, *Dimension de Hausdorff des attracteurs*, C. R. Acad. Sc. Paris **290** (1980), 1135–1138.

[49] J.-P. Eckmann and D. Ruelle, *Ergodic theory of chaos and strange attractors*, Rev. Modern Phys. **57** (1985), 617–656.

[50] H. Eggleston, *The fractional dimension of a set defined by decimal properties*, Quart. J. Math. Oxford Ser. **20** (1949), 31–36.

[51] K. Falconer, *The Hausdorff dimension of self-affine fractals*, Math. Proc. Cambridge Philos. Soc. **103** (1988), 339–350.

[52] K. Falconer, *A subadditive thermodynamic formalism for mixing repellers*, J. Phys. A: Math. Gen. **21** (1988), 1737–1742.

[53] K. Falconer, *Dimensions and measures of quasi self-similar sets*, Proc. Amer. Math. Soc. **106** (1989), 543–554.

[54] K. Falconer, *Bounded distortion and dimension for non-conformal repellers*, Math. Proc. Cambridge Philos. Soc. **115** (1994), 315–334.

[55] K. Falconer, *Techniques in Fractal Geometry*, John Wiley & Sons, 1997.

[56] K. Falconer, *Fractal Geometry. Mathematical Foundations and Applications*, John Wiley & Sons, 2003.

[57] A. Fan, D. Feng and J. Wu, *Recurrence, dimension and entropy*, J. London Math. Soc. (2) **64** (2001), 229–244.

[58] A. Fathi, M. Herman and J.-C. Yoccoz, *A proof of Pesin's stable manifold theorem*, in Geometric Dynamics (Rio de Janeiro, 1981), edited by J. Palis, Lecture Notes in Mathematics 1007, Springer, 1983, pp. 177–215.

[59] H. Federer, *Geometric Measure Theory*, Grundlehren der mathematischen Wissenschaften 153, Springer, 1969.

[60] D. Feng, *Lyapunov exponents for products of matrices and multifractal analysis. I. Positive matrices*, Israel J. Math. **138** (2003), 353–376.

[61] D. Feng, *The variational principle for products of non-negative matrices*, Nonlinearity **17** (2004), 447–457.

[62] D. Feng and K. Lau, *The pressure function for products of non-negative matrices*, Math. Res. Lett. **9** (2002), 363–378.

[63] D. Feng, K. Lau and J. Wu, *Ergodic limits on the conformal repellers*, Adv. Math. **169** (2002), 58–91.

[64] D. Feng and J. Wu, *The Hausdorff dimension of recurrent sets in symbolic spaces*, Nonlinearity **14** (2001), 81–85.

[65] D. Gatzouras and Y. Peres, *Invariant measures of full dimension for some expanding maps*, Ergodic Theory Dynam. Systems **17** (1997), 147–167.

[66] B. Gurevič, *Topological entropy of a countable Markov chain*, Soviet Math. Dokl. **10** (1969), 911–915.

[67] J. Hadamard, *Les surfaces à courbures opposées et leur lignes géodesiques*, J. Math. Pures Appl. **4** (1898), 27–73.

[68] T. Halsey, M. Jensen, L. Kadanoff, I. Procaccia and B. Shraiman, *Fractal measures and their singularities: the characterization of strange sets*, Phys. Rev. A (3) **34** (1986), 1141–1151; errata in **34** (1986), 1601.

[69] P. Hanus, R. Mauldin and M. Urbański, *Thermodynamic formalism and multifractal analysis of conformal infinite iterated function systems*, Acta Math. Hungar. **96** (2002), 27–98.

[70] G. Hardy and J. Littlewood, *Some problems on Diophantine approximations*, Acta Math. **37** (1914), 155–190.

[71] B. Hasselblatt, *Regularity of the Anosov splitting and of horospheric foliations*, Ergodic Theory Dynam. Systems **14** (1994), 645–666.

[72] B. Hasselblatt, *Hyperbolic dynamical systems*, in Handbook of Dynamical Systems 1A, edited by B. Hasselblatt and A. Katok, Elsevier, 2002.

[73] B. Hasselblatt and J. Schmeling, *Dimension product structure of hyperbolic sets*, Electron. Res. Announc. Amer. Math. Soc. **10** (2004), 88–96.

[74] B. Hasselblatt and J. Schmeling, *Dimension product structure of hyperbolic sets*, in Modern Dynamical Systems and Applications, Cambridge Univ. Press, Cambridge, 2004, pp. 331–345.

[75] F. Hofbauer, *Local dimension for piecewise monotonic maps on the interval*, Ergodic Theory Dynam. Systems **15** (1995),1119–1142.

[76] F. Hofbauer and P. Raith, *The Hausdorff dimension of an ergodic invariant measure for a piecewise monotonic map of the interval*, Canad. Math. Bull. **35** (1992), 84–98.

[77] H. Hu, *Box dimensions and topological pressure for some expanding maps*, Comm. Math. Phys. **191** (1998), 397–407.

[78] J. Hutchinson, *Fractals and sel-similarity*, Indiana Univ. Math. J. **30** (1981), 713–747.

[79] G. Iommi, *Multifractal analysis for countable Markov shifts*, Ergodic Theory Dynam. Systems **25** (2005), 1881–1907.

[80] G. Iommi and B. Skorulski, *Multifractal analysis for the exponential family*, Discrete Contin. Dyn. Syst. **16** (2006), 857–869.

[81] O. Jenkinson, *Rotation, entropy, and equilibrium states*, Trans. Amer. Math. Soc. **353** (2001), 3713–3739.

[82] A. Katok, *Bernoulli diffeomorphisms on surfaces*, Ann. of Math. (2) **110** (1979), 529–547.

[83] A. Katok, *Lyapunov exponents, entropy and periodic orbits for diffeomorphisms*, Inst. Hautes Études Sci. Publ. Math. **51** (1980), 137–173.

[84] A. Katok and B. Hasselblatt, *Introduction to the Modern Theory of Dynamical Systems*, Encyclopedia of Mathematics and Its Applications 54, Cambridge University Press, 1995.

[85] M. Keane, *Strongly mixing g-measures*, Invent. Math. **21** (1972), 309–324.

[86] G. Keller, *Equilibrium States in Ergodic Theory*, London Mathematical Society Student Texts 42, Cambridge University Press, 1998.

[87] M. Kesseböhmer and B. Stratmann, *A multifractal formalism for growth rates and applications to geometrically finite Kleinian groups*, Ergodic Theory Dynam. Systems **24** (2004), 141–170.

[88] M. Kesseböhmer and B. Stratmann, *A multifractal analysis for Stern-Brocot intervals, continued fractions and Diophantine growth rates*, J. Reine Angew. Math. **605** (2007), 133–163.

[89] F. Ledrappier, *Dimension of invariant measures*, in Proceedings of the Conference on Ergodic Theory and Related Topics, II (Georgenthal, 1986), Teubner-Texte, Math. 94, 1987, pp. 116–124.

[90] F. Ledrappier and M. Misiurewicz, *Dimension of invariant measures for maps with exponent zero*, Ergodic Theory Dynam. Systems **5** (1985), 595–610.

[91] F. Ledrappier and J.-M. Strelcyn, *A proof of the estimate from below in Pesin's entropy formula*, Ergodic Theory Dynam. Systems **2** (1982), 203–219.

[92] F. Ledrappier and L.-S. Young, *The metric entropy of diffeomorphisms I. Characterization of measures satisfying Pesin's entropy formula*, Ann. of Math. (2) **122** (1985), 509–539.

[93] F. Ledrappier and L.-S. Young, *The metric entropy of diffeomorphisms II. Relations between entropy, exponents and dimension*, Ann. of Math. (2) **122** (1985), 540–574.

[94] A. Lopes, *The dimension spectrum of the maximal measure*, SIAM J. Math. Anal. **20** (1989), 1243–1254.

[95] R. Mañé, *Ergodic Theory and Differentiable Dynamics*, Ergebnisse der Mathematik und ihrer Grenzgebiete; 3. Folge; 8, Springer, 1987.

[96] P. Mattila, *Geometry of Sets and Measures in Euclidean Spaces. Fractals and Rectifiability*, Cambridge Studies in Advanced Mathematics 44, Cambridge University Press, 1995.

[97] R. Mauldin and M. Urbański, *Dimensions and measures in infinite iterated function systems*, Proc. London Math. Soc. (3) **73** (1996), 105–154.

[98] R. Mauldin and M. Urbański, *Conformal iterated function systems with applications to the geometry of continued fractions*, Trans. Amer. Math. Soc. **351** (1999), 4995–5025.

[99] R. Mauldin and M. Urbański, *Parabolic iterated function systems*, Ergodic Theory Dynam. Systems **20** (2000), 1423–1447.

[100] H. McCluskey and A. Manning, *Hausdorff dimension for horseshoes*, Ergodic Theory Dynam. Systems **3** (1983), 251–260.

[101] P. Moran, *Additive functions of intervals and Hausdorff measure*, Proc. Cambridge Philos. Soc. **42** (1946), 15–23.

[102] M. Morse and G. Hedlund, *Symbolic dynamics*, Amer. J. Math. **60** (1938), 815–866.

[103] K. Nakaishi, *Multifractal formalism for some parabolic maps*, Ergodic Theory Dynam. Systems **20** (2000), 843–857.

[104] J. Neunhäuserer, *Number theoretical peculiarities in the dimension theory of dynamical systems*, Israel J. Math. **128** (2002), 267–283.

[105] E. Olivier, *Analyse multifractale de fonctions continues*, C. R. Acad. Sci. Paris **326** (1998), 1171–1174.

[106] E. Olivier, *Multifractal analysis in symbolic dynamics and distribution of pointwise dimension for g-measures*, Nonlinearity **12** (1999), 1571–1585.

[107] D. Ornstein and B. Weiss, *Entropy and data compression schemes*, IEEE Trans. Inform. Theory **39** (1993), 78–83.

[108] V. Oseledets, *A multiplicative ergodic theorem. Liapunov characteristic numbers for dynamical systems*, Trans. Moscow Math. Soc. **19** (1968), 197–221.

[109] J. Palis and F. Takens, *Hyperbolicity and Sensitive Chaotic Dynamics at Homoclinic Bifurcations*, Cambridge University Press, 1993.

[110] J. Palis and M. Viana, *On the continuity of Hausdorff dimension and limit capacity for horseshoes*, in Dynamical Systems (Valparaiso, 1986), edited by R. Bamón, R. Labarca and J. Palis, Lecture Notes in Mathematics 1331, Springer, 1988, pp. 150–160.

[111] W. Parry and M. Pollicott, *Zeta Functions and the Periodic Orbit Structure of Hyperbolic Dynamics*, Astérisque 187-188, 1990.

[112] Ya. Pesin, *Families of invariant manifolds corresponding to nonzero characteristic exponents*, Math. USSR-Izv. **10** (1976), 1261–1305.

[113] Ya. Pesin, *Characteristic exponents and smooth ergodic theory*, Russian Math. Surveys **32** (1977), 55–114.

[114] Ya. Pesin, *Geodesic flows on closed Riemannian manifolds without focal points*, Math. USSR-Izv. **11** (1977), 1195–1228.

[115] Ya. Pesin, *Dimension Theory in Dynamical Systems: Contemporary Views and Applications*, Chicago Lectures in Mathematics, Chicago University Press, 1997.

[116] Ya. Pesin and B. Pitskel', *Topological pressure and the variational principle for noncompact sets*, Functional Anal. Appl. **18** (1984), 307–318.

[117] Ya. Pesin and V. Sadovskaya, *Multifractal analysis of conformal axiom A flows*, Comm. Math. Phys. **216** (2001), 277–312.

[118] Ya. Pesin and H. Weiss, *On the dimension of deterministic and random Cantor-like sets, symbolic dynamics, and the Eckmann–Ruelle conjecture*, Comm. Math. Phys. **182** (1996), 105–153.

[119] Ya. Pesin and H. Weiss, *A multifractal analysis of Gibbs measures for conformal expanding maps and Markov Moran geometric constructions*, J. Statist. Phys. **86** (1997), 233–275.

[120] Ya. Pesin and H. Weiss, *The multifractal analysis of Gibbs measures: motivation, mathematical foundation, and examples*, Chaos **7** (1997), 89–106.

[121] C.-E. Pfister and W. Sullivan, *On the topological entropy of saturated set*, Ergodic Theory Dynam. Systems **27** (2007), 929–956.

[122] H. Poincaré, *Sur le problème des trois corps et les équations de la dynamique*, Acta Math. **13** (1890), 1–270.

[123] M. Pollicott and H. Weiss, *Multifractal analysis of Lyapunov exponent for continued fraction and Manneville-Pomeau transformations and applications to Diophantine approximation*, Comm. Math. Phys. **207** (1999), 145–171.

[124] F. Przytycki and M. Urbański, *On Hausdorff dimension of some fractal sets*, Studia Math. **93** (1989), 155–186.

[125] C. Pugh, *The $C^{1+\alpha}$ hypothesis in Pesin theory*, Inst. Hautes Études Sci. Publ. Math. **59** (1984), 143–161.

[126] C. Pugh and M. Shub, *Ergodic attractors*, Trans. Amer. Math. Soc. **312** (1989), 1–54.

[127] M. Rams, *Measures of maximal dimension for linear horseshoes*, Real Anal. Exchange **31** (2005/06), 55–62.

[128] D. Rand, *The singularity spectrum $f(\alpha)$ for cookie-cutters*, Ergodic Theory Dynam. Systems **9** (1989), 527–541.

[129] M. Ratner, *Markov partitions for Anosov flows on n-dimensional manifolds*, Israel J. Math. **15** (1973), 92–114.

[130] V. Rokhlin, *On the fundamental ideas of measure theory*, Amer. Math. Soc. Transl. **10** (1962), 1–52.

[131] D. Ruelle, *Statistical mechanics on a compact set with $\mathbb{Z}^\nu$ action satisfying expansiveness and specification*, Trans. Amer. Math. Soc. **185** (1973), 237–251.

[132] D. Ruelle, *Thermodynamic Formalism*, Encyclopedia of Mathematics and Its Applications 5, Addison-Wesley, 1978.

[133] D. Ruelle, *Ergodic theory of differentiable dynamical systems*, Inst. Hautes Études Sci. Publ. Math. **50** (1979), 27–58.

[134] D. Ruelle, *Repellers for real analytic maps*, Ergodic Theory Dynam. Systems **2** (1982), 99–107.

[135] O. Sarig, *Thermodynamic formalism for countable Markov shifts*, Ergodic Theory Dynam. Systems **19** (1999), 1565–1593.

[136] B. Saussol, *Recurrence rate in rapidly mixing dynamical systems*, Discrete Contin. Dyn. Syst. **15** (2006), 259–267.

[137] B. Saussol, S. Troubetzkoy and S. Vaienti, *Recurrence, dimensions, and Lyapunov exponents*, J. Statist. Phys. **106** (2002), 623–634.

[138] B. Saussol and J. Wu, *Recurrence spectrum in smooth dynamical systems*, Nonlinearity **16** (2003), 1991–2001.

[139] S. Savchenko, *Special flows constructed from countable topological Markov chains*, Funct. Anal. Appl. **32** (1998), 32–41.

[140] J. Schmeling, *Hölder continuity of the holonomy maps for hyperbolic sets. II*, Math. Nachr. **170** (1994), 211–225.

[141] J. Schmeling, *Symbolic dynamics for β-shifts and self-normal numbers*, Ergodic Theory Dynam. Systems **17** (1997), 675–694.

[142] J. Schmeling, *On the completeness of multifractal spectra*, Ergodic Theory Dynam. Systems **19** (1999), 1595–1616.

[143] J. Schmeling, *Entropy preservation under Markov coding*, J. Stat. Phys. **104** (2001), 799–815.

[144] J. Schmeling and R. Siegmund-Schultze, *Hölder continuity of the holonomy maps for hyperbolic sets I*, in Ergodic Theory and Related Topics III, Proc. Int. Conf. (Güstrow, 1990), edited by U. Krengel, K. Richter and V. Warstat, Lecture Notes in Mathematics 1514, Springer, 1992, pp. 174–191.

[145] J. Schmeling and H. Weiss, *An overview of the dimension theory of dynamical systems*, in Smooth Ergodic Theory and Its Applications (Seattle, 1999), edited by A. Katok, R. de la Llave, Ya. Pesin and H. Weiss, Proceedings of Symposia in Pure Mathematics 69, Amer. Math. Soc., 2001, pp. 429–488.

[146] M. Shereshevsky, *A complement to Young's theorem on measure dimension: the difference between lower and upper pointwise dimension*, Nonlinearity **4** (1991), 15–25.

[147] K. Simon, *Hausdorff dimension for noninvertible maps*, Ergodic Theory Dynam. Systems **13** (1993), 199–212.

[148] K. Simon, *The Hausdorff dimension of the Smale–Williams solenoid with different contraction coefficients*, Proc. Amer. Math. Soc. **125** (1997), 1221–1228.

[149] K. Simon and B. Solomyak, *Hausdorff dimension for horseshoes in $\mathbb{R}^3$*, Ergodic Theory Dynam. Systems **19** (1999), 1343–1363.

[150] D. Simpelaere, *Dimension spectrum of axiom A diffeomorphisms. II. Gibbs measures*, J. Statist. Phys. **76** (1994), 1359–1375.

[151] B. Solomyak, *Measure and dimension for some fractal families*, Math. Proc. Cambridge Philos. Soc. **124** (1998), 531–546.

[152] F. Takens, *Limit capacity and Hausdorff dimension of dynamically defined Cantor sets*, (Valparaiso 1986), edited by R. Bamón, R. Labarca and J. Palis, Lecture Notes in Mathematics 1331, Springer, 1988, 196–212.

[153] F. Takens and E. Verbitski, *Multifractal analysis of local entropies for expansive homeomorphisms with specification*, Comm. Math. Phys. **203** (1999), 593–612.

[154] F. Takens and E. Verbitski, *On the variational principle for the topological entropy of certain non-compact sets*, Ergodic Theory Dynam. Systems **23** (2003), 317–348.

[155] M. Urbański, *Measures and dimensions in conformal dynamics*, Bull. Amer. Math. Soc. (N.S.) **40** (2003), 281–321.

[156] M. Urbański, *Recurrence rates for loosely Markovdynamical systems*, J. Aust. Math. Soc. **82** (2007), 39–57.

[157] M. Urbański and A. Zdunik, *The finer geometry and dynamics of the hyperbolic exponential family*, Michigan Math. J. **51** (2003), 227–250.

[158] B. van der Waerden, *Algebra*, vol. 1, Springer, 1991.

[159] M. Viana, *Stochastic Dynamics of Deterministic Systems*, Brazilian Math. Colloquium, IMPA, 1997.

[160] J. von Neumann, *Proof of the quasi-ergodic hypothesis*, Proc. Nat. Acad. Sci. U.S.A. **18** (1932), 70–82.

[161] P. Walters, *A variational principle for the pressure of continuous transformations*, Amer. J. Math. **97** (1976), 937–971.

[162] P. Walters, *Equilibrium states for β-transformations and related transformations*, Math. Z. **159** (1978), 65–88.

[163] P. Walters, *An Introduction to Ergodic Theory*, Graduate Texts in Mathematics 79, Springer, 1981.

[164] C. Wolf, *On measures of maximal and full dimension for polynomial automorphisms of $\mathbb{C}^2$*, Trans. Amer. Math. Soc. **355** (2003), 3227–3239.

[165] L.-S. Young, *Dimension, entropy and Lyapunov exponents*, Ergodic Theory Dynam. Systems **2** (1982), 109–124.

[166] M. Yuri, *Multifractal analysis of weak Gibbs measures for intermittent systems*, Comm. Math. Phys. **230** (2002), 365–388.

Index

Progress in Mathematics (PM)

Edited by
Hyman Bass, University of Michigan, USA
Joseph Oesterlé, Institut Henri Poincaré, Université Paris VI, France
Alan Weinstein, University of California, Berkeley, USA

Progress in Mathematics is a series of books intended for professional mathematicians and scientists, encompassing all areas of pure mathematics. This distinguished series, which began in 1979, includes research level monographs, polished notes arising from seminars or lecture series, graduate level textbooks, and proceedings of focused and refereed conferences. It is designed as a vehicle for reporting ongoing research as well as expositions of particular subject areas.

BIRKHÄUSER

Progress in Mathematics (PM)

Edited by
Hyman Bass, University of Michigan, USA
Joseph Oesterlé, Institut Henri Poincaré, Université Paris VI, France
Alan Weinstein, University of California, Berkeley, USA

Progress in Mathematics is a series of books intended for professional mathematicians and scientists, encompassing all areas of pure mathematics. This distinguished series, which began in 1979, includes research level monographs, polished notes arising from seminars or lecture series, graduate level textbooks, and proceedings of focused and refereed conferences. It is designed as a vehicle for reporting ongoing research as well as expositions of particular subject areas.